UNIVERSITY OF STRATHCLYDE

30125 00640868 5

Books are to be returned on or before
the last date below.

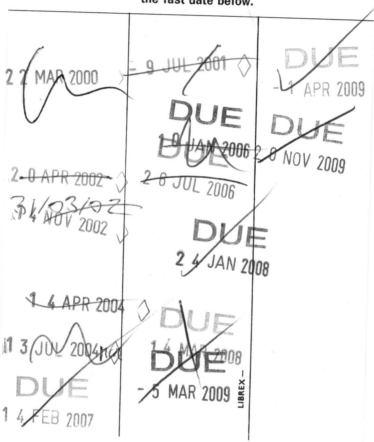

Environmental Engineering

Series Editors: U. Förstner, R. J. Murphy, W. H. Rulkens

Springer

Berlin
Heidelberg
New York
Barcelona
Budapest
Hong Kong
London
Milan
Paris
Santa Clara
Singapore
Tokyo

Günter Baumbach

Air Quality Control

Formation and Sources, Dispersion, Characteristics
and Impact of Air Pollutants –
Measuring Methods, Techniques for Reduction
of Emissions and Regulations for Air Quality Control

With the Assistence of
K. Baumann, F. Dröscher, A. Grauer, H. Gross, B. Steisslinger
and U. Vogt

With 244 Figures

Springer

Series Editors

Prof. Dr. U. Förstner
Arbeitsbereich Umweltschutztechnik
Technische Universität Hamburg-Harburg
Eißendorfer Straße 40
D-21073 Hamburg, Germany

Prof. Robert J. Murphy
Dept. of Civil Engineering and Mechanics
College of Engineering
University of South Florida
4202 East Fowler Avenue, ENG 118
Tampa, FL 33620-5350, USA

Prof. Dr. ir. W. H. Rulkens
Wageningen Agricultural University
Dept. of Environmental Technology
Bomenweg 2, P.O. Box 8129
NL-6700 EV Wageningen, The Netherlands

Author

Prof. Dr.-Ing. Günter Baumbach
Universität Stuttgart
Institut für Verfahrenstechnik
und Dampfkesselwesen
Pfaffenwaldring 23
D-70569 Stuttgart, Germany

Translator

Christina Grubinger-Rhodes
Epplestraße 9
D-70597 Stuttgart, Germany

Translated from the German Edition "Luftreinhaltung"

ISBN 3-540-57992-3 Springer-Verlag Berlin Heidelberg New York

Cataloging-in-Publication Data applied for
Die Deutsche Bibliothek - CIP-Einheitsaufnahme
Baumbach, Günter: Air Quality control : formation and sources, dispersion, characteristics and impact of air pollutants - measuring methods, techniques for reduction of emissions and regulations for air quality control / Günter Baumbach. With the assistance of K. Baumann... [Transl. Christina Grubinger-Rhodes]. - Berlin; Heidelberg; New York; Barcelona; Budapest; Hong Kong; London; Milan; Paris; Santa Clara; Singapore; Tokyo: Springer, 1996 (Environmental Engineering)   Dt. Ausg. u.d.T: Luftreinhaltung
   ISBN 3-540-57992-3

This work is subject to copyright. All rights are reserved, whether the whole or part of the material is concerned, specifically the rights of translation, reprinting, reuse of illustrations, recitation, broadcasting, reproduction on microfilm or in other ways, and storage in data banks. Duplication of this publication or parts thereof is permitted only under the provisions of the German Copyright Law of September 9, 1965, in its current version, and permission for use must always be obtained from Springer-Verlag. Violations are liable for prosecution act under German Copyright Law.

© Springer-Verlag Berlin Heidelberg 1996
Printed in Germany

The use of general descriptive names, registered names, trademarks, etc. in this publication does not imply, even in the absence of a specific statement, that such names are exempt from the relevant protective laws and regulations and therefore free for general use.

Typesetting: Data-conversion by M. Schillinger-Dietrich, Berlin
SPIN:10085939   61/3020-5 4 3 2 1 0 - Printed on acid-free paper

Preface

Air quality and air pollution control are tasks of international concern as, for one, air pollutants do not refrain from crossing borders and, for another, industrial plants and motor vehicles which emit air pollutants are in widespread use today. In a number of the world's expanding cities smog situations are a frequent occurrence due to the number and emission-intensity of air pollution sources. Polluted air causes annoyances and can, when it occurs in high concentrations in these cities, constitute a serious health hazard. How important clean air is to life becomes apparent when considering the fact that humans can do without food for up to 40 days, without air, however, only a few minutes.

The first step towards improving the air quality situation is the awareness that a sound environment is as much to be aspired for as the development of new technologies improving the standard of living. Technical progress should be judged especially by how environmentally benign, clean and noiseless its products are. Of these elements, clean air is of special concern to me. I hope that this book will awaken more interest in this matter and that it will lead to new impulses. Due to the increasing complexity of today's machinery and industrial processes science and technology can no longer do without highly specialized design engineers and operators. Environmental processes, however, are highly interdependent and interlinked. Environmental protection therefore requires interdisciplinary thought processes. This book attempts to make the reader aware of both the wide scope of air quality control and to direct his attention to specific problems arising in practical work.

During my work in foreign countries I have frequently been asked for an English translation of my German book "Luftreinhaltung". This idea was kindly received by the publishers Springer Verlag. For the English version several chapters were enlarged and international conditions included.

This book covers the causes of air pollutants, their dispersion and transformation in the atmosphere, their effects on humans, animals, plants and materials as well as emission reduction technologies. Measuring technology is a focal point as it plays a key role, be it in the detection of air pollutants or in the controlling and monitoring of measures for emission reduction. The book starts out with a historical overview and concludes with regulations on air quality control. Current problems such as SO_2 long-range transport, behavior of tropospheric ozone, recent forest damage, motor vehicle exhaust gases, polychlorinated dioxins and furans and reduction technologies are dealt with on the basis of my own research experience. Emission and behavior of radioactive substances have not been included as these represent fields by themselves and would exceed the scope of this work.

Air quality control being a wide field, it is inevitable that, depending on one's own experience, some facts can be interpreted differently. Hence, I would be grateful for new ideas, criticism or corrections.

This book is based on lectures held for students at the University of Stuttgart, Germany. It is primarily intended as a supplement to these lectures. Beyond that it is addressed to all those interested in air quality control. It is meant for engineers and practitioners working on emission reduction who wish to learn both more on the subject and more about the behavior and effects of pollutants. It is also for those affected by pollutants who wish to gain insight into the possibilities and problems of pollutant reduction.

It may not be expected that plants for air pollution control can be designed after reading this book. For in-depth treatment of individual fields special literature is recommended. Also, this book cannot replace practical experience in air quality control which can only be gathered by measuring air pollutants or by designing and operating air quality control units.

This book could not have been written without the assistance of a number of people. I want to express my thanks to the technical and scientific staff of the Institute of Process Engineering and Power Plant Technology of the University of Stuttgart, especially to the current and former colleagues of the Department for Air Pollution Control who have supported me in my work with their help and research activity. In particular, I would like to mention my co-authors Dr.-Ing. Frank Dröscher, Dr.-Ing. Harald Gross, Dr.-Ing. Bernd Steisslinger and Dipl.-Ing. Andreas Grauer. By preparing research results relevant for this book Dr.-Ing. Karsten Baumann and Dipl.-Ing. Ulrich Vogt have contributed a great deal to this work. My thanks are directed to the present and former directors of the Institute, Prof. Dr.-Ing. Klaus R. G. Hein, Prof. Dr. techn. Richard Doležal and Prof. Dr.-Ing. Rudolf Quack who made it possible for me to work in the field of air quality control and who stood by me in word and deed. I also want to thank Ms. Ursula Docter and Ms. Waltraut Wurster for writing the texts and tables and Mr. Femi Ogunsola for helping with the illustration translations. Ms. Christina Grubinger-Rhodes deserves special mention for the English translation. With painstaking care she has succeeded in producing this technically correct and highly readable English version. My special thanks are due to her.

Last but not least I would like to thank the publishers Springer Verlag for picking up this idea and realizing this book.

Contents

1	**General Overview**	1
1.1	Clean Air and Air Pollution	1
1.2	Historical Overview	2
1.3	Explanation of Terms	10
1.4.	Bibliography	13
2	**Origin and Sources of Air Pollution**	15
2.1	Emission of Pollutants Caused by Combustion Processes	15
2.1.1	Products of Complete and Incomplete Combustion	15
2.1.1.1	Carbon Monoxide	18
2.1.1.2	Hydrocarbons	18
2.1.1.3	Soot	23
2.1.1.4	Hydrocarbon Emissions in Different Combustion Processes	25
2.1.2	Sulfur Compounds	26
2.1.2.1	Sulfur in Coal	26
2.1.2.2	Sulfur in Fuel Oil	27
2.1.2.3	Sulfur in Natural Gas	28
2.1.2.4	Comparison of Sulfur Contents of Different Fuels	29
2.1.2.5	Combustion Products of Sulfurous Fuels	29
2.1.3	Oxides of Nitrogen	32
2.1.3.1	Origins of Oxides of Nitrogen	32
2.1.3.2	Nitrogen Oxide Emissions in Different Combustion Processes	35
2.1.4	Particles	36
2.1.4.1	Problem, Dependencies and Components	36
2.1.4.2	Soot and Particle Emissions during the Combustion of Liquid Fuels	39
2.1.4.3	Particle Emission in Industrial Coal Furnaces	41
2.1.4.4	Particle Emission in the Combustion of Lump Wood in Domestic Furnaces	43
2.1.5	Polychlorinated Dibenzodioxins and Dibenzofurans	46
2.1.5.1	Properties, Formation and Origin	46
2.1.5.2	Toxicity, Toxicity Equivalents and Threshold Values	48
2.1.5.3	Dioxin Sources	50
2.1.6	Exhaust Gases from Motorised Vehicles	51
2.1.6.1	Influences on their Formation	51
2.1.6.2	Exhaust Gas Emissions when Driving	57
2.1.6.3	Development of Motor Vehicle Emissions	60
2.2	Sources of Air Pollutants	62

2.2.1	Overview	62
2.2.2	Carbon Dioxide (CO_2)	64
2.2.3	Sulfur Oxides (So_x)	65
2.2.4	Nitrogen Oxides (No_x)	68
2.2.5	Carbon Monoxide (CO)	70
2.2.6	Volatile Organic Compounds (VOC)	72
2.3	Bibliography	74
3	**Air Pollutants in the Atmosphere**	**79**
3.1	Meteorological Influences on the Dispersion of Air Pollutants	79
3.1.1	Wind	80
3.1.2	Turbulence	80
3.1.3	Inversion	83
3.1.3.1	Inversion Types	84
3.1.3.2	Inversion Formation	85
3.1.4	Mixing Layer and Barrier Layers	87
3.1.5	Inversion Layers and Air Pollutants – Examples of the Dispersion of Pollutants	91
3.1.5.1	Widespread Dispersion of Air Pollutants: SO_2 Long-Range Transport	91
3.1.5.2	Short-Range Dispersion of Air Pollutants in Highland Valleys	96
3.2	Chemical Transformations of Pollutants in the Atmosphere	100
3.2.1	General Considerations	101
3.2.1.1	Atmosphere and Air Pollution	101
3.2.1.2	Calculation of Reaction Rates	105
3.2.2	Oxidation of SO_2	105
3.2.2.1	SO_2 Oxidation in the Gas Phase	106
3.2.2.2	SO_2 Conversion in Liquid Phase and on Solid Particles	106
3.2.3	Reactions of Nitrogen Oxides in the Atmosphere	110
3.2.3.1	NO Oxidation and the Formation of Ozone	111
3.2.3.2	Participation of Hydrocarbons in NO Oxidation	115
3.2.3.3	NO_2 Oxidation	118
3.2.3.4	NO_x and Acid Rain	119
3.2.4	Ozone in the Atmosphere	119
3.2.4.1	Ozone in the Stratosphere	119
3.2.4.2	Ozone in the Troposphere	121
3.2.5	Carbon Compounds	127
3.2.5.1	Organic Carbon Compounds	127
3.2.5.2	Inorganic Carbon Compounds	127
3.2.6	Particles in the Atmosphere	128
3.2.7	Precipitation Components	128
3.3	Distribution and Temporal Development of Air Pollutants in Unpolluted and Polluted Areas	130
3.3.1	Spatial Distribution	130
3.3.2	Temporal Trends in Air Quality	142
3.3.2.1	Carbon Compounds	142

3.3.2.2	Sulfur Dioxide (SO$_2$)	145
3.3.2.3	Suspended Particulate Matter	146
3.3.2.4	Nitrogen Oxides	150
3.3.2.5	Tropospheric Ozone (O$_3$)	150
3.4	Models of Pollutant Dispersion	156
3.4.1	Objective and Application of Mathematical Meteorological Simulation Models	156
3.4.2	Model Concepts	158
3.4.2.1	Flow and Turbulence Models	158
3.4.2.2	Modeling of Pollution Dispersion	160
3.4.3	Consideration of Chemical Transformations in Dispersion Models	166
3.4.4	Summary and Overview of Model Concepts	167
3.5	Bibliography	167

4 Effects of Air Pollution ... 174

4.1	General Considerations	174
4.1.1	The Range of Possible Types of Damage	174
4.1.2	The Path of Air Pollutants to the Location where they Become Effective	175
4.2	Climatic Changes Caused by Atmospheric Trace Substances	175
4.2.1	Temperature Increase	177
4.2.1.1	Destruction of the Stratospheric Ozone Layer	177
4.2.1.2	Greenhouse Effect of Infrared-Active Gases	179
4.2.2	Temperature Drop Due to Particle and Cloud Occurrence	180
4.2.3	Prognostic Difficulties	181
4.3	Effects on Materials	181
4.3.1	Mineral Building Materials	182
4.3.2	Metals	184
4.3.3	Other Materials	185
4.4	Effects on Vegetation	186
4.4.1	Plant Damage Caused by Air Pollution	188
4.4.1.1	Determination of Dose-Effect Relations	188
4.4.1.2	Damage Mechanisms and Profiles of Single Air Pollutants	189
4.4.2	Forest Damage	193
4.4.2.1	Damage Profiles	194
4.4.2.2	Assumed Mechanisms of Effect	195
4.5	Impact on Human Health	198
4.5.1	Possibilities and Difficulties of Recording Harmful Effects	199
4.5.2	Paths Air Pollutants Take in the Human Body	202
4.5.3	Effects of the Most Important Air Pollutants	202
4.6	Ambient Air Guidelines and Standards	202
4.6.1	Nature of WHO Guidelines	209
4.6.2	National Ambient Air Quality Standards	212
4.6.3	MIK Values of the Verein Deutscher Ingenieure (VDI = Association of German Engineers)	212

4.6.3	Smog Alarm Values	213
4.7	Bibliography	213
5	**Measuring Techniques for Recording Air Pollutants**	**216**
5.1	General Criteria	216
5.1.1	Applications of Measuring Techniques	216
5.1.2	Discontinuous or Continuous Measurements	216
5.1.3	Physical and Chemical Measuring Principles	219
5.1.4	Different Requirements for Emission and Air Quality Measurements	220
5.1.5	Emission Components to be Recorded	221
5.1.6	Ambient Air Components to be Recorded	222
5.2	Measuring Methods for Gaseous Pollutants	222
5.2.1	Photometry	225
5.2.1.1	IR Photometer	227
5.2.1.2	UV Photometer	230
5.2.1.3	Long-Path Photometry	233
5.2.2	Colorimetry	234
5.2.3	UV Fluorescence	235
5.2.4	Chemiluminescence	236
5.2.4.1	NO_x Measurement	236
5.2.4.2	O_3 Measurement	238
5.2.5	Flame Photometry	238
5.2.6	Flame Ionization	239
5.2.7	Conductometry	241
5.2.8	Amperometry	243
5.2.9	Coulometry	243
5.2.10.	Potentiometry	245
5.2.10.1	PH Measurement	245
5.2.10.2	HF and HCl Measurement with Ion-Sensitive Electrodes	246
5.2.10.3	O_2 Measurement with Solid-State Ion Conductor Zirconium Dioxide	246
5.2.11	Paramagnetic Oxygen Measurement	250
5.2.12	Measurement of Thermal Conductivity	250
5.2.13	Manual Analysis Methods	251
5.2.14	Chromatographic Techniques	251
5.2.14.1	Gas Chromatography	254
5.2.14.2	Gas Chromatography/Mass Spectrometry	256
5.2.14.3	High Performance Liquid and Ion Chromatography	257
5.2.14.4	Determination of Highly Toxic Organic Compounds	258
5.2.15	Method for Determining Odorous Substances – Olfactometry	261
5.3	Measuring Methods for Particulate Matter	262
5.3.1	Gravimetric Determination of Particulate Matter in Exhaust Gases	263
5.3.2	Continuous Registration of the Flue Gas Particle Concentrations	266
5.3.3	Determination of the Soot Number of Furnace Flue Gases	269
5.3.4	Determination of Grain Size Distribution of Particles	269
5.3.5	Measurement of Particle Sedimentation in Ambient Air	271
5.3.6	Measurement of Particle Concentration in the Air	274

5.3.6.1	Filter Sampling Instruments	274
5.3.6.2	Automatically Recording Particle Concentration Measuring Instruments	278
5.3.7	Determination of Chemical Composition of Particulates	280
5.4	Setting up of Measuring Equipment, Sampling Techniques and their Influences on the Accuracy of the Measurements	282
5.4.1	Emission Measurements in Furnaces and Process Equipment	282
5.4.1.1	Location of Sampling	282
5.4.1.2	Setting up of the Site of Measurement – Sampling System	284
5.4.1.3	Possible Faults during Sampling	286
5.4.2	Emission Measurements for Motor Vehicles	288
5.4.2.1	Exhaust Gas Sampling and Measurement According to the CVS Method	289
5.4.2.2	Various Driving Cycles	290
5.4.3	Ambient Air Measurements	292
5.4.3.1	Significance of the Locations of Measuring Sites	292
5.4.3.2	Sampling Systems in Measuring Stations	293
5.4.3.3	Set-up of Air Quality Measuring Stations – Example of a Forest Measuring Station	295
5.5	Calibration for Pollutant Measurements	298
5.5.1	Definitions	298
5.5.2	Calibration Gases	299
5.5.2.1	Static Methods for the Production of Calibration Gases	299
5.5.2.2	Dynamic Methods – the Mixing of Volume Flows	299
5.5.2.3	Example of a Calibration Gas Production	300
5.5.2.4	Difficulties in the Production of Calibration Gas	301
5.5.3	Significance of the Calibration	302
5.6	Accuracy of Measuring Methods and Measuring Instruments	302
5.6.1	Overview of Characteristic Values	302
5.6.2	Linearity of the Calibration Function and Sensitivity	307
5.6.3	Interference	308
5.6.4	Determination of the Efficiency of Measuring Methods by Ring Tests	310
5.6.5	Fault Consideration using the Example of a Complete Emission Measurement	312
5.7	Bibliography	314
6	**Evaluation of Air Pollution Measurements**	**322**
6.1	Determination of Pollutant Emissions	322
6.1.1	Detection of Pollutant Emissions from Concentration Measurements of Exhaust Gases	322
6.1.1.1	Emission Flows and Emission Factors	322
6.1.2	Calculation of Pollutant Emissions from Fuel Properties	324
6.1.2.1	Sulfur Dioxide	324
6.1.2.2	Nitrogen Oxides	325
6.1.2.3	Products of Incomplete Combustion (PIC)	325

6.1.2.4	Heavy Metal Emissions in Oil Furnaces	325
6.1.3	Registering Pollutant Emissions of an Area in Emission Inventories	326
6.1.3.1	Spatial Pollutant Distribution	326
6.1.3.2	Determination of Pollutant Emissions of Motor Vehicle Traffic	327
6.1.3.3	Pollutant Emissions of Domestic Furnaces and Small-Scale Industries	328
6.1.3.4	Pollutant Emissions from Industries	329
6.1.3.5	Summary of Annual Emissions	331
6.1.3.6	Temporal Pollutant Distribution	332
6.2	Evaluation and Graphic Representation of Air Quality Measurements	335
6.2.1	Temporal Resolution and Mean Value Formation	335
6.2.2	Summary and Graphic Representation of Measuring Data from Continuous Measurements	338
6.2.2.1	Unsmoothed Monthly Profiles	338
6.2.2.2	Mean Value Calculation for Smog Warnings	338
6.2.2.3	Diurnal Courses	338
6.2.2.4	Long-Term Annual Profiles	341
6.2.3	Frequency Distribution	343
6.2.4	Spatial Distribution of Pollutants	346
6.2.4.1	Method of Determination and Graphic Representation	346
6.2.5	Methods for the Investigation of Consistent Patterns in the Occurrence of Pollutants	348
6.2.5.1	Mean Diurnal Courses	348
6.2.5.2	Correlation Calculations	350
6.2.5.3	Pollutant Wind Roses: Pollutant-Wind Correlations	350
6.2.5.4	Abatement Curves	355
6.3	Bibliography	356
7	**Emission Control Technologies**	**360**
7.1	General Considerations	360
7.1.1	Process Modification	361
7.1.2	Emission Control for Furnaces	363
7.1.2.1	Products of Incomplete Combustion	364
7.1.2.2	Particulates	364
7.1.2.3	Nitrogen Oxides	365
7.1.2.4	Sulfur Dioxide	365
7.1.3	Efficiency of Exhaust Gas Purification Systems	367
7.2	Processes for the Removal of Particulate Matter from Exhaust Gases	369
7.2.1	Inertial Collectors	370
7.2.1.1	Inertial Force Particle Collectors	371
7.2.1.2	Centrifugal Force Particle Collectors	371
7.2.2	Wet Scrubbers	374
7.2.2.1	Principles of Wet Scrubbing	376
7.2.2.2	Different Types of Wet Scrubbers	379
7.2.3	Electrostatic Precipitators	383
7.2.3.1	Operating Principle	383

7.2.3.2	Mode of Operation	384
7.2.3.3	Precipitation Equation	389
7.2.3.4	Constructions	390
7.2.4	Filters	392
7.2.4.1	Tube Filters	393
7.2.4.2	Baghouse Filters	396
7.3	Nitrogen Oxide Reduction in Combustion Processes	397
7.3.1	Primary Measures in Furnaces	397
7.3.1.1	Reducing Air Excess	397
7.3.1.2	Stage Combustion, Multiple-Stage Mixing Burner and Top Air Nozzles	398
7.3.1.3	Slight Air Preheating	399
7.3.1.4	Reduction of the Volume-Specific Combustion Chamber Load	399
7.3.1.5	Flue-Gas Recirculation	400
7.3.1.6	Low-NO_x Burners	401
7.3.1.7	NO_x Reduction Potential of Primary Measures	401
7.3.2	Secondary Measures in Furnaces	402
7.3.2.1	Reduction Processes	403
7.3.2.2	Oxidation Processes	411
7.3.3	Catalytic Converter Technology for the Reduction of Nitrogen Oxides and Other Components in Automotive Exhaust Gases	413
7.4	Flue Gas Desulfurization	417
7.4.1	Dry Flue Gas Desulfurization	418
7.4.2	Semi-Dry Processes – Spray Absorption Technique	423
7.4.3	Wet Desulfurization Processes	426
7.4.3.1	Lime Scrubbing Processes	426
7.4.3.2	Other Wet Flue Gas Desulfurization Processes	434
7.5	Diagram of the Flue Gas Purification Units of a Power Plant	435
7.6	Present Situation of Flue Gas Purification in West German Power Plants	435
7.7	Removal of Organic Substances from Flue Gases	437
7.7.1	Overview of the Processes	437
7.7.2	Condensation	438
7.7.3	Absorption	439
7.7.3.1	Physical Absorption	440
7.7.3.2	Chemical Absorption	442
7.7.3.3	Biological Waste Air Purification – Bioscrubbing and Biofiltration	442
7.7.4	Adsorption	445
7.7.5	Combustion	450
7.7.5.1	Thermal Afterburning	450
7.7.5.2	Catalytic Afterburning	451
7.7.6	Membrane Processes	452
7.8	Bibliography	454

8	**Air Pollution Control Conventions, Laws and Regulations**	461
8.1	International Conventions	461
8.2	EU Directives	461
8.3	National Laws and Regulations	462
8.3.1	US Air Pollution Laws and Regulations	464
8.3.2	German Laws and Regulations	465
8.3.2.1	Plants Requiring Official Permission	466
8.3.2.2	Plants Exempted from Official Permission	470
8.3.2.3	Product-Related Air Pollution Prevention	470
8.3.2.4	Area-Related Air Pollution Prevention	472
8.4	Vehicle Exhaust Emissions	472
8.5	Bibliography	480
Subject Index		482

1 General Overview

1.1 Clean Air and Air Pollution

Generally, all substances changing the natural composition of air are considered pollutants. Gaseous components of naturally clean air are listed in Table 1.1. Apart from these substances, natural air can contain further components such as water vapor and traces of other gases, e.g., methane CH_4, ammonia NH_3, carbon monoxide CO and dinitric oxide N_2O as a result of decomposition processes, as well as low concentrations of ozone, possibly as a result of stratospheric penetration.

When discussing the origins of air pollution a distinction is made between natural and man-made (anthropogenic) sources. High levels of naturally occurring air pollutants, for example, during volcanic eruptions, during sand storms (deposits of Sahara sand have occasionally been observed in Central Europe!), during forest fires, during processes related to air-chemistry in thunderstorms and also from plant pollen. Odorous substances from blossoms are not part of the natural air composition either, though they can hardly be regarded as air pollutants.

Anthropogenic air pollutants will be primarily dealt with in this work. One differentiates between particulates and aerosols (fine dusts and extremely fine droplets distributed in the air) on one hand and gases on the other. Visible air pollutants are usually dusts or droplets, e.g., smoke, soot, oil mist; only few gases, however, are visible. The most important noxious gas recognizable by its color is the brown nitrogen dioxide (NO_2). Olfactory substances are another particularly conspicuous type of air pollutant. These are gases which are often perceptible even in extremely low concentrations.

Table 1. Natural composition of air

		Volume content in % related to dry air
Oxygen	(O_2)	20.93
Nitrogen	(N_2)	78.10
Argon	(Ar)	0.9325
Carbon dioxide	(CO_2)	0.03-0.04
Hydrogen	(H_2)	0.01
Neon	(Ne)	0.0018
Helium	(He)	0.0005
Krypton	(Kr)	0.0001
Xenon	(Xe)	0.000009

1.2 Historical Overview (F. Dröscher)

Anthropogenic air pollution had its beginnings when man started to use fire. Smoke, CO, CO_2 and organic gases are the pollutants resulting from this.

In nomadic and rural cultures pollution of indoor air was high, whereas in the densely populated cities of advanced civilisations the outdoor atmosphere also deteriorated. E.g., in the year 61 A.D. the Roman philosopher Pliny the Elder noted: "As soon as I had left Rome's heavy air and the stench of the smoking chimneys blowing all sorts of vapors and soot into the air behind, I felt a change in my condition" [1].

In pre-industrial times the most significant sources of air pollution were foundries, potteries as a trade with high energy consumption and the smoking of fish and

Fig. 1.1. Diagram of a dust chamber in the Ore Mountains (1556)[2], by courtesy of the VDI Commission Air Pollution Prevention

meat [1]. The smelting of ores released acid gas from roasting and dusts containing large amounts of heavy metals. The damage done to fields, meadows and fruit in the surroundings of the foundries of this era which were almost exclusively located in the highlands led to neighborhood conflicts between foundry people, landowners and farmers, especially in narrow valleys [2].

The development of effective flues in which coarse dust sedimented and was fed back into the process can safely be regarded as the beginning of environmental engineering in Germany (Fig. 1.1). These flues and dust chambers were mandatory in Bohemia and Saxony very early on.

With regard to air quality, the invention of the steam engine by James Watt in 1769 ushered in a new era. Within a few decades energy consumption soared. Smoke and ash became the main problem when coal was burned for the firing of boilers of stationary steam engines and locomotives as well as in home stoves. At the same time, production rates in almost all branches of trade increased and with them their emissions. The variety of air pollutants increased rapidly, particularly due to the growing chemical industry.

England and France, the vanguards of industrialization, were the first states whose governments and administrative authorities were forced to deal with complaints about increasing environmental damage. For this, the old trade regulations proved to be insufficient. New special regulations were constantly being issued and some plants had to fulfill certain requirements usually based on court orders regulating complaints. As early as 1810 France passed a national air pollution control act by decree. According to this law a total of 66 commercial activities were subject to official approval, and both the formal proceedings for the granting of permits as well as the successive stages of appeal were laid down. Trades were divided into three categories:
- factories which were on no account allowed in residential areas as they presented a fire hazard, had a negative effect on people's health or produced an offensive smell,
- factories which were tolerated in residential areas but had to be supervised,
- objectionable trades which were, however, subject to approval.

Approximately one third of the factories belonged to the first category.

In the ensuing period of rapid technological development – in 1845 as many as 307 branches of trade were subject to this law – the decree was supplemented and interpreted by export regulations and the supreme court [3].

Not only was France the first country to require approval for steam engines in 1823 (a practice adopted by Prussia in 1831), it was also the first to introduce regular checks for them by state-employed test engineers. In Germany, boiler operators took on this task from 1866 in voluntary cooperation with the technical inspection authorities.

While the majority of the population of the 19th century considered "smoking chimneys" (Fig. 1.2) the hallmark of technical progress and not least their livelihood, agriculture and forestry had to cope with the effects of air pollution. In the middle of the last century large sections of forests began dying in the surroundings of all industrial centres of Central Europe, in the Ore Mountains, in the Thuringian and Franco-

Fig. 1.2. Foundries and mines of the Baron von Rothschild near Moravian Ostrau [2], by courtesy of the VDI Commission Air Pollution Prevention

nian Forest and in Saxony and the Ruhr District. Mainly fir and spruce, the most sensitive coniferous trees, succumbed to this. Chemists and forestry scientists of the Bergakademie Freiberg and of the Forstakademie Tharandt in Saxony thoroughly examined the damage mechanisms at work. As early as 1850/1871 one of the founders of this smoke damage research, Stöckhardt, pointed out two ways in which today's so-called classical smoke damage takes effect[5]:
- by direct, acute poisoning from SO_2 through assimilating leaf organs,
- by indirect chronic poisoning of the soil due to dust containing heavy metals.

In this context, air quality analyses for SO_2 in air and for sulfate in rain water were developed, and critical SO_2 concentration threshold values were determined for plants and human beings [5].

Strongly increasing SO_2 emissions prompted Smith's examinations on "acid rain" [4] which were published in England in 1872.

After a certain understanding of the mechanisms of impact had been achieved, measures were taken. In the industrial centers (e.g., in the Ruhr District) the forestry services systematically replaced spruce forests by more smoke-resistant deciduous trees. At least in the close vicinity of polluting factories deciduous trees were planted in long rows as air quality protection, if necessary.

Measures for reducing harmful emissions were, however, given first priority. A series of innovations in industrial processing engineering led to the use of lower-emission raw materials, made possible higher yields or improved efficiency of burning processes and the utilization of by-products (e.g., the HCl yielded through condensation during soda production). Most frequently, however, reasons pertaining to the approval procedure rather than economic motives, were deciding factors for the reduction of pollutants.

With the help of powerful blowers exhaust gases carrying acid gases and dust could be conducted into "deacidifying plants" (dry adsorption on lime, absorption in scrubbing towers), cyclones or fabric filters.

In the particularly exposed valleys of the highlands foundries and steam engine operators, however, used blowers also for the dilution of flue gases in the chimney or made do with high stacks. At the turn of the century the Freiberger Hüttenwerke boasted the world's highest stack with a height of 144 m.

Another example of the relocation of environmentally harmful substances was described by Wislicenus in 1908 [5]:

> "Particularly worth mentioning is the example of a Saxon ultramarine factory (Schindlers Blaufarbwerke near Bockau in the Ore Mountains) which only a few years ago was extremely dangerous to the surrounding forest but today has a lot of success by combining deacidification with the dilution by air.
>
> First of all, exhaust gases are pushed through large chambers filled with limestone with the help of a fan. In this way the originally high level of acidity is considerably reduced but not totally eliminated. Instead of permitting the remaining gases to escape through the stack into the air, they are passed through a long wooden channel filled with birch brushwood to another washing arrangement where they are passed over large amounts of water rushing over mill-wheels. They then finally pass through numerous cracks of a large chamber made of loose wooden planks into the air. Only a fine mist containing a little sulfuric acid or SO_3 escapes. Its harmfulness cannot be compared with the original sulfuric acid gases."

It was realized that another means of reducing emissions was to give the people operating the machines a sound education. This led to the setting up of schools for furnacemen.

All through the 19th century air pollution was merely considered a problem pertaining to neighborhood law. Workers' health played a secondary role in this period of economic liberalism and manpower surplus. In the end it was the social legislation of the German Empire which provided new impulses in this respect, as it made the employer share the financial risk involved in the ill-health of his employee.

By the middle of the 20th century industrialisation in North America and Europe had been well-established. Myriad technical innovations and developments influenced emissions. Electrical energy displaced the mechanical energy of the steam engine as an energy resource. Energy production and the resulting emissions were concentrated in central power plants with steam turbines.

With the further development of the dust removal techniques of the 19th century and the invention of the electrostatic precipitator, dust technology became established in the twenties. In 1926 a technical committee for "Dust Techniques" was founded within the VDI (Verein Deutscher Ingenieure – Association of German Engineers).

Simultaneously, the development of theoretical industrial chemical engineering permitted the planning of apparati and equipment and particularly the construction of effective exhaust gas purification plants. The two World Wars during which the economy was diverted to wartime production had a detrimental effect on the preservation of clean air. During the war and postwar years measures for emission reduction were neglected in favor of production.

In this context the strivings for self-sufficiency of the Third Reich played a special role, as scarce products were substituted to a particularly high degree by available raw materials. E.g., the expensive process of extracting gasoline from mineral coal was developed at that time.

Not only did the war economy block civilian plans, it also took up the major share of the existing development capacities. Moreover, the development of chemical, biological and atomic weapons of this era created tremendous, new potential hazards for man and his environment. This development has continued incessantly from post-war times up to the present.

However, by striving to protect mankind against these weapons, new branches of science have been founded: meteorology of air pollution and toxicology, both of which have decisively influenced our present knowledge of the distribution and effects of air pollutants.

In the USA which was the least affected by the effects of the two World Wars, great technological changes which were effected in Europe only decades later were brought about. The rapid increase of automobile traffic deserves special mention. Unburned hydrocarbons and nitric oxides from car exhaust gases led to the notorious photosmog episodes in sunny Los Angeles as early as the forties. From the twenties on the natural gas pipeline net started to spread in the south and west of the USA. Gas heating gradually replaced traditional coal and oil furnaces even in the coal areas of the eastern states of America.

In the meantime mainly the European centres of coal consumption were plagued by intermittent periods of the so-called London smog caused by the accumulation of sulfur dioxide and air-borne particles during low-exchange atmospheric conditions with high humidity. Table 1.2 compiles the most important episodes of London smog described in literature along with the epidemiological data collected at the time. A direct consequence of the smog episode of 1962 in the Ruhr District was the institution of a smog warning service with the help of many measuring stations in 1963.

After the years of reconstruction in Europe and the recession in America, the fifties saw a boost in industrialisation the world over, which also swept through the agrarian Southern and Eastern European countries and Japan.

Primary energy consumption continued to increase exponentially (s. Fig. 1.3c) while practically unchanged technology made the air quality situation in the centers of heavy industry increasingly critical. There was a definite need for legislative measures. Fuel quality and discharge restrictions were first regulated in clean air preservation laws in the USA (1955) and in England (Clean Air Act, 1956), to be followed later by all the industrial nations.

In the Federal Republic of Germany the trade regulation was changed accordingly in 1959. This paved the way for both imposing conditions on new plants which were subject to approval as well as for later changes of the operating permits of old plants. The first version of the general regulation TA Luft (Technical Directive On Air Pollution Control) took effect in 1964.

Table 1.2. Chronology of the most important episodes of London smog [6-10]

Date		Place	Casualties	Cases of sickness	SO_2 max. 24 h-values mg/m^3
1873	12/09-12/11	London			
1880	01/26-01/29	London	1000		
1892	12/28-12/30	London			
1909		Glasgow	80-176		
1930	12/01-12/05	Maas Valley (Belgium)	63	6000	25
1948	10/26-10/30	Donora (USA)	20	6000	1.6
1948	11/26-12/01	London	700-800		
1950	11/24	Poza Rica (Mexico)[a]	22	320	
1952	12/05-12/09	London	4000		3.8
1953	11/15-11/24	New York	250		2.2
1956	01/03-01/06	London	1000		1.5
1957	12/02-12/05	London	700-800		1.6
1958		New York			
1959	01/26-01/31	London	200-250		0.8
1962	12/05-12/10	London	700		3.3
1962	12/03-12/07	Ruhr Area (FRG)			5.0
1963	01/07-01/22	London	700		
1963	01/09-02/12	New York	200-400		1.3
1966	11/23-11/25	New York	168		
1979	01/17	Ruhr Area (FRG)	–		approx. 0.9
1982	01/22	Stuttgart (FRG)	–		0.57
1985	01/16-01/20	Ruhr Area (FRG)	–		0.85

[a] Hydrogen sulfide eruption in a natural gas processing plant during an inversion weather situation

The guidelines of the VDI-Commission "Reinhaltung der Luft" (preservation of clean air) which was founded in 1957 have subsequently supplemented these regulations and have a normative character.

Both the general introduction of dust removal technology and also high stacks successfully led to the positive result that "the sky over the Ruhr District became blue again" in the seventies.

In 1970, the "Year of the Environment", the time had come for a new beginning in environmental awareness in Germany. The Federal Government passed an immediate program for environmental protection [11, 12]. The first measure was a change of the Basic Constitutional Law for environmental protection with legislative authority passed from the federal states to the federal government. This laid the basis for the Federal Air Pollution Control Act passed in 1974. Based on this law, numerous administrative regulations were passed, e.g., for the inspection of domestic heating and for the desulfurization of light fuel oil. Further regulations based on this law are

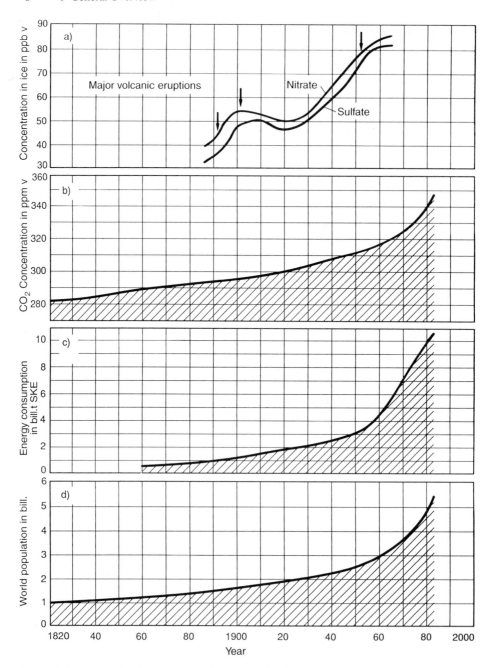

Fig. 1.3. Long-term development of environment-related parameters on a global scale.
a. nitrate and sulfate content of glacier ice in Southern Greenland (representative of the northern hemisphere), compensating curve 1885-1978 [13]; **b.** background concentration of CO_2: 1820-1960 obtained through an analysis of air bubbles in the arctic ice, from 1960 onward direct measurements on the Mauna-Loa, Hawaii [14]; **c.** global consumption of primary energy: [15, 16]; **d.** world population since 1960, UN estimates [17, 18]

general administrative regulations as, e.g., a new Technical Directive On Air Pollution Control (TA Luft). Besides, the law requires that plans for the preservation of clean air and smog regulations be set up in high-pollution areas. As a result of these smog regulations Berlin and the Ruhr District have had several smog alarms, as did Stuttgart in 1982. More stringent threshold values led to more frequent smog alarms in 1985, e.g., in the Ruhr District, in Hesse, Berlin, Northeast Bavaria and in the greater Nuremberg area.

Owing to the change to low-sulfur fuels in the past years and to the higher efficiency of domestic heating furnaces, SO_2 pollution has declined in the cities whereas oxides of nitrogen have become one of the main pollutants due to the sharp increase in automobile traffic.

The widespread occurrence of a new type of forest damage all over Central Europe from 1981/82 again led to a surge of even more stringent legislative measures for the preservation of clean air: 1983 saw a partial re-enactment of the Technical Directive On Air Pollution Control (TA Luft) when the Regulation for Large-Scale Furnaces was passed. In 1985 more stringent exhaust levels were introduced for internal combustion engines, and in 1986 the Technical Directive On Air Pollution Control (TA Luft) was supplemented by requirements for numerous plants. Ozone (O_3) formed by nitrogen oxides and hydrocarbons when exposed to UV light entered the discussion as a secondary pollutant. As early as 1972 Scandinavian scientists at a UNO conference in Stockholm had already pointed out that atmospheric pollution crosses national borders. They attributed the acidification of many thousands of lakes in Northern Europe to acid-forming SO_2 emissions transported there from Great Britain and Central Europe.

As is apparent from the analysis of air bubbles in glacier ice and more recently from direct measurements, the constantly increasing use of fossil fuels has long since led to a perceptible increase of carbon dioxide concentrations in the atmosphere in all parts of the world (s. Fig. 1.3b).

Like carbon dioxide, other products of fossil combustion disperse in the atmosphere over large areas. Their concentrations have increased almost proportionally with world energy consumption (Fig. 1.3c). This development is illustrated by Fig. 1.3a by glacier water analyses in Southern Greenland. The concentration of sulfate (nitrate) formed secondarily from sulfur dioxide (or nitric oxide) fluctuated within a certain range up until approx. 1940 (1950). Since then it has been rising sharply.

So far little attention has been paid to the role of the many organic air pollutants which have been released into the atmosphere for decades, whether gradually, as in the case of pesticides and freons, or spectacularly, as in the dioxine catastrophe of Seveso (1976) and the methylene isocyanite catastrophe in Bhopal (1984), or even entirely unnoticed. It is only in the rarest of cases that the effect of these organic compounds has been explained. E.g., even today it is still unclear which role is played by organic compounds in the rising frequency of cancer in industrial nations or the forming of ozone over large areas. A similar statement can be made for the long-term effects of radioactive gases and aerosols on the environment. The reactor accident of Chernobyl (1986), however, threw light on the hazards of the steadily increasing use of nuclear power plants.

Faced with the many open questions concerning the long-term effects of air pollutants on the entirety of the ecological system, it is imperative to find responsible compromises between keeping production processes economical on one hand and conserving the environment on the other.

The more the world population increases and with it, inevitably, the amount of supplies needed (Fig. 1.3d), the greater this challenge will be.

1.3 Explanation of Terms

Air pollutants – harmful (noxious) substances

Whereas all substances changing the natural composition of air are called pollutants, substances are only termed harmful or noxious if they have a harmful effect.

Emission

Emissions are those pollutants, such as gases and dusts, released into the environment by industrial plants, automobiles or products. In other fields, too, one may talk of emissions: sounds, rays, heat, tremors etc. can also be emitted.

Air quality

The degree of air pollution.

Transmission

The transmission of air pollutants from the air to another medium or object. The dispersion of air pollutants between emission and deposition is called transmission. Besides physical dilution air pollutants can also go through a chemical transformation during transmission. Fig. 1.4 shows these correlations in a diagram.

With high emission sources the path from emission to deposition is very long: hence, air pollutants are diluted before they reach their final destination where they take effect. A high source altitude also permits dispersion of the pollutants over a large area. In this case one speaks of so-called remote transports. With low source altitudes, e.g., automobile emissions, air pollutants can reach the human respiratory system by the shortest route.

Smog

Smoke + fog = smog. High concentrations of pollutants combined with fog are termed smog.

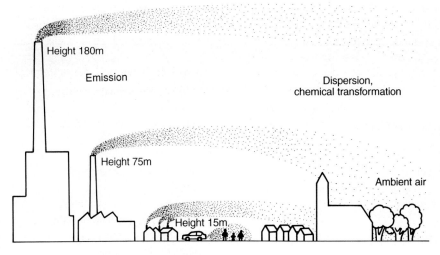

Fig. 1.4. Path of air pollutants from points of emission to deposition

Exhaust gas

Unutilized gas released into the environment by industrial plants, automobiles etc.. In general, it represents the carrier for the air pollutants emitted.

Flue gas

Furnace exhaust gas that becomes visible owing to fine particles (smoke). Even if the exhaust gas contains virtually no particles, one generally speaks of flue gas where furnaces are concerned.

Vent air

Exhaust gas whose carrier gas consists of air, e.g., workplace ventilation, where pollutants carried by air are released into the environment via a stack.

Emission sources

Emission sources are industrial plants, automobiles etc. which generate air pollutants and release them into the environment. One talks of point sources if air pollutants are released in high concentrations at one location, e.g. by industrial stacks. Numerous small sources make a plane source, e.g., leakages in pipes and fittings in large-scale industrial plants, e.g., refineries. House chimneys of residential areas can also be called a plane source. Streets with a high traffic volume, e.g., represent line sources.

Table 1.3. Units of measurement in air pollution control (s. also [19])

Parameter	Unit	Description
Emission		
Mass concentration c	mg/m^3 (at 0 °C, 1013 mbar)	Pollutant mass per cubic meter exhaust gas in normal condition, usually dry
Volume concentration c_v	cm^3/m^3, also ppm v $1\ cm^3/m^3 = 1$ ppm v	Pollutant volume per exhaust gas volume, parts per million
Emission mass flow	kg/h t/a	Emitted pollutant mass per unit of time
Emission factor	mg/kg kg/t	Emitted pollutant mass per mass of fired fuel or produced quantity
Emission factor	kg/TJ	Kilograms per terajoule; emitted pollutant mass per thermal power (in furnaces)
Emission factor	g/km	Emitted pollutant mass per kilometer driven (in motor vehicles)
Air quality concentrations		
Mass concentration c	mg/m^3 $\mu g/m^3$ 1000 µg=1 mg	Pollutant mass per cubic meter of air
Volume concentration c_v (mixing ratio)	cm^3/m^3, also ppm v ppb ($\mu l/m^3$)	Pollutant volume per air volume; parts per million; parts per billion
Pollutant concentration in rain water	mg/l µg/l	Pollutant mass per liter of rain water
Pollutant deposition (e.g., deposited particulate matter, but also wet precipitation and gases)	$mg/m^2/day$ $\mu g/m^2/day$	Deposited pollutant mass per surface area and time
Pollutant dose	$mg/m^3 \cdot h$	Concentration times time ($c \cdot t$)
Dose of effect	µg/kg	Received (effective) pollutant mass per acceptor (recipient) mass

Units of measurement

Table 1.3 shows an overview of the units of measurement used in air pollution control.

Conversion of volume concentration in mass concentration.

Noxious gas concentrations are often measured as volume concentrations. Most of the time, however, mass concentrations are of interest. The conversion is carried out as follows:

$$1 \text{ mg/m}^3 = 1 \text{ cm}^3/\text{m}^3 \cdot \rho = 1 \cdot \frac{cm^3 \ pollutant \cdot mg \ pollutant}{m^3 \ air \cdot cm^3 \ pollutant}$$

ρ = gas density

$$\rho = \frac{molar \ mass}{molar \ volume} \text{ g/l or mg/cm}^3$$

$$\rho = \frac{molar \ mass}{22.4} \text{ mg/cm}^3 \text{ (at 0 °C, 1 013 mbar)}$$

$$\rho = \frac{molar \ mass}{24} \text{ mg/cm}^3 \text{ (at 20 °C, 1 013 mbar)}$$

1.4. Bibliography

1 Stern, A. et al.: Fundamentals of air pollution. 2nd Ed., Orlando: Academic Press 1984
2 Spiegelberg, F.: Reinhaltung der Luft im Wandel der Zeit. VDI-Komm. Reinhaltung der Luft, Düsseldorf: VDI 1984
3 Mieck, I.: Luftverunreinigungen und Immissionsschutz in Frankreich und Preußen zur Zeit der frühen Industrialisierung. Technikgeschichte 48 (1981) 239–251
4 Smith, R.A.: Air and rain. The beginnings of a chemical climatology. London: Longmans, Green & Co 1872
5 Wislicenus, H. (Hrsg.): Waldsterben im 19. Jahrhundert, Sammlung von Abhandlungen über Abgase und Rauchschäden. Reprintausgabe Parey, Berlin, 1908–1916/Einf. von d. VDI-Komm. Reinhaltung der Luft. Düsseldorf: VDI 1985
6 Leithe, W.: Die Analyse der Luft und ihre Verunreinigungen. Stuttgart: Wissenschaftliche Verlagsgesellschaft 1974
7 Dreyhaupt, F.J.: Smog-Alarm, Umwelt 8/86, 522–526
8 Giebel, J.; Bach, R.-W.: Ursachenanalyse der Immissionsbelastung während der Smogsituation am 17.1.1979. Schriftenreihe der Landesanstalt für Immissionsschutz des Landes Nordrhein-Westfalen, H. 17, 60–73. Essen: W. Giradet 1979
9 Baumüller, J.; Reuter, U.; Hoffmann, U.: Analyse der Smog-Situation in Stuttgart – Januar 1982. Landeshauptstadt Stuttgart, Chemisches Untersuchungsamt, Mitteilung Nr. 4/1982
10 Külske, S.; Pfeffer, H.-U.: Smoglage vom 16.–20. Januar 1985 an Rhein und Ruhr. Staub-Reinhaltung der Luft 45 (1985) Nr. 3, 136/141
11 Bundesministerium des Innern: Umweltprogramm der Bundesregierung, Referat für Öffentlichkeitsarbeit des Bundesinnenministeriums Bd. 9, Bonn 1971
12 Bundesministerium des Innern: Umweltplanung, Materialien zum Umweltprogramm der Bundesregierung 1971, zu Bundestags-Drucksache VI/2710, 6. Wahlperiode, Bonn, 1971

13 Neftel, A. et al.: Sulphate and nitrate concentrations in snow from South Greenland 1895–1978. Nature 314 (1985) 611
14 Graßl, H.: Klimaveränderung durch Spurenstoffe. Energiewirtschaftliche Tagesfragen 37 (1987) 127–133
15 Schilling, H.D.; Hildebrandt, R.: Die Entwicklung des Verbrauchs an Primärenergieträgern und an elektrischer Energie in der Welt, in den USA und in Deutschland, Bd. 6 Rohstoffwirtschaft International. Essen: Glückauf 1977
16 N.N.: Energy statistics yearbook. UN-Dept. of Int. Econ. and Soc. Aff., New York 1960–1985
17 N.N.: The prospects of world urbanization. UN-Publ. No. 87133, New York 1987
18 Meadows, D.: Die Grenzen des Wachstums. Stuttgart: Deutsche Verlagsanstalt 1972
19 Umweltbundesamt: Was Sie schon immer über Luftreinhaltung wissen wollten. Stuttgart: W. Kohlhammer 1986

2 Origin and Sources of Air Pollution

man-made

The most important anthropogenic source groups of air pollution are industrial furnaces and industrial processes, traffic, small-scale businesses and domestic furnaces as well as special sources such as animal confinement systems, spray cans etc.

A major proportion of the pollutants caused in these different areas has its origin in combustion processes, either in industrial and domestic furnaces or in traffic from combustion engines and aircraft engines. For this reason pollution formation during combustion processes will be looked into more closely in the following chapter.

2.1 Emission of Pollutants Caused by Combustion Processes

When burning fossil fuels for generating heat or power a wide variety of air polluting substances can be created. Type and amount of the air pollutants emitted depend on the type of combustion process, on the fuel used and on processing of combustion. For the area of furnaces these influencing factors and the possible emissions of air pollutants with their interdependence are diagrammatically shown in Fig. 2.1. In a general sense the same is true for combustion engines; however, other fuels – gasoline and diesel – containing fewer impurities are used. The following overview describes the individual pollutant groups with their corresponding origins.

2.1.1 Products of Complete and Incomplete Combustion

The fossil fuels gas, gasoline, fuel oils and coals mainly consist of hydrocarbon compounds with varying C/H ratios.

During complete combustion the carbon in the fuel reacts with oxygen to become carbon dioxide, the hydrogen to become water according to the following gross reaction equation:

$$C_nH_m + \left(n + \frac{m}{4}\right)O_2 \rightarrow nCO_2 + \frac{m}{2}H_2O. \tag{2.1}$$

Depending on the type of combustion process and fuel used, technical processes need a certain amount of excess oxygen or air for complete combustion. Lack of air causes incomplete combustion with higher pollutant emissions.

16   2 Origin and Sources of Air Pollution

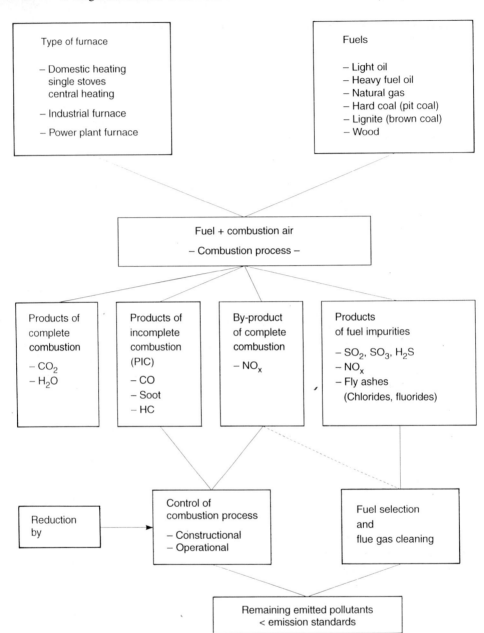

Fig. 2.1. Parameters influencing pollutant emission spectrum and level using the example of different types of furnaces

2.1 Emission of Pollutants Caused by Combustion Processes

Excess of air during combustion processes is defined as follows:

$$\lambda = \frac{actual\ amount\ of\ air\ A}{stoichiometrically\ required\ amount\ of\ air\ A_{min}}. \tag{2.2}$$

The minimum amount of air A_{min} can be determined by applying the relevant combustion calculations. The same is true for the theoretical amount of exhaust gas V_{min} [1].

As the actual amount of combustion air A can in practice only be determined with difficulty, excess air is determined alternatively from the remaining oxygen content O_2 in the exhaust gases or from the carbon dioxide content CO_2; complete combustion is, however, a precondition for this:

$$\lambda = 1 + \frac{V_{min}}{L_{min}} \cdot \frac{O_2}{21 - O_2}. \tag{2.3}$$

If pure carbon is burned completely, only CO_2 results, whereby one mol CO_2 is formed from a mol O_2 [equation (2.1): n molecules $O_2 \rightarrow$ n molecules CO_2]. In this case the exhaust gas volume V_{min} is equal to the air volume A_{min}. However, if water vapor is also formed, then V_{min} is no longer equal to A_{min}. Where fuels with a high C/H ratio are concerned $V_{min} = A_{min}$ can be assumed in estimated calculations. In this case the following applies:

$$\lambda = \frac{21}{21 - O_2}. \tag{2.4}$$

During complete combustion the maximum possible reduction of CO_2 is effected exclusively via the dilution of the exhaust gases with excess air. Thus, the CO_2 content can also be used to determine excess air. For this, however, the maximum possible CO_2 content $CO_{2\ max}$ must be known. During stoichiometrical combustion of pure carbon n molecules O_2 become n molecules CO_2; thus the $CO_{2\ max}$ content amounts to 20.93 percentage of volume.

In fuels consisting of hydrocarbons the $CO_{2\ max}$ will be lower depending on the hydrogen; for light fuel oil, e.g., it amounts to 15.4 percentage of volume (Vol. %).

The following equations are valid for determining the excess air λ from the CO_2 content of the exhaust gases:

$$\lambda = 1 + \frac{V_{min}}{L_{min}} \cdot \frac{CO_{2\ max} - CO_2}{CO_2} \tag{2.5}$$

or estimated with the simplified assumption $V_{min} = A_{min}$:

$$\lambda = \frac{CO_{2\ max}}{CO_2}. \tag{2.6}$$

Excess air is an important factor for the low pollution conduct of combustion processes. It will play a part in many of the following considerations. Lack of air causes incomplete combustion. Apart from this factor, the following causes can impair the completeness of the combustion:
- inadequate mixing of fuel and combustion air (local lack of air),
- insufficient pulverizing of solid fuels or atomizing of liquid fuels,
- sudden cooling of the flame gases through combustion chamber walls,
- residence time at high temperatures too short,
- flames burning in lifted condition, escaping of intermediate products from under the flame base.

During incomplete combustion the following pollutants can be formed and reach the environment in the form of emissions: carbon monoxide (CO), hydrocarbons (C_nH_m), oxidized hydrocarbons, and others such as odorous substances and soot (products of incomplete combustion: PIC).

2.1.1.1 Carbon Monoxide

Carbon monoxide (CO) is the intermediate product in the combustion process of carbon to CO_2. CO oxidation requires a so-called ignition temperature of at least 990 K, and for a complete burn-up a certain residence time at this temperature is required. If combustion temperature and residence time in the flame are insufficient, or if there is a lack of air, part of the CO can escape into the exhaust gas.

Using oil furnaces as an example, the dependency of CO emission on air or oxygen excess during combustion is shown in Fig. 2.2. As can be seen, there is a dramatic increase of the exhaust gases' CO content when there is not enough air.

How high the minimum air excess has to be so that combustion is as complete as possible depends on the combustion temperature, on the residence time and on how well mixed the fuel-air mixture is.

The dependency of CO emission on temperature is shown in Fig. 2.3. A room stove with an oil evaporation burner is used as an example. As can be seen, there is a clear CO emission increase in the lower temperature range.

In manually stoked furnaces, e.g., in wood-burning central heating boilers, burning conditions are changing continuously. After charging there is an initial intensive fire with high temperatures. Most of the time, however, combustion air is not sufficient for CO and hydrocarbon emissions to be released. In the later stages of combustion the amount of fuel is reduced and excess air increases. Simultaneously, the combustion temperature decreases steadily causing CO emissions to rise again. A combustion process like the one described is shown in Fig. 2.4.

2.1.1.2 Hydrocarbons

If hydrocarbons are not totally oxidized during combustion, a variety of substances can appear in the exhaust gas, e.g., alcohols, aldehydes or organic acids. Hydrocar-

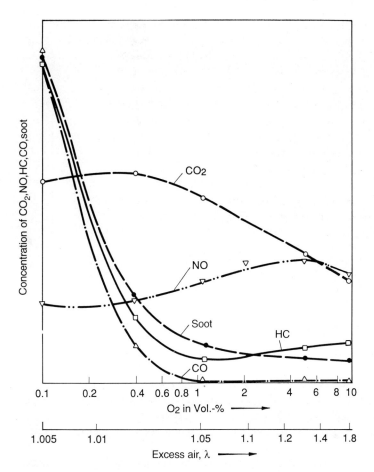

Fig.2.2. Typical course of pollutant emissions in the exhaust flue gas of oil furnaces versus excess air [2]

bons, for instance, can be oxidized to carbon dioxide and water via the following more or less stable oxidation states which are also possible emissions components.

$CH_4 \rightarrow$	$CH_3OH \rightarrow$	$HCHO \rightarrow$	$HCOOH \rightarrow$	$CO \rightarrow$	CO_2, H_2O
methane (hydro-carbon)	methanol (alcohol)	form aldehyde	formic acid	carbon monoxide	carbon-dioxide water

During incomplete combustion or insufficient mixing of fuel and air in the flame, part of the fuel can escape unburned along with the exhaust gas [4, 5]. In contrast to that, if there is a lack of air, thermal decomposition (pyrolysis) can set in. This decomposition process either takes place via the reaction of a partial oxidation

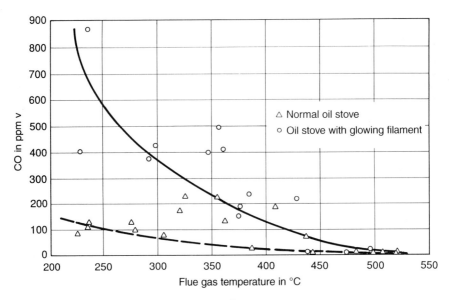

Fig. 2.3. CO emission of an oil stove versus exhaust gas temperature

described above or, e.g., it leads to the formation of new hydrocarbons not originally contained in the fuel via the separation of hydrogen atoms. This is also how aromatic (e.g., benzene, toluene, xylenes (BTX)) and polycyclic aromatic hydrocarbons (PAH), some of which are known to be carcinogenic, are formed. Fig. 2.5 shows a diagram of BTX and PAH formation according to Badger [5, 6].

Using this type of polycyclic formation, the reaction sequence can be summarized as follows:
1. Addition of smaller aliphatic compounds and cyclization into hydroaromatic hydrocarbons with medium-sized molecules,
2. transformation of hydroaromatic hydrocarbons into fully aromatic hydrocarbons,
3. formation of larger polycyclic aromatic hydrocarbons from smaller ones.

Investigations [7] show that apart from the type of the pyrolyzed material, the composition of pyrolytic products depends strongly on the combustion temperature.

Fig. 2.6 shows the structural formula of the well-known 3,4-benzopyrene and of a nitrated polycyclic aromatic hydrocarbon. Aromatic and polycyclic aromatic hydrocarbons occur as pyrolytic products in cigarette smoke and in automobile and furnace exhaust gases [8]. They are known for their carcinogenic effect. Oxidized and nitrated PAH [7], e.g., in the exhaust gases of diesel engines, are also significant.

As an example, the hydrocarbon emissions of a central heating boiler with an atomizer oil burner are shown with the help of a gas chromatogram of the exhaust gas in Fig. 2.7. As a comparison a gas chromatogram of the burned light fuel oil is shown (Fig. 2.7a).

On the right-hand side of the exhaust gas chromatogram unburned fuel oil components can be observed. New substances created during pyrolsis are to be found on the left, especially lower aromatic hydrocarbons (BTX).

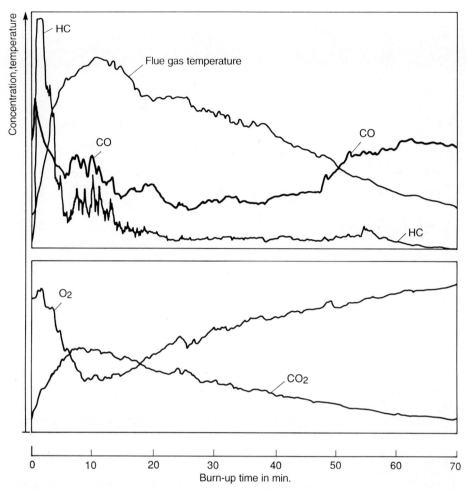

Fig. 2.4. CO and H emission of a wood-fired central heating boiler during a burn down from fresh charge to embers [3]

Apart from a faulty burner setting, the processes of turning the burner on and off cause emissions of organic substances in oil furnaces [9, 10].

Hydrocarbon emissions of furnaces are of note mainly owing to their disagreeable odour. It is not easy to assign odorous emissions directly to the hydrocarbon emissions determined by gas chromatography as the substances combinations are complicated. Even substances at low, hardly measurable concentrations can contribute considerably to emission with disagreeable odours. Nevertheless, one can generally establish a connection between the emissions of unburned substances – CO, C_nH_m and soot – and the odor intensity of the exhaust gas, s. Fig. 2.8: If the exhaust gas contains many products of incomplete combustion, then the odor intensity of the exhaust gas is high [11].

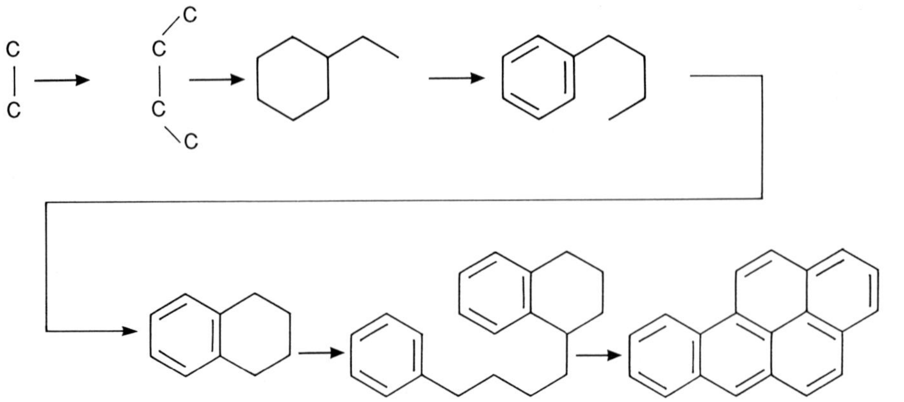

Fig. 2.5. Reaction diagram for the formation of polycyclic aromatic hydrocarbons (PAH) during combustion processes

3,4 Benzopyrene

Nitrated PAH: 1-Nitro-pyrene

Polycyclic aromatic hydrocarbons (PAH)

Fig. 2.6. Structural formulas of a polycyclic aromatic hydrocarbon (3,4 benzopyrene) and of a nitrated PAH (1-nitro-pyrene) [7, 8]

Formaldehyde will be used here as an example for the emission of oxidized hydrocarbons during combustion processes.

As an example Fig. 2.9 shows the formaldehyde emission of a central heating wood boiler depending on the CO content of the exhaust gas. One can see a definite correlation. If carbon monoxide occurs during the incomplete combustion of the wood, then one must also assume the presence of formaldehyde emissions.

During combustion processes which are particularly prone to incomplete combustion, e.g., wood stoves in private homes or engine combustion in automobiles, oxidated hydrocarbons occur to a greater extent [12, 13], they manifest themselves by characteristic odorous emissions.

Table 2.1 shows the emission rates of different aldehydes in milligram/mile measured in three different automobiles. The values were obtained with the US Test 75.

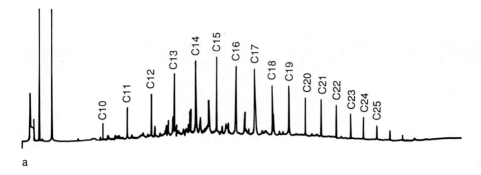
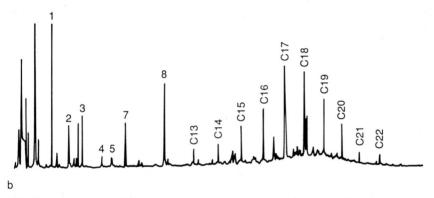

Fig. 2.7. Hydrocarbon emissions of a central heating boiler with an atomizer oil burner, the C_i represent alkanes with their corresponding carbon number. **a** gas chromatogram of the light fuel oil burned; **b** gas chromatogram of the organic substances contained in the flue gas. Burner setting: slight lack of air; the peak numbers mean: *1* toluene, *2* ethylbenzene, *3* xylenes, *4* propylbenzene, *5* ethyl-methylbenzene, *7* trimethylbenzene, *8* naphthalene

2.1.1.3 Soot

If fuels containing hydrocarbons are heated in the absence of oxygen, thermal decomposition processes ensue. In the course of these processes the hydrogen is split off and soot can be the resulting final product. It consists of agglomerates of elementary carbon and, in part, of hydrocarbons. For chain-like initial substances one may conceive of the formation of soot via acetylenes (hydrocarbons with a triple bond) and polyacetylenes. Ultimately, soot particles can be formed via ring closures, aromatic compounds and polycyclic aromatic hydrocarbons [14, 15]. S. also Fig. 2.5.

Soot formation is favored by lack of oxygen in the flame base, e.g., by inadequate mixing of fuel and air and by high temperatures in this phase. It depends not only on burning conditions but also on the fuel used. If there is a lack of air fuels with a high C/H ratio are rather more prone to soot formation than fuels with a low

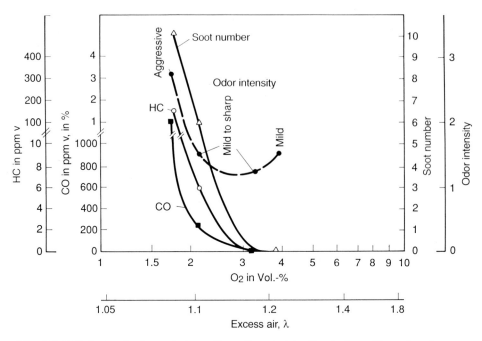

Fig. 2.8. Pollutant and odor emissions of an oil central heating boiler with modern burner versus excess air

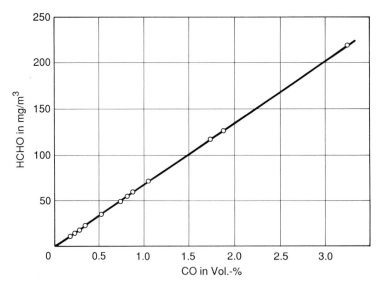

Fig. 2.9. Correlation between formaldehyde and carbon monoxide emissions in a wood-fired central heating boiler

Table 2.1. Emission rates of different aldehydes and ketones as determined by US-test 75 in three different automobiles (acc. to [13]

Components	Without Catalytic Converter Automobile no.3		Three-way Catalytic Converter Automobile no.6		Diesel Automobile no.16	
	mg/mi	%[a]	mg/mi	%[a]	mg/mi	%[a]
Formaldehyde	38	60.4	4	47.6	18	41
Acetaldehyde	10	15.9	2	23.8	11	25.1
Acrolein	2	3.2	1	12	3	6.8
Acetone	3	4.8	1	12	5	11.4
Propionaldehyde	< 0.5	< 0.8	< 0.1	< 1.2	3	6.8
MEK + Isobutyric aldehyde	3	4.8	< 0.1	< 1.2	2	4.6
Crotonic aldehyde	< 0.2	< 0.3	< 0.1	< 1.2	1	2.3
Benzaldehyde	6	9.6	< 0.1	< 1.2	0.8	1.8
Total aldehydes (DNPH)	63	100	8.4	100	44	100

[a] proportion of total aldehydes, determined by the DNPH method

C/H ratio. Thus, when burning natural gas which consists largely of methane (CH_4), only a slight soot formation will occur even with a lack of air, whereas with an acetylene flame (ethine, C_2H_2) large flakes of soot will be released immediately when there is a lack of air (s. autogenous welder).

One can see by the color of the flame whether soot is being formed in this flame. The burning soot particles have a radiation behavior resembling a black body. Depending on the temperature the flame has a dark yellow to a light yellow hue. On the other hand, flames without soot formation and without soot combustion release a pure gas radiation, e.g., with a transparent blue hue.

Depending on the conditions in the flame (temperature, mixture and excess of air) the resulting soot burns out more or less completely. As an example for oil flames Fig. 2.2 also shows soot emissions as a function of air excess. It can be seen that when there is a lack of air soot emissions increase sharply. Soot formation in the flame base is the precondition of this type of behavior. There are oil flames, so-called blue burning flames, which do not emit any soot even with a lack of air but merely CO and hydrocarbons [2].

2.1.1.4 Hydrocarbon Emissions in Different Combustion Processes

Efforts have been made to show that in combustion processes emissions of incomplete combustion products – CO, hydrocarbons and soot – depend on a variety of factors. Quite generally it can be said that the more stable the flame burns and the

better it is set and supervised, the lower these emissions are. Different combustion processes producing incomplete combustion products can be listed in approximately the following increasing order of magnitude:
1. large industrial furnaces with control and supervising equipment,
2. small industrial furnaces with control and supervising equipment,
3. domestic furnaces with oil or gas blowpipe burners,
4. automobiles (with catalytic converters),
5. domestic furnaces with non-stationary combustion conditions: single stoves for oil, coal, wood or solid fuel central boilers,
6. automobiles without catalytic converters.

The technical developments aim at improving and regulating the individual combustion processes, so that combustion can be effected as completely as possible and CO, C_nH_m and soot emissions are minimized. Exhaust gases of automobiles will be looked into more closely in a later section.

Quantifying products of incomplete combustion is hardly possible, particularly where furnaces are concerned, as emissions depend on the momentary setting and operation of the different combustion processes. If, despite this, numerical data are to be found in the literature, then these values were based on individual furnaces and cannot necessarily be applied to all other furnaces.

2.1.2 Sulfur Compounds

Sulfur dioxide is formed during the combustion of fossil fuels containing sulfur compounds. When these fossil fuels – coal, mineral oil and natural gas – were formed, nitrogen compounds and also sulfur compounds found their way into these fuels via the amino acids which are the fundamental components of the plant protein. Methionine and cysteine will be shown here as two examples of sulfurous amino acids :

$$\begin{array}{cc} \text{COOH} & \text{COOH} \\ | & | \\ \text{H}_2\text{N}-\text{C}-\text{H} & \text{H}_2\text{N}-\text{C}-\text{H} \\ | & | \\ \text{CH}_2-\text{CH}_2-\text{SCH}_3 & \text{CH}_2\text{SH} \\ \\ \text{methionine} & \text{cysteine} \end{array}$$

2.1.2.1 Sulfur in Coal

Depending on the transformation stage of the fuels, the organic sulfur components, too, were transformed into different compounds. The older the substances, the more the organic compounds were mineralized. In the oldest fossil fuel, mineral coal,

Table 2.2. Mean composition of mineral coals and lignites (acc. to [16])

Fuel	Carbon C mass-%	Hydrogen H mass-%	Oxygen Nitrogen O + N (N=1% assumed) mass-%	Sulfur S mass-%	Water H_2O mass-%	Ash mass-%	Heating Value h_u kJ/kg
Westphalian Min. Coal	79	4.5	7	1	2.5	6	31 400
Saar Min. Coal	74	4.5	10	1	3.5	7	29 300
Silesian Min. Coal	71	4.5	12.5	0.5	5	6.5	27 630
Saxon Min. Coal	70	4	9.5	1	8	7.5	27 215
Bavarian Min. Coal	53	4	12	5	9	17	21 770
Coke	84	0.8	3.4	1	1.8	9	29 310
Crude Lignite							
Lower Rhine	23	1.9	12.1	1	59.3	2.7	8 120
Upper Palatinate	25.3	2	12	1	52.7	7	8 540
Bohemia	52	4.2	13	1	24	6	20 100
Saxony	40	3	11	2	37	7	15 070
Rhine Lignite, Lignite Briquets	54.5	4.2	20.4	0.4	15	5	20 100
Middle German Lignite Briquets	52	4.3	16	2	17	9	20 100

sulfur compounds can in the most extreme of cases be present entirely in anorganic form, e.g., as pyrite sulfur, as sulfides or sulfates. The higher the content of volatile components, i.e., hydrocarbons, in the coal, the bigger the content of organic sulfur compounds.

Table 2.2 shows the composition of different coals. The varying levels of sulfur are evident.

2.1.2.2 Sulfur in Fuel Oil

In the mineral oils where carbon appears exclusively in the form of hydrocarbons, the sulfur compounds correspondingly occur in an organically bound state, e.g., in the form of mercaptanes. Crude oils have a varying sulfur content, depending on their origin, s. Fig. 2.10.

Organic sulfur compounds have substantially higher boiling points than their corresponding pure hydrocarbons, e.g.:

C_2H_5—SH $\qquad$ C_2H_6
ethyl mercaptane $\qquad$ ethane
boiling point: 35 °C $\qquad$ boiling point: –88.6 °C.

For this reason sulfur compounds accumulate in the heavier fractions during refinement; in mineral oil products sulfur content increases accordingly in the following order:

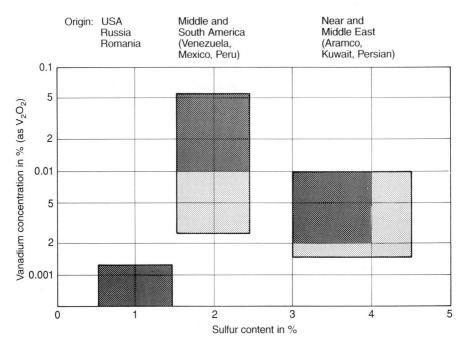

Fig. 2.10. Sulfur and vanadium content of crude oils of different origins [16]

gas, liquid gas→	crude gasoline→	intermediate distillate→	atmospheric
CH_4 to C_4H_{10}		e.g., kerosine, light fuel oil, Diesel	residue heavy fuel oil
up to 0 °C very little sulfur	up to 200 °C up to 0.05 % sulfur	175 to 350 °C 0.2 % to 0.8 % sulfur	>350 °C 1 to >3 % sulfur

Besides the selection during the refining process the sulfur content of the different fractions stems primarily from the sulfur content of the crude oil, s. Table 2.3. In the FRG the sulfur content of light fuel oil and diesel must not be in excess of 0.2 % [17]. This sulfur content is either kept low from the outset owing to the low sulfur content of the crude oil, or it is obtained by mixing various oils of different origin or, in the least desirable case, by partially desulfurizing the fuel oil [18].

Heavy fuel oils can be obtained with differing degrees of sulfur content; sulfur content can range from 0.7 (in special cases 0.3) and 1.0 up to >2 %. Oils with a low sulfur content (0.7 and 1.0 %) are expensive and less easily available than the ones with a higher sulfur content.

2.1.2.3 Sulfur in Natural Gas

In natural and synthetic gaseous fuels sulfur occurs in the form of hydrogen sulfide (H_2S).

Table 2.3. Overview of the sulfur content of different fuels after their technical purification (acc. to [16-20] and information from gas suppliers)

Fuel	Sulfur Content, Mass Ratio in %
Natural gas	0.0005-0.02
Liquefied gas, $C_3H_8+C_4H_{10}$	not detectable
Gasoline	0.001-0.06
Light fuel oil, diesel	up to 1; in Germany limited to 0.20
Heavy fuel oil	0.7 to >2
Wood, pure	not detectable
Wood bark	<0.15
Mineral coal, depending on its origin	$\begin{cases} 0.5-2 \\ 0.8-1 \text{ mean value} \end{cases}$
Coke	0.6-1
Rhine lignite	1
Saxon lignite	2

As it is both a health hazard and causes corrosion in gas pipes, H_2S is to a large extent removed from fuel gases through alkaline gas washing or aqueous solutions of amines even before it is distributed. The bound hydrogen sulfide will then be expelled by heating and is processed further in the Claus process. Hereby H_2S is first oxidized to SO_2, which is subsequently used as an oxidizing agent for more H_2S [16]:

$$2H_2S + 3O_2 \rightarrow 2H_2O + 2SO_2 \tag{2.7}$$

$$2H_2S + SO_2 \rightarrow 2H_2O + 3S. \tag{2.8}$$

Accordingly, natural gas contains only very small amounts of sulfur compounds, unless high-sulfur gases are used for purposes of combustion, e.g., directly at the locations of distribution.

2.1.2.4 Comparison of Sulfur Contents of Different Fuels

Table 2.3 compiles the sulfur contents of different fuels after their technical purification.

2.1.2.5 Combustion Products of Sulfurous Fuels

Sulfur dioxide SO_2

If the sulfur contained in fuel is burned completely, sulfur dioxide is formed, e.g., according to the following reaction:

$$CH_3-SH+3O_2 \rightarrow SO_2+CO_2+2H_2O \qquad (2.9)$$

SO_2 is a colorless gas with a pungent odor; it can be detected in the air from approx. 0.6-1 mg/m^3 by its smell.

Hydrogen sulfide H_2S

During incomplete combustion, e.g., when there is a lack of air, elementary sulfur (S) or hydrogen sulfide (H$_2$S) can, depending on how high the temperature is, be formed under reducing conditions from the fuel's sulfur compounds:

$$CH_3-SH+0,5O_2 \rightarrow H_2S+HCHO \qquad (2.10)$$

$$2H_2S+O_2 \rightarrow 2H_2O+2S. \qquad (2.11)$$

In most combustion processes, the reduced sulfur compounds in exhaust gas are insignificant. It is only where smouldering fires are concerned, e.g., when brown-coal briquettes in domestic stoves are burned with the air flap closed for a low flame, that H$_2$S formation can certainly be of significance. H$_2$S emissions can be recognized by their characteristic rotten egg smell. The smell detection threshold is very low at approx. 0.002 mg/m^3 [21].

Even if H$_2$S emissions play an insignificant role in the sulfur balance of combustion, they can lead to annoying odors even in the smallest of amounts. The smell of H$_2$S is quite common in cities where brown-coal furnaces are widespread.

In recent times H$_2$S emissions have been observed in automobiles with catalytic converters: When running on excessively rich mixtures, e.g., when the vehicle is fully loaded, sulfur compounds in the catalytic converter can be reduced to H$_2$S and thus lead to emissions with a disagreeable smell. The emission of H$_2$S is influenced by the gasoline's sulfur content, even if it is low.

Sulfur trioxide SO_3

During combustion or in the flue gas channels of furnaces part of the SO$_2$ can be oxidized to SO$_3$. In combination with water vapor sulfuric acid (H$_2$SO$_4$) in the form of aerosol is formed from SO$_3$. Due to this chemism, SO$_3$ is an even stronger irritant than SO$_2$. The smell threshold value is at approx. 0.6 mg/m^3 [22].

According to the law of mass action, quite high SO$_2$-SO$_3$ transformation rates result [23] during the cooling of the gases in a temperature range of 800-400 °C with sufficient excess of air. However, the reaction speed constants of the possible reactions are not great enough for the theoretically possible SO$_3$ content to form during the rapid cooling of the flue gases on their way through the flue gas channels. The balances are most frequently frozen at a very low SO$_3$ content. If measurable SO$_3$ concentrations occur at all in the flue gases, it is usually due to the presence of catalytic materials in the exhaust channels which accelerate SO$_2$-SO$_3$ transformation. Vanadium pentoxide is known for its effect here, e.g., as particles in the flue gases of heavy oil furnaces.

2.1 Emission of Pollutants Caused by Combustion Processes

Depending on the formation conditions (with catalytic particles and excess air being present during combustion), the SO_3 content, e.g., in the flue gas of heavy oil flames can range between approx. 5-60 mg/m^3 [24]. This corresponds to a SO_3 percentage of less than 1 % to approx. 4 % of the total SO_x emission.

Up to now SO_3 formation was mainly taken notice of because sulfuric acid forming with water vapor led to corrosion in the flue gas channels of furnaces. Under certain conditions "sulfuric acid rain" can sometimes occur in the vicinity of stacks of individual problematic furnaces.

Sulfur emission rate

When determining SO_2 emissions of combustion processes, the question as to what extent the sulfur which enters into the combustion with the fuel reappears in the exhaust gases as SO_2 arises. Sulfur not emitted as SO_2 can either be bound in the ash (in coal furnaces) or be emitted as SO_3 or as H_2S (negligible). Binding in the ash depends on the combustion temperature and the properties of the ash (e.g., alkalinity or alkaline earth content). In coal fires a higher SO_3 formation would favor sulfur binding into the ash.

Various investigations have provided the sulfur dioxide emission rates compiled in Table 2.4 as a measure of the percentage of the fuel sulfur in gaseous combustion products (as SO_2). For this, the SO_2 emission rate was determined either by comparing SO_2 emission measurements and calculations of the SO_2 emissions from the fuel sulfur or from the sulfur content of the ash.

It may be assumed that oil furnaces have a sulfur emission of 100 %. With coal furnaces, however, and depending on the combustion conditions, sulfur emission can go down to 90-60 % due to the binding into the ash.

Table 2.4. Sulfur dioxide emission proportion ηSO_2 of the fuel sulfur content

Furnace	ηSO_2 (mean values)	References
Lignite		[22]
grate firing and pulverized fuel firing	$\begin{cases} 0.8 - 0.9 \\ approx.\ 0.5 \\ 0.6 - 0.8 \end{cases}$	[25] [26]
Mineral coal	0.95	[22]
pulverized fuel firing	1.0	[26]
grate firing	0.8	[26, 27]
Lignite briquets	0.9-0.95	[28]
Fuel oil	1.0	[22, 29]

2.1.3 Oxides of Nitrogen

2.1.3.1 Origins of Oxides of Nitrogen

Oxides of nitrogen (NO_x) are formed during combustion processes at high temperatures through oxidation of the nitrogen in the combustion air and through the combustion of the fuel-bound nitrogen. Primarily nitrogen monoxide NO is formed, whereas the more poisonous nitrogen dioxide is formed only after the combustion when there is an adequate oxygen content in the exhaust gases and finally in the atmosphere. In combustion processes involving a high air excess, e.g., in gas turbines or in diesel engines with low load, considerable NO_2 emissions must be expected. As all NO is ultimately transformed into NO_2, NO_x emissions ($NO+NO_2$) are indicated as NO_2, threshold values are also listed as NO_2.

There are three mechanisms of NO formation depending on the particular conditions of temperature and concentration, on the residence time and the type of fuel: thermal NO, prompt NO and fuel NO.

Thermal NO

The mechanism of the so-called thermal NO formation was discovered by Zeldovic. The concentration of oxygen atoms available during or after combustion is responsible for this type of NO formation. At over 1300 °C and with rising temperatures the O concentration increases considerably as a result of the dissociation of O_2. Consequently, the NO formation speed also increases. The two following reactions take place in high oxygen areas (excess of O_2!) of the flame or in the post reaction zone:

$$O + N_2 \rightleftarrows NO + N \tag{2.12}$$

$$N + O_2 \rightleftarrows NO + O. \tag{2.13}$$

In fuel-rich zones of the flame the following reaction takes place mainly at temperatures over 1300 °C:

$$N + OH \rightleftarrows NO + H. \tag{2.14}$$

The amount of NO formed is influenced by the following factors:
- air-fuel ratio in the reaction zone which influences the concentration of atomic oxygen. NO emissions generally decrease with decreasing air-fuel ratio;
- temperature in the reaction zone: apart from O_2 dissociation the reaction itself (2.12) is highly temperature-dependent, so that both influences lead to a clear increase in NO formation with rising temperatures [30];
- the gases' residence time in the reaction zone at a maximum temperature or mixing speed after the reaction with cooler reaction products: the shorter the residence time, the smaller the NO formation.

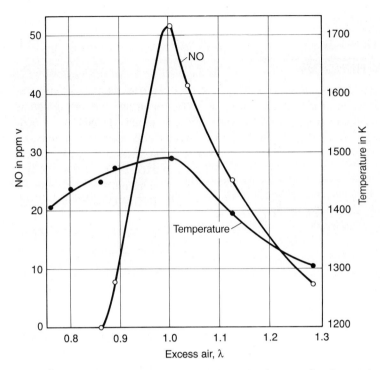

Fig. 2.11. Influence of excess air on NO formation in an enclosed natural gas premix flame (acc. to [30])

The influence of air excess and the resulting combustion temperature are illustrated in Fig 2.11 on the basis of results measured in a laboratory natural gas premix flame. Even at a slowly rising temperature one can see a dramatic increase of the NO emission with increasing air. When the flame is cooled by the excess air NO emission again decreases distinctly.

We are looking at an ideal laboratory flame here where complete combustion occurs at $\lambda=1$. In technical combustion processes complete combustion only occurs when there is a certain excess of air, so that the NO_x maximum also lies in a slight air excess range.

The dependency of the NO emission on air excess practically runs counter to the emission of unburned products, so that when there is a minimum of CO and C_nH_m emissions there is a maximum output of NO. (Compare Fig. 2.2)

Prompt NO

In the low-oxygen area of flames, NO can be formed via fuel radicals (e.g., CH) with molecular nitrogen – so-called prompt NO. This phenomenon was discovered by Fenimore. The following reactions are believed to lead to the formation of NO:

$$CH + N_2 \rightleftarrows HCN + N \qquad (2.15)$$

$$C + N_2 \rightleftarrows CN + N \qquad (2.16)$$

$$CN + H_2 \rightleftarrows HCN + H \qquad (2.17)$$

$$CN + H_2O \rightleftarrows HCN + OH \qquad (2.18)$$

$$HCN, CN + O \rightarrow NO + R. \qquad (2.19)$$

R = organic residue

The CN radical arising from fuel-nitrogen compounds is transformed further into HCN. Furthermore, atomic nitrogen is released. The cyanogen compounds and the nitrogen atoms can then be oxidized to NO. The temperature range where prompt NO formation sets in is shown in Fig. 2.13.

It is generally assumed that prompt NO formation in technical flames is of secondary significance [32]. In the fuel-rich area of flames, however, this type of NO formation can gain a certain significance via, among others, CN and N formation through heterogenous reactions with soot particles, s. equation (2.16), e.g., in oil atomizer burners in small heating boilers. By avoiding soot formation, e.g., by the so-called blue burning flames, NO formation can be reduced [2] in this case, too.

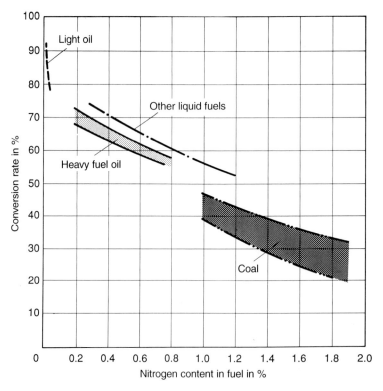

Fig. 2.12. Transformation of fuel nitrogen into NO, showing three fuels as examples [31]

Fuel NO

Fuel oils as well as coal contain organic nitrogenous compounds such as amines, R—NH_2, and amides, R—CO—NH_2, nitrogenous compounds such as nitrobenzene, $C_6H_5NO_2$, heterocyclic compounds such as pyridine, C_5H_5N, etc.. Fuel oil contains 0.1-0.6 % nitrogen, coal 0.8-1.5 %. Oxidation of these nitrogenous compounds is less temperature-dependent occuring even at lower temperatures (Fig. 2.13) and increasing with a higher excess of air:

$$\text{z.B.} \quad NH_2 + 1/2\,O_2 \rightarrow NO + H_2. \tag{2.19}$$

However, during combustion only part of the fuel nitrogen is transformed into NO. In fuel-rich areas of the flame not only fuel nitrogen but also thermally formed NO (e.g., in coke particles) can be reduced to molecular nitrogen [33]:

$$NH_3 \xrightarrow{C} \tfrac{1}{2}\,N_2 + 1\tfrac{1}{2}\,H_2 \tag{2.20}$$

$$NO + CO \xrightarrow{C} \tfrac{1}{2}\,N_2 + CO_2 \tag{2.21}$$

$$HCN + 2\,NO \rightarrow 1\tfrac{1}{2}\,N_2 + CO_2 + \tfrac{1}{2}\,H_2. \tag{2.22}$$

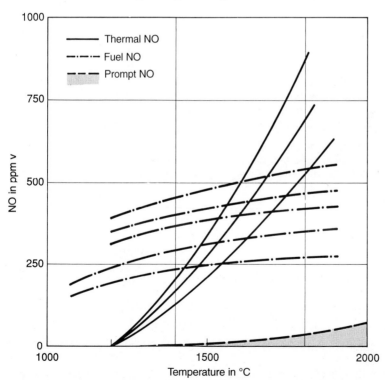

Fig. 2.13. Diagram of NO formation versus combustion temperature (acc. to [31]); parameter for thermal NO: O_2 excess

The lower the fuel's bound nitrogen content, the higher the transformation rate to NO, s. Fig. 2.12. Nevertheless, fuels with high N contents have higher NO_x emissions than those with low ones.

The difference in fuel nitrogen content is the reason why coal furnaces have, in principle, a higher NO emission than oil furnaces, and these again a higher emission than gas furnaces. NO formation from fuel nitrogen, however, is strongly overlapped by thermal NO formation. Even with a low fuel nitrogen content high NO emissions can occur, e.g., in automobiles.

Fig. 2.13 illustrates the influence of the combustion temperature on NO formation according to different formation mechanisms. As can be seen, fuel and prompt NO formation are much less temperature-dependent than thermal NO formation.

2.1.3.2 Nitrogen Oxide Emissions in Different Combustion Processes

Large-scale industrial furnaces, e.g., power plant furnaces, emit nitrogen oxides to a particularly high extent. In these cases NO_x emissions depend on the one hand on the type of combustion, particularly on how high combustion temperatures are and how long the residence time at these temperatures is. On the other hand, NO_x emissions are determined by the thermal load which has a retroactive effect on the furnace-heat capacity and thus on the combustion temperature. Fig. 2.14 illustrates the NO_x emission ranges of large-scale coal furnaces. As can be seen, the highest NO_x emissions occur in furnaces with high temperatures, in the smelting chamber (fluid cinder outlet) and cyclone furnaces. In dry furnaces (dry cinder outlet) of different designs and construction NO_x emissions are distinctly lower.

Table 2.5 compares the nitrogen oxide emissions of different combustion processes. Emissions are listed as concentration values in exhaust gas, as emission per amount of fuel burned and as emission per thermal capacity of fuel. So far, a comparison of the emissions of the different source groups is not yet possible in every respect. For further information, emissions in mass per unit of time, or in the case of automobiles, emissions in mass per unit of distance covered, as well as size and number of furnaces/automobiles and their respective fuel consumptions must be taken into account (see also Chap.6).

2.1.4 Particles

2.1.4.1 Problem, Dependencies and Components

When furnace chimneys or automobile exhaust pipes emit smoke, then this is visible due to the particles or droplets suspended in the exhaust gases. The fine particles, in combination with the carrier gas – exhaust gas or air – also called aerosol, are of particular significance as:
- they have a small mass but a large surface; this is why exhaust gases are visible even in small particle-mass concentrations; the layer of smog over large cities also stems from fine particles or drops which do not sediment;

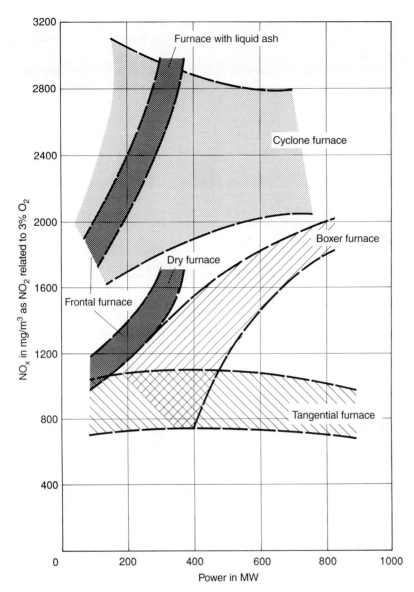

Fig. 2.14. NO_x emissions of different industrial furnaces (coal) in the USA without NO reduction measures (acc. to [34])

- they are respirable, they can have a toxic effect in the respiratory system itself or they can transport toxic substances into the lungs due to their absorptive properties,
- many particle precipitators have a low collection efficiency for fine particles, i.e., residual dust after such precipitators consists mainly of fine particles.

A great percentage of the particulate air pollutants have their origins in combustion processes with particle formation depending on different factors:

- on the type of fuel: fuel gases, gasolines, fuel oils, wood, coal,
- on the type of fuel preparation, e.g., atomizing or vaporization of liquid fuels, chopping up or pulverization of solid fuels,
- on the conditions of combustion, such as flame temperatures, mixing of fuel and air and on the amount of oxygen available.

Table 2.5. Nitrogen oxide emissions of various technical combustion processes without application of special reduction measures

Combustion process	NO_x as NO_2			References
	Concentration[a] in the exhaust gas mg/m^3	Emission per fuel g/kg	Emission per thermal capacity of fuel mg/kWh	
Power plant furnaces				
Mineral coal pulverized fuel firing				
melting furnaces	1200...3000	13 ...30	1400...3600	[19,31,34]
dry furnaces	700...1800	8 ...20	900...2300	[19,31,34,35]
Lignite	600...1000	4 ... 8	850...1400	[31,36]
Industrial grate furnaces	150.....650	2 ... 9	200... 950	[19,27]
Fluidized bed furnaces				
stationary	100...1000	1.2...11.6	140...1400	[37,38,39]
circulating	80.....300	0.9... 3.5	100... 410	[40]
Industrial oil furnaces	300...1100	3.5...13	300...1200	[2,19,41,42]
Industrial gas furnaces	100.....800	0.4... 3.4[b]	85... 700	[19]
Domestic furnaces				
oil fan burner	80... 250	1 ... 3	80... 260	[2,43,44]
gas fan burner	60....170	0.2... 0.7[b]	50... 150	[43]
Gas, atmospheric burner	100....200	0.4... 0.9[b]	85... 170	[45]
Motor vehicles[c]				
automobiles with Otto engines	1000...8000	10 ...84	900...7000	[46,47]
idling	20... 50	0.2... 0.6	18... 50	[46,47]
with catalytic converter and λ control	40... 400	0.4... 4	35... 350	
Automobiles and trucks with diesel engines	400...3000	12 ...40	1000...3500	[46,48,49]
idling	20... 50	0.8... 2	70... 180	

[a] The exhaust gas concentrations listed in some cases refer to varying air excess numbers (=varying exhaust gas dilution); the resulting deviations are within the ranges of the given spans.
[b] g/m^3 gas
[c] In the case of motor vehicles pollutant emissions are usually indicated in mass per distance driven. For the comparison of exhaust gas concentrations of different combustion processes carried out here only sources containing information on pollutant concentrations could be used, e.g., from roller dynamometer investigations. NO_x emissions usually indicated in ppm v were converted into mg/m^3 with the help of NO_2 density (NO_2=2.05 mg/cm^3).

The particles can also consist of the following components:
- soot,
- high-molecular, condensed hydrocarbons, e.g., tar; these substances can also be adsorbed to the soot particles,
- ash particles, e.g., oxides of metals,
- also unburned carbon particles in solid fuels.

2.1.4.2 Soot and Particle Emissions during the Combustion of Liquid Fuels

Soot formation has already been dealt with in Chap. 2.1.1. As soot is formed during the gaseous phase, soot formation can, in principle, also occur during the combustion of gases. However, soot emissions during the combustion of liquid and also solid fuels are of greater significance. As has already been mentioned, not every fuel is equally prone to soot formation. Soot consists of very fine particles agglomerating into larger ones after the flame. As a general rule, the more complete the combustion, the lower the soot emission, but also the finer the particles will be. In the combustion of heavy fuel oil particle emission can be reduced by so-called fuel oil additives which help to reduce soot formation in the flame and to improve the burning. At the same time these additives shift particle-size range to finer particles in the exhaust gas [50], s. Fig. 2.15. It shows the size distribution of the emitted particles of heavy fuel oil combustion with and without additives. Approx. 25 % of the dust particles emitted during the combustion without additives are mineral ash components, the re-

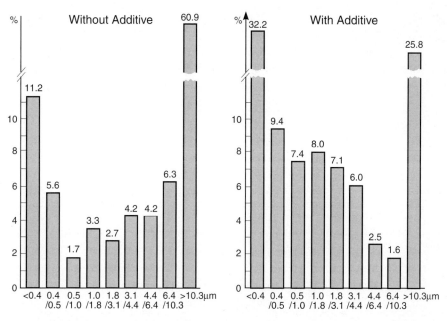

Fig. 2.15. Particle-size spectrum of the emitted flue dust of heavy fuel oil combustion in a power plant furnace without and with additive (from cascade-impactor measurements) [50, 51]

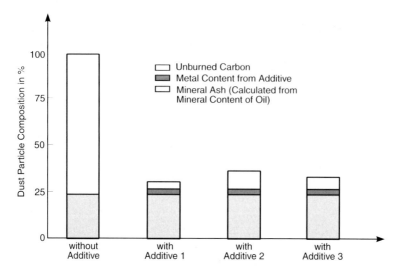

Fig. 2.16. Example of flue gas particle composition of heavy fuel oil combustion in a power plant furnace without and with additive [50, 51]

mainder is soot, or flue coke, see Fig. 2.16. The additive improves the burn-out process to such an extent, that virtually no unburned components in the exhaust gas remain [50]. The main metal components in the particles of heavy oil furnaces are, as can be seen from Fig. 2.16, iron, nickel, cadmium and vanadium.

Fig. 2.17 illustrates the particle-size distribution in a small modern light oil furnace. With a generally low concentration level (e.g., 0.18 mg/m^3) the emitted parti-

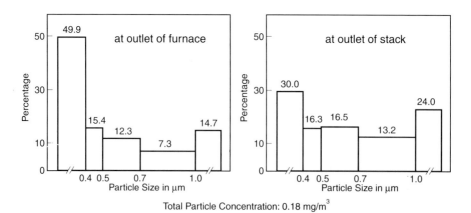

Fig. 2 17. Particle-size spectrum of the emitted flue dust of light fuel oil combustion in a modern domestic boiler (from cascade-impactor measurements)

cles are very fine. The particle content below 0.4 µm aerodynamic diameter amounts to approx. 50 % at the outlet end of the furnaces. However, a shifting to larger particles occurs as early as in the chimney, thus indicating agglomeration of soot particles in the exhaust channels.

2.1.4.3 Particle Emissions in Industrial Coal Furnaces [52]

In Germany coal is the primary fuel used in fossil-fuel fired powerplants. It is first pulverized and then fed into the furnace in dust form.

Apart from its main components of carbon, hydrogen, oxygen, nitrogen and sulfur, coal consists from 5 to 40 % of mineral components which are not burned and will be emitted at the end of the combustion process as particles. These inorganic additions are either present as mineral inclusions varying in size between the fraction of a micrometer and several micrometers, or they occur in the shape of atoms bound to organic residue. The coal dust present after comminution with a grain size of between approx. 10 to 70 µm can be divided into four categories: coal particles containing hardly any minerals, coal particles with mineral inclusions, mineral particles consisting of different minerals but containing hardly any coal and monomineralic particles such as pyrite or quartz.

During the combustion of coal dust the volatile components are released first, while the mineral components remain almost completely in the carbon skeleton for the time being, s. Fig. 2.18. Ash can be formed from this in different ways. Larger ash particles in the range of between 1 and 20 µm are formed through the agglomeration of mineral inclusions on the surface of burning coke particles. During combustion in coal dust furnaces under typical conditions the predominant share of the mineral components conglomerates, thus forming spherical ash droplets on the coke surface. If the coke particle remained intact during combustion, exactly one ash particle would remain per burning coke particle. Due to coke fragmentation every coke

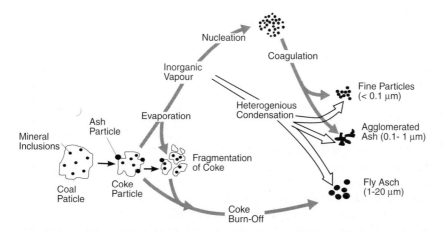

Fig. 2.18. Possible ash transformation processes during pulverized coal combustion [52]

particle typically generates 3-5 larger ash particles with a diameter of 10-20 μm and a larger number of ash particles with a diameter of 1-3 μm (see Fig. 2.18, lower arrow).

Smaller particles in the range below 0.1 μm are formed during the vaporization and subsequent recondensation or desublimation of a small part of the mineral components in the boundary layer surrounding the burning grain of coal (see Fig. 2.18, center arrow). Under conditions in coal dust furnaces at flame temperatures of 1400-1600 °C approx. 1 % of the mineral components typically vaporizes, leading to volatile inorganic vapors such as Na, As, Sb, Fe, Mg and the suboxide SiO. Due to the reducing atmosphere in the boundary layer surrounding the burning particle, not metal oxides but suboxides or metals are formed. After having diffused from the grain boundary layer into the higher oxygen surroundings, they can be oxidized and subsequently, depending on the saturation pressure of the vapors, undergo a homogenous nucleation, causing the formation of a large number of extremely fine ash particles (see Fig. 2.28, upper arrow). Alkaline vapors can in turn condense on these and due to coagulating processes, individual fine dust particles can grow together to form larger aggregates in a diameter range of 0.1-1 μm. Thus, the resulting shape of the aerosol particle size distribution depends on the complex interaction of the competing processes of homogenous nucleation, heterogenous condensation and growth via coagulation.

The size distribution of the individual mineral inclusions in the coal grain represents the upper (fine) limit of the flue ash particle size distribution assuming that no coalescence of the mineral inclusions takes place during combustion. The lower (rough) limit is represented on the assumption that exactly one ash particle is generated from each coal particle, i.e., that a complete coalescence of all mineral inclusions takes place in each coal particle. Hence, one tendency of the coal particle, namely to fragmentate during combustion, leads to a size distribution of the flue ash particles which is closer to the size distribution of the mineral inclusions in the coal.

In coal dust combustion the fine particle percentage of the total flue ash before filtering (raw gas) amounts to approx. 1-2 %; after filtering, the percentage of fine particles in pure gas dominates. The grain range of the flue ash of coal dust furnaces (pure gas) is between 0.01 and 100 μm, most particles being smaller than 20 μm. Different authors have observed a bimodal distribution of the pure gas dust with peaks between 0.1 and 0.5 μm and between 1 and 5 μm. Over 90 % of the particles exhibit a spherical structure [53]. The resulting flue ash grain size distribution is influenced by the comminution of the coal dust, by the temperatures and by the residence time in the fire.

Pure gas and raw gas particles consist 85-90 % of SiO_2 (quartz), Al_2O_3 (pure clay), Fe_2O_3 (hematite) and K_2O. For the main components listed, the particle composition of the filtered and the pure gas particle hardly shows any differences. Apart from the main components pure gas particles have a multitude of trace elements, with heavy metals playing an important role [54].

As there is no theory so far covering the degree of volatility of trace elements, the enrichment behavior in the flue ash must be determined by measurements. If one looks at the steam pressure curves of the elements As, Cd, Cr, Ni, Sb, Co, Hg and Zn in the temperature range of dust furnaces, one finds that apart from Co, Ni and Cr, all elements and their chlorides have boiling points below 1600 °C. Accordingly, they should all vaporize during combustion and partially accumulate in the flue dust. However, this phenomenon could only be partially confirmed by practical experiments. Neither can the varying tendencies of accumulation be attributed to the different boiling points of the metal oxides [55].

Whereas part of the trace elements leaving the furnace chamber in the shape of predominently gaseous oxides already condense on the walls and fittings of the flue gas channels, most of the trace elements condense during the cooling of the flue gas to temperatures of between 130 and 140 °C. Analyses of flue dust behind electric filters show a distinct decrease in the trace element concentration of the electric filter dust at rising temperatures. One exception is the element Hg which does not accumulate by condensation in flue dust due to its usually low concentration in coal, but, if at all, can find its way into flue ash in small amounts via adsorption and sublimation processes.

The increasing concentration of trace elements in flue ash with diminishing fine particle diameter can be explained by the fact that the accumulation takes place on the surface and that the surface-to-volume ratio is in inverse proportion to the diameter, so that the trace element concentration increases with the diminishing size of the particle.

Owing to their accumulation behavior, trace elements can be divided into one of three classes. Group I contains the non-volatile elements with high boiling points and low steam pressures such as, among others, Al, Cs, Fe. Group II contains the medium volatile elements such as, e.g., As, Co, Cr, Pb and Zn and Group III contains the highly volatile elements such as Br, Cl, F, Hg and Se which are completely gaseous after combustion and do not condense within the furnace due to their low dew points. These last elements are either scrubbed in the flue gas desulfurization plant or emitted as gas.

2.1.4.4 Particle Emission in the Combustion of Lump Wood in Domestic Furnaces

A wood fire in an open fireplace, in a chimney stove or in a tiled stove is still the epitome of cosy warmth. In well-wooded areas wood is still used to a certain extent as a basic means for generating room heat. Moreover, in the wood working industry, residual wood is generally burned, and most of the time the energy released in the process is utilized. When lump wood is burnt, however, particles consisting of soot, tar, ash or unburned carbon are released to a particularly high extent. That is why wood furnaces frequently have much higher particle emissions which can amount to 10 000 mg/m^3 [56]. The significance of particle emissions from wood furnaces, particularly of very small particles, has so far been grossly underestimated. One investigation, e.g., has proved that for the area of Vancouver and Portland approx. 52 % of the respirable particles emitted in January 1978 originated in domestic wood stoves [57]. But even in well-wooded areas of Germany, particularly in valleys, the burning of wood can be the cause of considerable annoyance and high aerosol concentrations in the ambient air. Fig. 2.19 shows the dense haze stemming from some wood furnaces over a village.

Fig. 2.19. Wood combustion flue gases lay a smoke-screen over a whole village during a winter surface inversion situation (Photo: Günter Baumbach)

Incomplete combustion frequently occurs in domestic wood fires. Causes are uneven, instationary burning conditions with inadequate mixing of fuel and air due to a local or general lack of air, and combustion temperatures which are periodically too low. This results in soot formation and emission. In many cases the volatile wood components do not burn properly but merely vaporize or pyrolyze partially and then condense in the exhaust gases. The condensate consists of high-molecular hydrocarbons, among others a large number and high concentration of polycyclic aromatic hydrocarbons (PAH) [56] which not only settle on heat exchange surfaces but also on emitted particles, particularly on soot. Thus, the emitted particles do not only contain elementary carbon (soot) but also a large share of organically bound carbon. As a general example Table 2.6 shows carbon emission values during the combustion of wood as compared to other emissions.

The emitted particles not only contain carbon and hydrocarbons but also, depending on the type of wood burned, various trace elements. An overview of these can be seen in Table 2.7. However, the values listed are merely guiding values, as the concentration of trace elements in wood depends strongly on the location, the conditions of growth and the age of the trees, among other factors.

With the exception of the ash particles the size distribution of the particles emitted by wood fires is largely independent of the type of furnace, its conditions of operation and the type of wood burned. According to investigations, the diameters of the emitted particles, if they originate from incomplete combustion, are below 10 μm in size, the biggest part of the particles and thus the great mass (roughly up to 40 %) even clearly below 1 μm. The majority of the results available suggest that the size range is located from between <0.1 up to approx. 0.2-0.3 μm [56]. To illustrate this, Fig. 2.20 shows a size distribution of flue gas particles of wood combustion. Merely

Table 2.6. Typical carbon emissions from different emission sources related to the amount of fuel burnt (acc. to [58])

Fuel/Process	Organic Carbon in mg/kg	Elementary Carbon in mg/kg
Natural gas heating	0.4	0.2
Motor vehicles with Otto engine		
without catalytic converter	40	14
with catalytic converter	14	11
Diesel engine	710	2100
Wood combustion	4000	1900

Table 2.7. Mean concentrations of trace elements in exhaust gases of wood-fired furnaces operated with North American woods in %, relative to the total particle mass [59]

Element	Pine	Oak
Al	0.45	0.27
Cl	2.87	1.04
Fe	0.03	0.01
K	11.61	5.84
P	0.19	0.12
Pb	0.14	0.05
Rb	0.02	0.01
Si	0.55	0.28
S	2.22	1.90
Zn	0.15	0.05

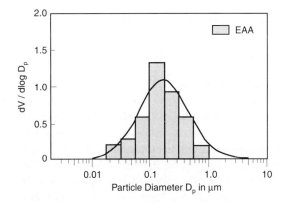

Fig. 2.20. Example of particle-size distribution of exhaust-gas particles from a wood furnace (acc. to [58])

particles ≤1 μm were taken into account in this distribution; larger particles were previously separated in a cascade impactor (approx. 10-17 % of the entire particle mass). One can see that the smoke formed during the combustion of wood contains extraordinarily small particles, thus representing a particular health risk because of its respirability and its polycyclic aromatic hydrocarbon content.

2.1.5 Polychlorinated Dibenzodioxins and Dibenzofurans

2.1.5.1 Properties, Formation and Origin

Since the 1976 catastrophe of Seveso in Northern Italy the public has become extremely aware of polychlorinated dibenzodioxins. In the wake of this accident methods of analyses were steadily improved so as to be able to detect even the slightest of concentrations. Improved measuring technology revealed that chlorinated dioxins and furans not only occur at the assumed places of origin but practically everywhere.

Polychlorinated dibenzodioxins (PCDD) and polychlorinated benzofurans (PCDF) are compound classes of aromatic ethers, i.e., phenyl rings which are bound with two and one oxygen atom respectively and substituted with many different chlorine atoms. Fig. 2.21 shows the structure formulas of polychlorinated dibenzodioxins, dibenzofurans and, to be able to compare, those of biphenyls. Dioxins have 75 congeners (compounds of one substance class), furans 135 which can occur in 8 homolog groups each (compounds with the same amount of chlorine atoms). Chlorine substitutions are possible in positions 1,2,3,4 and 6,7,8,9 (s. Fig. 2.21). Nomenclature is provided in Table 2.8. In Seveso dioxin chlorine atoms are located at positions 2,3,7 and 8: 2,3,7,8-tetrachlorodibenzodioxin (TCDD). Amount and position of the chlorine atoms influence the chemical properties and thus the toxicity (poisonous effect) of the substances:

- Water solubility decreases drastically with an increasing degree of chloridization and can generally be classified as slight.
- Solubility in fat increases with increasing chloridization and is about 4 powers of ten higher than solubility in water.
- The boiling point is between 300 °C and 400 °C; it rises with increasing chloridization; steam pressure and with it volatility thus decrease.

Polychlorinated dibenzodioxins and dibenzofurans are formed during thermal processes, e.g., in the exhaust gases of refuse incineration plants, in particular owing to the following mechanisms:

- Formation from precursor compounds (predioxins), e.g., from chlorinated benzenes, phenols, biphenyls or chlorinated biphenylether.
- Formation from non-chlorinated organic substances and chlorine (de-novo-synthesis).
- Incomplete combustion of substances already containing dioxins.

It is a well-known fact that in refuse incineration plants hardly any PCDD and PCDF are to be found at the end of the combustion. They are formed at a later stage in the exhaust gases in a temperature range between 250 and 350 °C [61]. At higher temperatures a dechlorination takes place, i.e., the share of low chlorine PCDD and

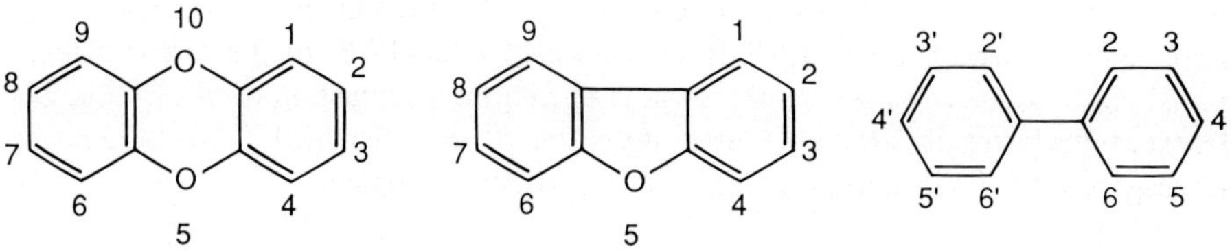

Dibenzodioxin Dibenzofurane Biphenyl

Polychlorinated aromatic compounds

Fig. 2.21. Chemical structure and numeration of the polychlorinated dibenzodioxins, -furans and biphenyls [60]

Table 2.8. Nomenclature of polychlorinated dibenzodioxins and -furans

Number of Chlorine Atoms	Dibenzodioxin	Number of Isomers	Dibenzofuran	Number of Isomers
1	monochloro-(MCDD)	2	monochloro-(MCDF)	4
2	dichloro-(DCDD)	10	dichloro-(DCDF)	16
3	trichloro-(TrCDD)	14	trichloro-(TrCDF)	28
4	tetrachloro-(TeCDD)	22	tetrachloro-(TeCDF)	38
5	pentachloro-(PeCDD)	14	pentachloro-(PeCDF)	28
6	hexachloro-(HxCDD)	10	hexachloro-(HxCDF)	16
7	heptachloro-(HpCDD)	2	heptachloro-(HpCDF)	4
8	octachloro-(OCDD)	1	octachloro-(OCDF)	1

PCDF homologs increases. PCDD and PCDF are relatively stable thermally, if the temperature rises, decomposition increases via the separation of the aromatic ether compound. In the presence of oxygen and at high temperatures the molecules can also be decomposed oxidatively. Fig. 2.22 qualitatively shows the opposite development of dioxin formation and decomposition and the dioxin concentration resulting from it.

It was possible to prove that for the formation of dioxin in incineration exhaust gases, the initial reaction is the formation of chlorine from copper or other metal chlorides or from metal oxides reduced with HCl. The so-called Deacon process [62]:

$$\begin{array}{lcl} CuCl_2 + \tfrac{1}{2}\,O_2 & \xrightarrow{300\ °C} & CuO + Cl_2 \\ CuO + 2HCl & \longrightarrow & CuCl_2 + H_2O \\ \hline 2\,HCl + \tfrac{1}{2}O_2 & \longrightarrow & H_2O + Cl_2 \end{array}$$

Accordingly, copper, e.g., takes part in the reactions, but is released again at the end; it can therefore be regarded as the catalyst for dioxin formation.

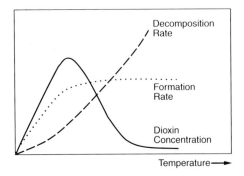

Fig. 2.22. Formation and decomposition of chlorinated dioxins in filter particulate when oxygen is present (acc. to [61])

2.1.5.2 Toxicity, Toxicity Equivalents and Threshold Values

One acute effect of the 2,3,7,8-TCDD is evident in chloric acne, still not all people are affected to the same extent by this skin disease [63]. The other acute toxicity becomes evident when chlorine is taken in with food. In animal experiments the poisoning takes a slow and non-specific development and leads to death in approx. two to three weeks. During this time the animals lose up to 50% of their body weight ("wasting syndrome"). However, no specific cause of death can be found [62]. Table 2.9 shows the relative toxicity of 2,3,7,8-TCDD in comparison with other toxins. It is worth mentioning here, that as far as acute toxicity is concerned 2,3,7,8-TCDD is the most poisonous substance to occur so far in connection with chemical production processes. Only bacterial toxins have considerably lower LD_{50} values yet.

In the Seveso accident pregnant women were considered a special risk group. However, despite highly intensive investigations no indications of an elevated abortion or stillbirth rate or an increased occurrence of macrostructural abnormalities in new-born children could be detected [63]. Neither could a mutagenic effect be observed even with the 2,3,7,8-TCDD dosages relatively relevant for longterm human

Table 2.9. Relative toxicity ($<D_{50}$-dose) of some selected toxic substances (2,3,7,8-tcdd=1) [62]

Substance	Relative Toxicity
Botulinum Toxin A	0.00003
Tetanus Toxin	0.0001
Diphtheria Toxin	0.3
2,3,7,8-TCDD	1
Saxitoxin	9
Tetrodotoxin	8-20
Curare	500
Strychnine	500
Muscarine	1 100
Sodium Cyanide	10 000

exposure. In addition, no significant binding of TCDD to DNA could be proven in vivo (rat liver) [64]. However, a carcinogenic potential could be proven in rats, with dioxins working as a promotor causing – from a certain threshold value given in a daily dosage – the formation of monooxygenases in the animals' liver which in turn functioned as a tumor initiator. Having an initiator potential of its own is considered improbable [64] or even impossible [65]. What is of importance is that a threshold value can be indicated for which no effects need be feared. As this threshold value is two to three powers of ten higher than the daily absorbed amount of the average human load [66] and as rats accumulate ten times more PCDD and PCDF in their livers than human beings, a hazard for man via this route can be excluded, as long as he is not exposed to considerably higher concentrations.

To be able to classify the toxicity (poisonous effect) of dioxin and furan emissions containing different isomers, so-called toxicity equivalents (TE) were introduced. TE concentrations are calculated such that each dioxin and furan important for total toxicity is assigned a factor according to its relative toxicity in relation to the 2,3,7,8-TCDD (Seveso dioxin). 2,3,7,8-TCDD, as the most poisonous representative of this substance class, is assigned the factor 1, the other isomers accordingly factors ≤1. If these individual concentrations are assessed with their factors and then added up, a concentration value is obtained as if 2,3,7,8-TCDD were the only substance considered. This facilitates the evaluation of emissions and loads, as even with varying distributions of congeners, toxicity can be rated with a value. Unfortunately, different institutions use different TE factors, Table 2.10 gives an overview of these.

Table 2.10. Toxicity equivalents (TE factors) of different international agencies

Congener	NATO CCMS 1988 [67]	Switzerland 1988 [68]	Scandinavia 1988 [69]	BGA (FRG) 1985 [70]	EPA (USA) 1985 [71]	Eadon 1982 [71]
2,3,7,8-TCDD	1	1	1	1	1	1
1,2,3,7,8-PCDD	0.5	0.4	0.5	0.1	0.2	1
1,2,3,4,7,8-HCDD	0.1	0.1	0.1	0.1	0.04	0.03
1,2,3,6,7,8-HCDD	0.1	0.1	0.1	0.1	0.04	0.03
1,2,3,7,8,9-HCDD	0.1	0.1	0.1	0.1	0.04	0.03
1,2,3,4,6,7,8-HCDD	0.01	0.01	0.01	0.01	0	0
OCDD	0.001	0.001	0.001	0.001	0	0
2,3,7,8-TCDF	0.1	0.1	0.1	0.1	0.1	0.33
1,2,3,7,8-PCDF	0.05	0.01	0.01	0.1	0.1	0.33
2,3,4,7,8-PCDF	0.5	0.4	0.5	0.1	0.1	0.33
1,2,3,4,7,8-HCDF	0.1	0.1	0.1	0.1	0.01	0.01
1,2,3,6,7,8-HCDF	0.1	0.1	0.1	0.1	0.01	0.01
1,2,3,7,8,9-HCDF	0.1	0.1	0.1	0.1	0.01	0.01
2,3,4,6,7,8-HCDF	0.1	0.1	0.1	0.1	0.01	0.01
1,2,3,4,6,7,8-HCDF	0.01	0.01	0.01	0.01	0.001	0
1,2,3,4,7,8,9-HCDF	0.01	0.01	0.01	0.01	0.001	0
OCDF	0.001	0.001	0.001	0.001	0	0
non 2,3,7,8-substituted	0	0	0	0.001–0.01	0–0.01	0

One can see that the toxicity of the individual isomers is assessed quite dissimilarly by some of the different institutions and that these ratings have also varied over the years. Owing to these different ratings, toxicity equivalents must always be provided with source references, otherwise they are not comparable. In more recent publications and when maximum values are determined, e.g., in the Refuse Incineration Regulation of the Federal Republic of Germany [72], the international calculation mode according to NATO/CCMS is applied.

There are, for instance, expert recommendations of the German Federal Health Office (BGA) which enable assessments regarding the necessity of regenerating polluted soil [73, 74]. These values were recommended by the working group "Dioxin" of German Federal and State governments for inclusion as standard values for applicable legal codes and regulations, Baden-Wuerttemberg being the first federal state to introduce these regulations in January 1992 [75]:

≤ng/TE/kg	for unrestricted agricultural use
5-40ng TE/kg	use restricted by cultivation recommendations
≥ 40ng TE/kg	no agricultural use
≥ 100 ng TE/kg	soil exchange in playgrounds
≥ 1000 ng TE/kg	soil exchange in housing areas
≥ 10.000 ng TE/kg	need for regeneration outside of housing areas.

In North Rhine-Westphalia the standard value for outdoor air is listed as 3 pg/m^3 for 2,3,7,8-TCDD (standard immission value).

In the 1990 Regulation for Refuse Incineration Plants [2] a PCDD/PCDF maximum emission value of 0.1 ng TE/m^3 was stipulated for the exhaust gases of refuse incineration plants which must be adhered to immediately in new plants and in old plants by the 1st of December 1993.

2.1.5.3 Dioxin Sources

Table 2.11 shows the estimated contributions of individual PCDD/PCDF sources to the annual load in the old states of Germany and an extrapolation for the total emission by each source during the last 20 years [61]. Obviously, only the sources known up to this point are listed here. The amounts found in the environment, however, are far higher than the contribution of the sources known so far. According to a Swedish extrapolation [76], e.g., dioxin deposition is 25-30 times higher than the emissions of the known sources. Therefore quantitatively relevant emittents are still being looked for.

Based on individual measuring results, e.g., wood fires in private homes or commercial enterprises, metal-melting plants, low temperature carbonization of wires and similar processes must also be considered as further sources of dioxin and furan [77].

In refuse incineration the larger part of the PCDD and PCDF occurs in filter dust, the remainder mainly in the pure gas; the slag contains almost no PCDD and PCDF. If the exhaust gases which are fed back into the combustion process for the destruction of the PCDD and PCDF are scrubbed catalytically [78] or via activated carbon

Table 2.11. Quantitative estimate of different PCDD/PCDF sources in TE (BGA=German Federal Health Office) for the old states of Germany (1989) (acc. to [61])

Sources	Calculation basis	TE/year	TE/20 years
Garbage incinerators	10 ng TE/m^3 exhaust gas	0.4 kg	8 kg
	13 ng TE/g flue dust	3.1 kg	62 kg
Hospital garbage incinerators	15 ng TE/m^3 exhaust gas	0.0015 kg	0.03 kg
Hazardous waste incinerators	0.5 ng TE/m^3	0.001 kg	0.02 kg
Automobile exhaust	0.002 ng TE/g gasoline	0.05 kg	1 kg
Recycling	estimated	0.4 kg	0.4 kg
PCP	300 or 2400 ng TE/g	1.3 kg	26 kg
PCB	open usage		90 kg
	1500 ng TE/g		
	closed cycle		90 kg

[79], and if subsequently the filter dusts are treated thermally under the exclusion of air [80], then fewer PCDD and PCDF escape from such a modern refuse incineration plant than were previously fed into it with the refuse. These modern refuse incineration plants can therefore be regarded as dioxin decreasers rather than dioxin sources [78, 81].

2.1.6 Exhaust Gases from Motorised Vehicles

2.1.6.1 Influences on their Formation

Most of the automobiles in street traffic are driven by combustion engines – Otto or Diesel engines. In these engines an intermittent, unstationary combustion takes place in the combustion chambers; with every other piston stroke the sucked-in air/fuel mixture ignites, expands due to the released oxidation heat and drives the piston downward. The rising piston subsequently presses the exhaust gas out of the cylinder. How completely this combustion is carried out and how high exhaust gas emissions are at the best possible fuel exploitation depends on numerous influencing factors which are represented in Fig. 2.23 as examples.

One of the main problems is caused by the fact that automobile engines are not operated with a constant load and a constant number of revolutions. It is hardly possible to achieve minimal exhaust gas emissions with a maximum exploitation of fuel at every point of engine operation. The dependencies of the air excess λ and of the exhaust gas emissions on the engine speed and torque are illustrated in engine performance curve characteristics. As an example for a four-stroke automobile Otto engine Fig. 2.24 shows the dependency of the air excess on revolutions and mean effective pressure, the last-mentioned being a measure for load. Figs. 2.25 and 2.26 illustrate the dependencies for the component carbon monoxide (CO) and for nitrogen oxide (NO) respectively. Analogous to the emission behavior of oil furnaces (see Fig. 2.2) emissions of CO and hydrocarbons increase when there is a lack of air (rich range $\lambda < 1$), see Fig. 2.27, whereas nitrogen oxides are relatively low when there is

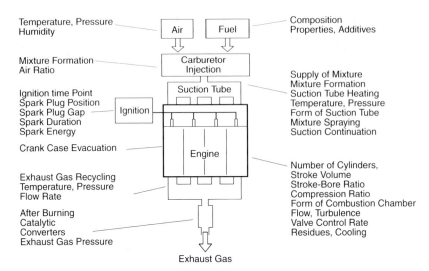

Fig. 2.23. Factors influencing exhaust gas composition in combustion engines with spark ignition [82]

a lack of air. As we are mainly dealing with thermal NO here, nitrogen oxides (see Fig. 2.11) are at their maximum during the most intense stage of combustion, i.e., at λ just above 1.

One can see from the engine performance characteristics that the engine runs on a rich mixture ($\lambda < 1$) when it is idling (lowest number of revolutions, lowest load). Accordingly, CO and HC (hydrocarbon) emissions rise in this case. Nitrogen oxides

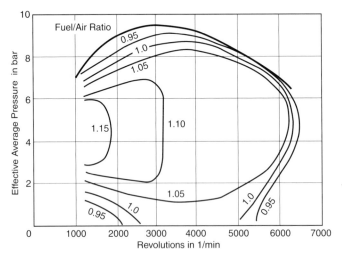

Fig. 2.24. Engine performance curves of a 4-cylinder-4-stroke spark-ignition engine of an automobile during stationary operating conditions: effective mean pressure (as measure of load) versus revolutions during different air ratios λ [46]

2.1 Emission of Pollutants Caused by Combustion Processes   53

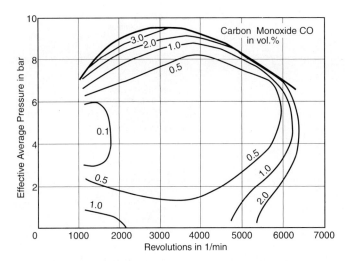

Fig. 2.25. Engine performance curves of a 4-cylinder-4-stroke spark-ignition engine showing lines of constant carbon monoxide concentrations [46]

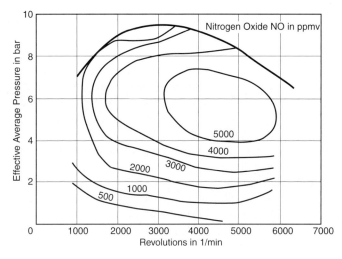

Fig. 2.26. Engine performance curves of a 4-cylinder-4-stroke spark-ignition engine showing lines of constant nitrogen oxide concentrations [46]

are at a minimum in this range. At a relatively low number of revolutions the engine has its highest performance in the rich range ($\lambda < 1$). Accordingly, CO and HC emissions rise in this operation phase. Due to this enrichment nitrogen oxides are not at their maximum at full load, see Fig. 2.27, but at approx. 2/3 load and 3/4 of the number of revolutions.

The following further causes can increase exhaust gas emissions in Otto engines [46, 48, 84]:

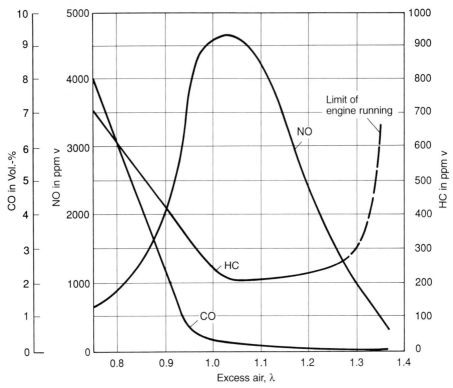

Fig. 2.27. Exhaust gas emissions of an automobile spark-ignition engine versus excess air λ [83]

- If the engine is offered an excessively rich mixture in cold starts (in carburetor engines: choke!) it invariably leads to emissions of unburned substances. λ-controlled injection engines generally do not receive an excessively rich mixture, but even with $\lambda=1$, CO and hydrocarbons can be found in the exhaust gases which should really be oxidized subsequently by the catalytic converter. But if the catalytic converter has not yet warmed up, then the substances escape with the exhaust gases. Developments are therefore at present aimed at minimizing cold start emissions of the engines.
- If the flame is extinguished in cracks existing in the engine due to its construction, CO and HC emissions are increased [84, 85].
- Poor maintenance and wear and tear after long-term performance of the engines generally lead to increased CO and hydrocarbon emissions, e.g., a poorly adjusted ingnition increases exhaust gas emissions [84]. Worn piston rings and valve guides lead to increased, often incomplete combustion of engine oil. Thus, distinctly visible blue smoke with strong odoros emissions can occur in old engines especially during downhill driving [86].
- In automobiles with catalytic converters operating with λ-control[1] nitrogen oxide emissions can, e.g., be drastically increased when the lambda probe is faulty and

---
[1]Catalytic converter technology is described in chapter 7.3.3

the engine is running on a lean basic setting. With a rich setting, however, CO and HC emissions will be increased drastically [87].
- In automobiles with catalytic converters operating without λ-control operating conditions are sometimes lean and sometimes rich. Accordingly, in some ranges the catalytic converter does not decrease nitrogen oxide emissions, while in other ranges there is no decrease of hydrocarbon and carbon monoxide emissions. New compounds distinguished by strong odors or irritating effects, e.g., hydrogen sulfide or aldehydes may be formed as a result of incomplete oxidation or reduction processes in the catalytic converter.
- In automobiles with catalytic converters operating with λ control the mixture is enriched to improve performance when there is a full load (see engine performance characteristics, Fig. 2.24). For this, the λ control is turned off. In this case partial oxidation or reduction processes with strong odoros emissions also occur.

The exhaust gas components emitted by automobiles do not only consist of carbon monoxide, nitrogen oxides and hydrocarbons in general. Hydrocarbons, e.g., can have a variety of compositions; in addition other trace substances such as cyanide, ammonia, particles etc. occur in the exhaust gases. Fig. 2.28 shows an exemplified overview of the most important air pollutants occurring in the exhaust gases of pas-

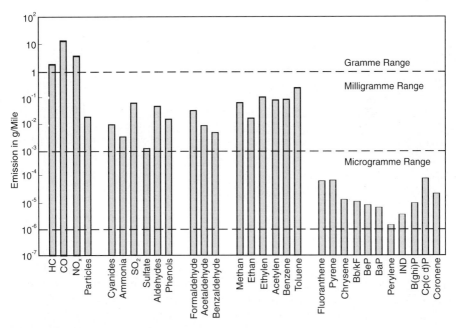

Fig. 2.28. Overview of emission values (log. scale) of 4 spark-ignition automobiles without catalytic converter, determined by 3 different US tests [87].
Bb/kF = sum of benzo(b)- and benzo(k)fluoranthene; BeP = benzo(e)pyrene;
BaP = benzo(a)pyrene; IND = indeno(1,2,3-cd)pyrene; B(ghi)P = benzo(ghi)perylene;
Cp(cd)P = cyclopenta-(cd)pyrene

56   2 Origin and Sources of Air Pollution

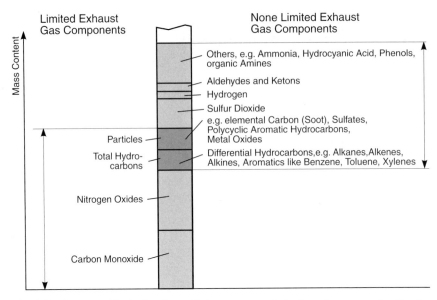

Fig. 2.29. Content of limited and non-limited compounds in the exhaust gas of a diesel engine [87]

senger cars with internal combustion engines without catalytic converters along with their orders of magnitude.

When there is a lack of air, diesel engines are primarily prone to emissions of soot and odoros substances, but less so to CO and hydrocarbon emissions. Fig. 2.29 illustrates which components are to be expected in the exhaust gas of a diesel engine. One distinguishes between the limited (CO, NO_x, HC and particles) and the manifold non-limited components which can constitute up to a third of exhaust gas emissions (without CO_2 and H_2O).

Apart from the air-fuel ratio, soot formation in diesel engines is dependent on pressure and temperature, on the combustion process, on start of injection, injection ending and ignition delay. Particularly an inadequate mixing which is influenced by the factors listed above leads – at high temperatures and via cracking processes – to the formation of soot [84, 86].

In diesel engines performance is not controlled via a throttle valve, as is the case in internal combustion engines, but via the amount of fuel injected. The amount of air sucked in is determined by the engine revolutions and the supply. Accordingly, when idling and when there is a partial load, vehicles generally operate with a great excess of air. If the injected amount of fuel is too high as against the amount of air sucked in ($\lambda < 1.2$-1.4, depending on the factors listed above) at full load and revolutions in the low to medium range, then soot formation increases dramatically, see performance characteristics of a diesel engine, Fig. 2.30. This is why diesel engines are prone to increased soot emissions at full loads, e.g., when driving uphill, with inadequate mixing and when large amounts of fuel are injected.

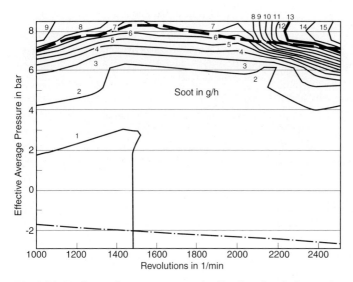

Fig. 2.30. Engine performance curves of a diesel engine during stationary operating conditions showing lines of constant soot emission [88]

In diesel engines *blue and white smoke* stem from condensation aerosols of unburned hydrocarbons. This can occur during a cold start or warm-up phase [86]. Soot emission during cold starts is also a well-known occurrence.

2.1.6.2 Exhaust Gas Emissions when Driving

Exhaust gas emissions calculated on the basis of stationary engine performance characteristics can only be a good measurment for air pollution if driving behavior and the route driven can be determined precisely, i.e., if revolutions and loads as well as transmission losses are known. In addition, the number of non-stationary working points of the engine (e.g., shifting gears and speedy accelerations) may not be too large, as the engine performance characteristics only show exhaust gas emissions during stationary operation. In individual cases emission behavior of the vehicles has been calculated with the engine performance characteristics on certain roads [89]. However, engine performance characteristics and transmission losses are often not known and would have to be determined by time-consuming and costly experiments. Moreover, performance characteristics change in older engines. For these reasons and also because it is relatively elaborate this process has rarely been used to determine emissions.

To determine exhaust gas emissions in automobiles, e.g., for classification in certain pollution emission classes, previously defined cycles are driven with the automobiles on roller dynamometers during which exhaust gas emissions are measured. Though this measuring process is standardized the driving cycles can vary greatly depending on the individual country. These test methods will be described more closely in chapter 5.4.2.

These tests serve to determine exhaust gas emissions of individual automobiles in a standardized driving cycle so that they can be compared, and they also serve to establish whether the prescribed maximum values are exceeded or not (s. also Chap. 8.3). To determine emissions of individual street blocks, however, e.g., for emission inventories or for the assessment of air pollution in the vicinity of heavy traffic streets, emission factors which apply to certain kinds of driving behavior and certain automobile collectives are necessary. In Germany the Technical Control Board (TÜV) of the Rhineland on behalf of the Federal Environmental Agency has, e.g., determined the driving behavior in different so-called driving modi and with these driving modi determined the exhaust gas emissions of a representative fleet of automobiles on roller dynamometers at regular intervals and published them [49, 88, 90, 91].

Based on these driving modi, the dependency on speed, e.g., of the exhaust gas emissions was determined. They are shown on Fig. 2.30 with nitrogen oxides as an example. It may be observed that NO_x emissions rise with increasing speeds owing to the increasing thermal load of the engines at higher speeds when thermal NO formation increases dramatically. The speed dependency of NO_x emissions illustrated in Fig. 2.31 does not apply at full loads. As pointed out earlier, at that point NO_x emissions would decrease again due to the enrichment of the mixture.

The driving modi on the roller dynamometers reflect driving behavior and resulting exhaust gas emissions in an idealized and simplified form and with reference to a certain automobile collective. If it is necessary to learn more about the emissions of a certain section of a street, e.g., to learn about the effects of traffic control measures, then there are basically two possibilities to investigate this more closely:

1. By measuring speeds in the flow of traffic (in-traffic driving) driving behavior is investigated in the relevant street section. This driving behavior is then reproduced on the roller dynamometer with individual automobiles or an automobile collective and emissions are measured. This method was used in 1985, e.g., in the large-scale experiment for the speed limit of 100 km/h [92].

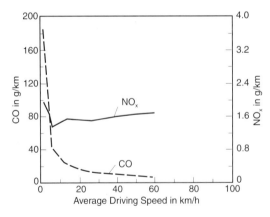

Fig. 2.31. Nitrogen oxide emission versus driving speed on a plane roadway – emissions of an automobile collective of the reference year 1985 (acc.to [90])

2. It is also possible to measure exhaust gas emissions directly in the automobiles driving along in traffic in the sections in question. This requires a special measuring technique developed for this purpose. Measuring concepts of this type were

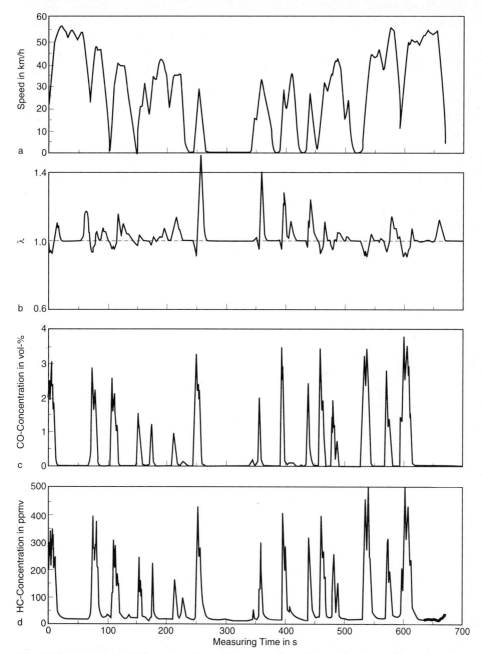

Fig. 2.32. Real urban driving example: **a** driving speed profile; **b** air ratio; **c** CO and **d** HC concentrations. Automobile equipped with three-way catalytic converter and λ control

developed by the company VW [93] and by the working group Luftreinhaltung (Clean Air) of the University of Stuttgart [94]. These so-called on-board measurements have the advantage that emissions can be determined directly during genuine dynamic operation of the automobile, e.g., during acceleration and deceleration, during uphill and downhill driving a.s.o..

2.1.6.3 Development of Motor Vehicle Emissions

Motor vehicle engines have undergone a constant development in the past years to improve specific performance and to lower fuel consumption and exhaust gas emissions. This development can be seen clearly from Fig. 2.33 where, as an example, CO, HC and NO_x emissions of different, commonly used German passenger cars, which were determined in a European test, are compared for the years of construction of 1978/79, 1982/83 and 1991 (with and without catalytic converters). At the end of the seventies minimizing fuel consumption and with it the reduction of CO and hydrocarbon emissions had priority, so that vehicles of the model years 1982/83 showed distinctly lower values in these exhaust gas components than vehicles of the years 1978/79. In all shown vehicle types, except for Opel, NO_x emissions increased for the time being due to improved combustion and lower CO and HC emissions. In the newer models (1991) CO values could be even further lowered while simultaneously NO_x emissions were also reduced. However, these engine-related measures did not suffice to comply with the new EC and US threshold values (s. Chap.8.3). For this reason, catalyzer technology was introduced in the eighties which again drastically reduced exhaust gas emissions (s. Fig. 2.33, right side).

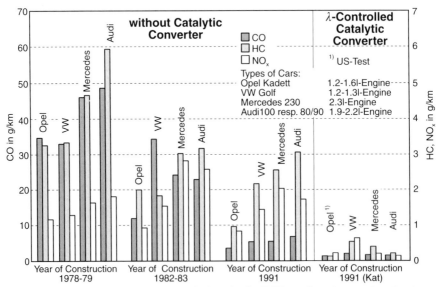

Fig. 2.33. Reduction of exhaust gas emissions in the ECE test based on engine development and catalytic converter technology in different automobiles (acc. to data taken from [95] and [96])

Apart from changes in modes of transportation and other measures, fuel consumption of motor vehicles must be reduced even further so that emissions of the greenhouse gas CO_2 can be lowered in the future. Basically, diesel engines are inherently better suited to this than gasoline engines; as far as these engines are concerned minimizing noise, particle (soot) and odoros substance emissions represent the main tasks. But in internal combustion engines, too, the fuel saving potential has not been completely exploited. Gasoline engines with direct injection promise a fuel consumption reduction to nearly the level of comparable diesel engines [97].

Table 2.12. Possible pollutant emissions of some important industrial processes

Industrial process	Emission sources	Gaseous or particulate emissions, to a varying extent, depending on process optimization and exhaust gas purification
Coking plant	Chamber oven	Particulate, CO, NH_3, H_2S, C_nH_m
Cement Production	Quarries, grinding mills, loading facilities	Particulate of varying compositions
	Rotary kiln	Particulate, NO_x, generally little SO_2, CO, H_2S
Glass production	Glass melting furnaces or tanks	Particulate (saline and other), NO_x, SO_2 (oil furnaces)
Pig iron mining	Calcining and sintering plants	Particulate containing heavy metals, SO_2, NO_x, CO, HCl, HF
	Blast furnace	Particulate (Pb, Zn, Cd, As) H_2S, HCN, CO
Steel manufacturing	Top blown basic	Fine particulate (brown smoke) oxygen converter consisting of iron oxides, CO
Casting	Coupola furnace	Particulates, CO
	Core production	Organic substances, odors
Mineral oil production (refineries)	Furnaces	SO_2, NO_x, Particulates
	Separator and transformer plants, storage, conveyance, loading	Various C_nH_m, possibly H_2S
Pulp production	Furnaces for the incineration of the sulfite pulping agents	SO_2, particulate
	digester, pulp washing	SO_2, odors
Sulfuric acid production	Residual gases of SO_2 catalysis	SO_2, SO_3
Nitric acid production	Adsorption plants	NO, NO_2
Fertilizer production	Mixing vessel	Raw material particulate, Fluorine and chlorine compounds, NH_3, ammonium salts
Sugar industry	Furnaces (partly heavy fuel oil)	SO_2, NO_x, particulate

2.2 Sources of Air Pollutants

2.2.1 Overview

Combustion processes are the main sources of anthropogenic air pollution. They are used for generating energy in industrial and power plant furnaces, in vehicular combustion engines, in airplane engines and in domestic heaters. There are a number of industrial processes which can also be sources of air pollution, with each industrial branch having its own problems with keeping the air clean. A description of the numerous processes with their specific emissions cannot be undertaken in this work: The interested reader is referred to special literature on this subject. Many environmentally relevant plants and their specific air pollution prevention measures have been, e.g., described in the German VDI guidelines [98]. For exhaust purification the book *Air Pollution Control Equipment* is mentioned here [99]. Some important industrial processes and their emissions are listed in Table 2.12 as examples. The main

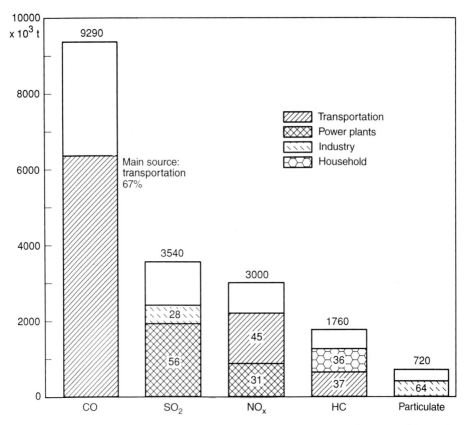

Fig. 2.34. Annual emissions from different air pollution sources in the old states of the FRG for the reference year 1982 (acc. to [100] – empty fields: other sources)

sources of the main individual air pollution compounds (except CO_2) are represented in Fig. 2.34. The proportion of individual sources contributing to the total emissions of the individual compounds and their absolute contribution are shown using the example of the Federal Republic of Germany for the reference year 1982. Naturally, many things have changed since then. We will therefore deal with the temporal development of emissions in more detail in the next sections. The main emission components and main source groups shown in Fig. 2.34 can, however, be regarded as typical of many industrial countries which have not installed exhaust gas purification equipment to any extended degree.

In developing countries, though, with less industrialization and motorization, the technical processes applied have not been optimized yet, so that their specific emissions are frequently higher. Combustion processes of, e.g., motor vehicles, thus often emit more products of incomplete combustion (CO, hydrocarbons, soot) than the same processes in the industrial states of North America, Europe and Japan. As an example, Fig. 2.35 shows the specific emissions (per capita) of an African country as compared to that of Germany. Total emissions of the typically consumption-related compounds CO_2, SO_2 and NO_x are much lower in the developing countries than in the industrial ones. This becomes particularly obvious when using the example of CO_2 emissions which will be dealt with in the next section.

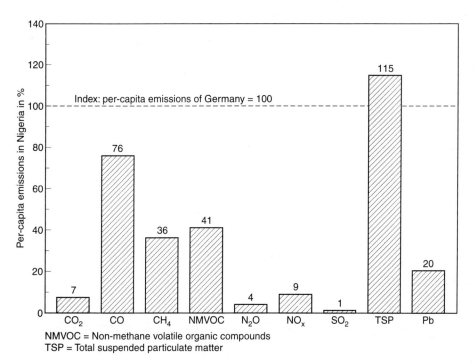

Fig. 2.35. Per-capita emissions in the African country of Nigeria as compared to Germany for the reference year 1990 [101]

2.2.2 Carbon Dioxide (CO_2)

As a result of the constantly increasing consumption of energy emissions of CO_2, the main product of combustion, rise exponentially. Fig. 2.36 shows the estimated rise of global CO_2 emissions. These emissions are mainly caused by the combustion of solid, liquid and gaseous fossil fuels, by the flaring of gases in oil production and in cement manufacturing. In 1991, approximately 6,200 million metric tons of carbon were emitted into the atmosphere compared to 93 million metric tons of carbon in 1860 [102].

Fig. 2.37 shows the national estimates of the 20 highest CO_2 emitting countries in 1991 compared with their emissions in 1950. The data are presented in descending

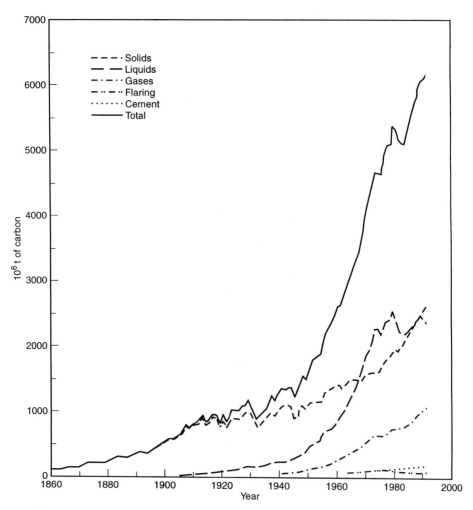

Fig. 2.36. Global CO_2 emissions from fossil fuel burning, cement production and gas flaring, 1860-1991 [102]

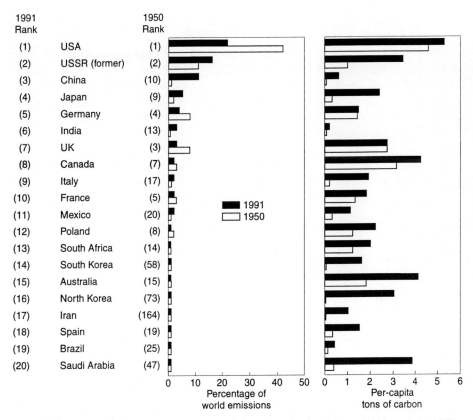

Fig. 2.37. Ranking of the 20 highest CO_2-emitting countries in 1991 and their rank in 1950, contributions to global emissions and per-capita CO_2 emission rates [102]

order with the highest emitting country, the United States, presented first. These 20 countries contributed just over 81 % of all the 1991 world CO_2 emissions from fossil fuel consumption. In the right part of Fig. 2.37 the CO_2 emissions per capita and year are shown. It can be seen that the North American countries USA and Canada have the highest specific CO_2 emissions followed by Australia, Saudi Arabia, the former USSR and Germany. The countries with a high population, China and India, produce very low amounts of CO_2 per capita.

2.2.3 Sulfur Oxides (SO_x)

Sulfur oxides, mainly in the form of sulfur dioxide (SO_2), are emitted during the combustion of sulfurous fuels, primarily coal and mineral oils. In industrial countries the main sources are industrial and power plant furnaces, where the fuels highest in sulfur are used. In many countries, however, sulfurous coal or fuel oils are burned at home for heating purposes and diesel fuel can contain up to 0.8 % of sulfur. The sulfur dioxide emissions estimated for 1985 of the 10 highest emitters in Europe and the USA are shown in Fig. 2.38 in absolute numbers and in relation to the surface

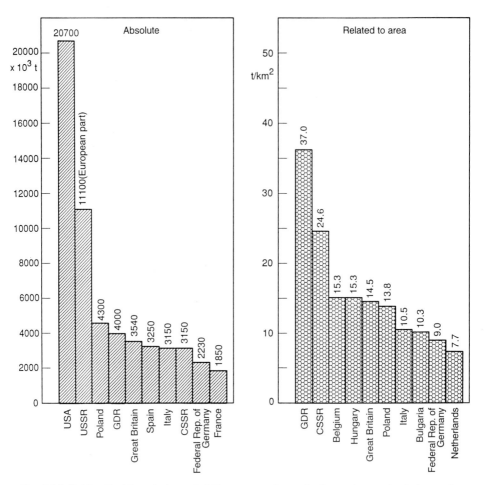

Fig. 2.38. Sulfur dioxide emissions of different countries – absolute values and relative to the surface area of the country shown, reference year 1985 (acc. to [100])

area of each country. It can be seen, that the former GDR and the Eastern European countries CSSR, Poland and Hungary relatively and absolutely emitted a lot of SO_2. As a result of German unification, the breaking up of the USSR and due to increased application of desulfurization measures emission conditions have undergone drastic changes in the meantime, particularly in Germany. The SO_2 development for the old states of Germany is shown in Fig. 2.39. SO_2 emissions have been greatly reduced in the past years owing to the installation of desulfurization equipment in large-scale furnaces, to fuel changes in medium-sized industrial furnaces, mainly from heavy fuel oil to natural gas, and due to the reduction of the sulfur content in light fuel oil and diesel fuel. The development of SO_x emissions in the USA is represented in Fig. 2.40. In this country, too, there is a decreasing tendency in the past 10 years. But even before this period a reduction was achieved by installing flue gas desulfurization plants.

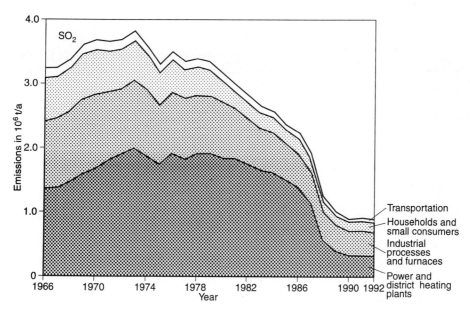

Fig. 2.39. Trend of SO_2 emissions in the old states of the FRG from different sources (acc. to [103])

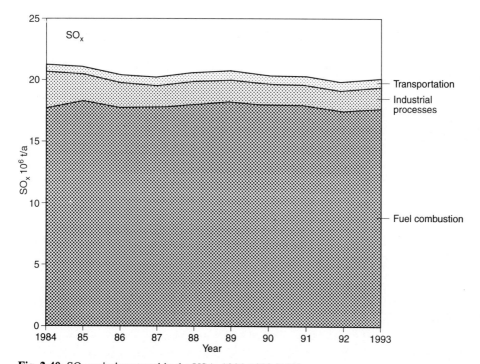

Fig. 2.40. SO_x emissions trend in the USA, 1984-1993 [104]

2.2.4 Nitrogen Oxides (NO_x)

With SO_x emissons the development of NO_x emissions is represented using the examples of Germany (old states) and USA, s. Figs. 2.41 and 2.42. While fuel combustion is the main emission source of SO_2, it is motor-vehicle traffic that causes the main emissions of nitrogen oxides. Up until approx. 1986 NO_x emissions increased distinctly in Germany. The NO_x decrease apparent from 1986 on was achieved mainly through emission reductions in power plant furnaces. In traffic the NO_x reduction achieved by automobiles with catalytic converters was initially offset by the growing numbers of vehicles. In the meantime, however, the number of automobiles having a catalytic converter with a lambda sonde (stringent US regulation, s. Chap. 8) has grown from 22 % in the year 1991 to 38 % in 1994 [105] in Germany. In the USA there have not been any great changes in NO_x emissions in the past ten years. In this large country fuel combustion is responsible for 50 % of NO_x emissions and transportation 45 %. While fuel combustion emissions (primarily from coal-fired electric utilities) were three percent higher in 1993 than in 1984, NO_x emissions from

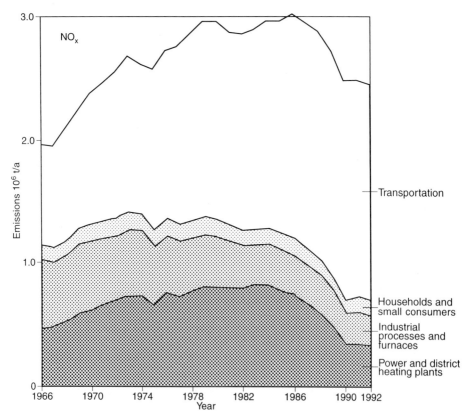

Fig. 2.41. Trend for nitrogen oxide emissions (NO_x, calculated as NO_2) in the old states of the FRG (from different sources) (acc. to [103])

2.2 Sources of Air Pollutants   69

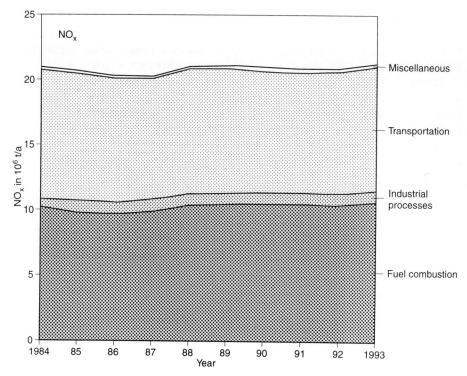

Fig. 2.42. NO$_x$ emissions trend in the USA, 1984-1993 [104]

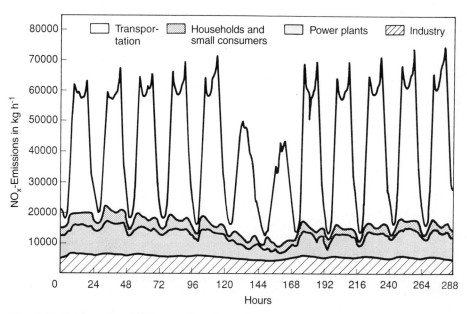

Fig. 2.43. Course of total NO$_x$ emissions in the German state of Baden-Württemberg, FRG, during a period of two weeks from March 18 – 29, 1988 [106]

highway vehicles decreased by 11 % during this 10-year period. Thus, total 1993 NO_x emissions were one percent higher than 1984 emissions [104]. In the USA the NO_x emissions from transportation have not decreased as much during the last ten years because of the efficiency of prevention measures before that time.

Nitrogen oxides are not emitted at a constant temporal rate. As, in Germany, a large part of the emissions is caused by automobiles, emissions basically follow the diurnal course of traffic. Using the example of a period of fourteen days Fig. 2.43 graphically shows the recorded nitrogen oxide emissions from all sources in the entire state of Baden-Wuerttemberg. It is obvious that the course follows a time-of-day pattern. The two low peaks in the middle show the emissions of a weekend with distinctly less traffic.

2.2.5. Carbon Monoxide (CO)

Carbon monoxide forms during incomplete combustion. Accordingly, it is released by internal combustion engines and turbines, by domestic furnaces, by industrial

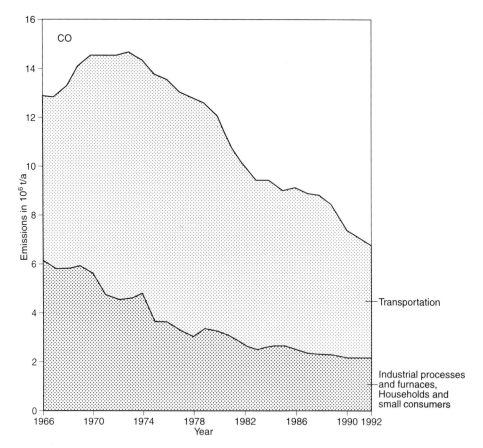

Fig. 2.44. Trend of CO emissions in the old states of the FRG from different sources (acc. to [103])

furnaces, particularly by special processes with substoichiometric operation, e.g., in coking plants and foundries, and to a high degree by unplanned and uncontrolled fires, e.g., by bush fires or when burning straw on fields. CO emissions have been greatly reduced in the industrial countries through improved combustion processes and control as can be seen in Figs. 2.44 (for Germany) and 2.45 (for the USA). It is obvious, that in these countries transportation contributes considerably to CO emissions. The emission-reducing effect of the low-emison vehicles sold in the past years has not yet had a real bearing on it, as it takes more than 10 years for an industrial country to renew a major part of its automotive pool. In the future, however, CO emissions will drop further with the increasing use of low-emission vehicles.

Developing countries lag behind in their improvements of combustion processes and thus to CO reduction. Due to this aspect as well as to bush and savanna burning processes specific (per capita) CO (as also VOC and particulate matter emissions) is relatively high, s. Fig. 2.35.

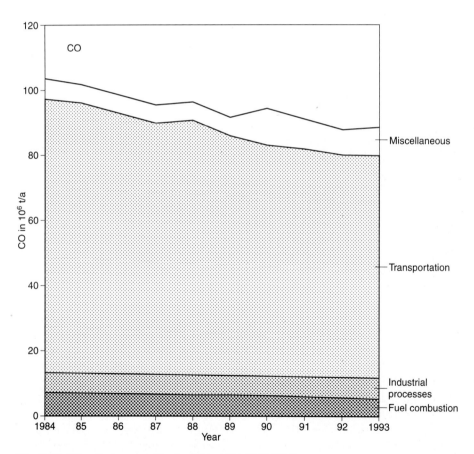

Fig. 2.45. CO emissions trend in the USA, 1984-1993 [104]

2.2.6 Volatile Organic Compounds (VOC)

Emissions of volatile organic compounds (VOC) occur in various forms and from a variety of sources. In automobile transportation, one of the main VOC sources, mainly aliphatic and aromatic hydrocarbons are released from the exhaust pipes of vehicles. A large number of aromatic compounds are considered carcinogenic, e.g., benzene and different PAH (polycyclic aromatic hydrocarbons), others contribute to ozone formation in the troposphere, particularly alkenes and aromatic compounds. Diesel vehicles, but also vehicles with catalytic converters running on a too-rich mixture and engines burning motor oil substances produce an intense smell and are an emission problem. Moreover, many industrial processes also release strong odors. Odorous emissions are, e.g., released during the production of food and semiluxuries. Odor-intensive substances are formed by protein degradation during the production of meat flavoring agents in spice factories. Emissions containing fat and water vapor develop in canned meat factories, e.g., when immersing the meat in hot fat or broth. In starch factories strong-smelling vapors are produced and coffee and cocoa roasters give off the well-known odorants consisting of aldehydes, organic acids and other hydrocarbons. Further producers of odors are malt factories, beer breweries and smoking plants. Odor emitters processing high-protein material of animal origin are fish meal factories, bone processing plants and suet melting plants. The following substances can be considered as odorants: ammonia and amines as alkaline nitrogenous substances, hydrogen sulfide and mercaptanes as sulfurous compounds, carbonyl compounds, fatty and amino acids. Odor emissions also emanate from intensive livestock breeding, from different waste disposal plants for sludge conditioning and composting. In animal carcass processing plants where carcasses of sick and condemned animals, of those killed in accidents and also slaughtering waste are processed small fragments having an intensive, nauseating odor are formed from protein particles as a result of rotting and decomposition processes.

In the industry adding organic solvents is a preliminary step to processsing and treating many materials, e.g., in the surface treatment of paint shops, in the printing industry and in the chemical and pharmaceutical industry. Dry cleaners and metal degreasing plants, e.g., work with halogenated hydrocarbons. Volatile organic substances partially vaporize during the process and are released into the ambient air. In metal processing plants organic substances are frequently released, too, e.g., by vaporizing oil in hardening and electrolytic refining plants, in metals melting and in foundries during the production of cores etc.

In plastic and textile processing monomers (formaldehyde, styrol), solvents, decomposition products of thermal molding and propellants from foaming processes are released, depending on the task concerned. For instance, during the spinning process in cellulose (viscose) fiber production the solvent carbon disulfide (CS_2) evaporates from the thread taking shape. The air required for this leaves the vent laden with carbon disulfide [107].

All estimates concerning the emission of organic substances are variable, as the substances released in combustion processes and from other sources depend on many factors. Thus, measuring results of emissions cannot be generally applied to all cases. The estimates formed nevertheless provide an initial overview and show where air

Table 2.13. Estimated, most important emissions of volatile organic compounds (VOC) in Baden-Wuerttemberg, Germany, in 1985 (acc. to [108])

Source	VOC emissions	
	kt/a	%
Furnaces	15	3
Automotive traffic		
exhaust gases	153	32
vaporized gasoline	38	8
Refineries	5	1
Gasoline storage and loading	13	3
Use of solvents	114	24
Gas distribution network	26	5
Dumps	10	2
Livestock breeding	70	15
Forests	48	10
Total	474	100

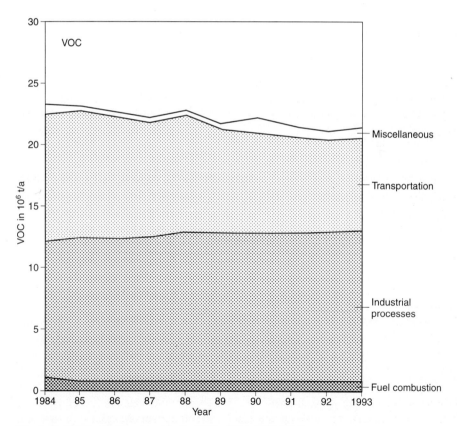

Fig. 2.46. VOC emissions trend in the USA, 1984-1993 [104]

purification measures have priority. Obermeier et al. have, for instance, estimated the main emissions of organic substances from a compilation of many data for the German state of Baden-Wuerttemberg, s. Table 2.13. The substance groups in the individual source areas have been broken down further [108].

The development of VOC emissions is represented in Fig. 2.46 using the example of the USA. Due to stringent emission reduction measures there has been a distinct decline in traffic emissions in the last few years, but in industrial processes VOC emissions have shown a tendency to rise.

2.3 Bibliography

1 Günther, R.: Verbrennung und Feuerungen. Berlin: Springer 1984
2 Baumbach, G.: Emissionen organischer Schadstoffe von Ölfeuerungen. Dissertation, Universität Stuttgart 1977
3 Bauer, A.; Baumbach, G.: Beeinflussung der Schadstoff-Emissionen bei der Holzverbrennung in Zentralheizkesseln. Fortschr. Ber. VDI Z. Reihe 15 Nr. 31, 1984
4 Baumbach, G.; Street, D.; Zayouna, A.: Gas-Chromatographische Bestimmung organischer Spurenstoffe in Ölfeuerungsabgasen mit einer Dünnfilm-Kapillare. Staub-Reinhalt. der Luft 39 (1979) Nr. 5, S. 87/91
5 Weller, L.; Straub, D.; Baumbach, G.: Gaschromatographische Bestimmung der Kohlenwasserstoffverbindungen in Feuerungsabgasen. Fortschr. Ber. VDI Z Reihe 15, Nr. 44, 1986
6 Badger, G.M. et al.: Nature 187 (1960) 663 f
7 Wenz, H.W.: Untersuchungen zur Bildung von höhermolekularen Kohlenwasserstoffen in brennerstabilisierten Flammen unterschiedlicher Brennstoffe und Gemischzusammensetzungen. Dissertation, TU Darmstadt 1983
8 Grimmer, D. et al.: Luftqualitätskriterien für ausgewählte polyzyklische aromatische Kohlenwasserstoffe; Umweltbundesamt Berichte 1/79. Berlin: Erich Schmidt 1979
9 Baumbach, G.: An- und Abfahremissionen bei Ölbrennern kleiner Leistung. Oel+Gasfeuerung 10 (1979) 598/600; Schornsteinfeger 12 (1979), Der Rauchfangkehrer 12 (1979)
10 Kremer, H.: Schadstoffemission von Ölbrennern niedriger Leistung mit Druckzerstäubung; VDI Ber. 246 (1975)
11 Baumbach, G.: Bestimmung der Geruchsemissionen von Ölfeuerungen. VDI Ber. 561 (1986) 199/226
12 Schriever, E.; Marutzky, R.; Merkel, P.: Emissionen bei der Verbrennung von Holz in Kleinfeuerungsanlagen, Staub-Reinhalt. Luft 43 (1983) Nr. 2, S. 62/65
13 Lies, K.-H.; Postulka, A.; Gring, H.; Hartung, A.: Aldehyde emissions from passenger cars. Staub-Reinhalt. Luft 46 (1986) Nr. 3, S. 136 – 139
14 Wagner, H.G.: Homogene Verbrennungsreaktionen, VDI Ber. 146 (1970) 5 – 9
15 Haynes, B.S.; Wagner, H.G.: Soot formation. Progr. Energy Combust. Sci., Vol. 7 (1981) 229 – 273
16 Netz, H.: Betriebstaschenbuch Wärme. Gräfelfing: Technischer Verlag Resch 1974
17 3. Verordnung zur Durchführung des Bundes-Immissionsschutzgesetzes (VO über Schwefelgehalt von leichtem Heizöl und Dieselkraftstoff – 3. BImSchV) vom 15.1.1975, BGBl. I, S. 264 ff. geändert durch Gesetz vom 18.2.1986, BGBl. I, S. 265 ff. und durch VO vom 14.12.1987, BGBl. I, S. 2671 ff
18 Deutsche BP AG: Mineraloel-Story, Druckschrift, Hamburg 1981
19 TÜV Rheinland; Umweltbundesamt: Emissionsfaktoren für Luftverunreinigungen. Materialien 2/80, Berlin: Erich Schmidt 1980
20 Stiftung Warentest: Umweltprobleme trotz guter Qualität (Benzintest), test 4/87, S. 334 – 339

21 Dollnick, H.W.O.; Thiele, V.; Drawert, F.: Olfaktometrie von Schwefelwasserstoff, n-Butanol, Isoamylalkohol, Propionsäure und Dibutylamin. Staub-Reinhalt. Luft 48 (1988) H. 9, S. 325 – 331
22 Mohry, H.; Riedel, H.G.: Reinhaltung der Luft. Leipzig: VEB Deutscher Verlag für Grundstoffindustrie 1981
23 Zentgraf, K.M.: Beitrag zur SO_2-Messung in Rauchgasen und zur Rauchgasentschwefelung mit Verbindungen der Erdalkalimetalle. Fortschr. Ber. VDI Z, Reihe 3, Nr. 22, 1967
24 Struschka, M.; Baumbach, G.: Meßtechnik für die Säuretaupunktmessung in Feuerungsabgasen, Fortschr. Ber. VDI Z Reihe 19, Nr. 2, 1986
25 Speich, R.: Schwefelbilanzuntersuchungen an Braunkohlenkesselanlagen. Braunkohle (1965) H. 9, S. 364/371
26 Stratmann, H.: Schwefelbilanz-Untersuchungen bei Steinkohlen- und Braunkohlenfeuerungen. Mitteilungen der VGB, H. 52 (1958) 23/29
27 Vollner, P.: Untersuchungen der SO_2-Emissions-Verringerung an Rostfeuerungen durch trockene Kalkzugabe. Studienarbeit am Institut für Verfahrenstechnik und Dampfkesselwesen der Universität Stuttgart, 1983
28 Verfuß, F.; Marutzki, D.: Laboruntersuchungen zur Entschwefelungswirkung von Brikettzusätzen, Schlußbericht. Bergbauforschung GmbH, Essen 1985
29 Kuttler, W.: Schwefeldioxid-Emissionen und -Immissionen beim Heizkraftwerk der Universität Stuttgart. Studienarbeit am Institut für Verfahrenstechnik und Dampfkesselwesen der Universität Stuttgart, 1981
30 Kremer, H.; Schulz, W.; Zellkowski, J.: NO_x-Entstehung in Feuerungen, NO_x-Minderung bei Feuerungen. VGB-Schrift TB 310, Technische Vereinigung der Großkraftwerksbetreiber e.V., Essen 1984
31 Technische Vereinigung der Großkraftwerksbetreiber e.V.: NO_x-Bildung und NO_x-Minderung bei Dampferzeugern für fossile Brennstoffe. VGB-Handbuch B 301, Essen 1986
32 Leukel, W.: Schadstoffbildung bei industriellen Verbrennungsanlagen und primäre Minderungsmaßnahmen. Bericht des 1. TECFLAM-Seminars „Schadstoffe bei Verbrennungsvorgängen", TECFLAM, DFVLR-Stuttgart 1985
33 De Soete, G.G.: Heterogene Stickstoffreduzierung an festen Partikeln. VDI Ber. Nr. 498 (1983) S. 171 – 176
34 Technische Vereinigung der Großkraftwerksbetreiber e.V.: Minderungstechnologie für NO_x-Emissionen steinkohlengefeuerter Großkraftwerke. VGB-Techn.-wiss. Berichte „Wärmekraftwerke", H. 8, TW 301, Essen 1980, Kap. 2 NO_x-Emissionen, S. 83 ff
35 Schuster, H.: Primäre NO_x-Minderungsmaßnahmen. in: Heft NO_x-Minderung bei Feuerungen, VGB-TB 310, S. 43/63, VGB Technische Vereinigung der Großkraftwerksbetreiber, Essen 1985
36 Rennert, K.D.: Möglichkeiten der Stickstoffoxidreduzierung in Feuerräumen; Fachreport Rauchgasreinigung. Düsseldorf: VDI 2/86
37 Gauld, D.W.; Pragnell, R.J.: Combustion Parameters Affecting NO_x Emissions in Atmospheric Fluidized Bed Combustion. Proc. 4th Intern. Fluidized Combustion Conference. London: The Institute of Energy 1988
38 Braun, A.; Renz, U.; Drischel, J.; Köser, H.J.K.: N_2O- und NO_x-Emissionen einer Wirbelschichtfeuerung. VGB-Konferenz „Wirbelschichtsysteme 1990", Essen, 1990
39 Braun, A.; Renz, U.: Vergleichende Untersuchungen der Emissionen von Stickoxiden aus stationären Wirbelschichtfeuerungen. VDI Berichte Nr. 922, S. 597/608, Düsseldorf: VDI Verlag 1991
40 Mjörnell, M.; Hallström, C.; Karlsson, M.; Leckner, B.: Emissions from a Circulating Fluidized Bed Boiler II. Chalmers University of Technology, Report A89–180, Göteborg, 1989
41 Babcok, ASR-Brenner, Versuchs- und Betriebsergebnisse, Firmendruckschrift, Deutsche Babcock Werke AG, Oberhausen 1985
42 Diehl, P.: Möglichkeiten zur Verminderung der NO_x-Emissionen bei schweröl- und erdgasbefeuerten industriellen Großfeuerungsanlagen. Diplomarbeit, Institut für Verfahrenstechnik und Dampfkesselwesen der Universität Stuttgart 1983
43 Kremer, H.; Otto, D.: NO_x-Emission von Heizungsanlagen mit Öl- und Gasbrennern. gaswärme-international 30 (1981) H. 1, S. 41/47

44 Weishaupt: Saubere Umwelt durch Öl und Gas, Druckschrift Nr. 888, Max Weishaupt GmbH, 7959 Schwendi, 1985
45 Pfeiffer, R.: Ruhrgas Essen: Minderung von Stickstoffoxid-Emissionen bei gasgefeuerten Anlagen. Vortrag am Institut für Kernenergetik der Universität Stuttgart, 29.11.1985
46 Beier, R. u. a.: Verdrängungsmaschinen, Teil II Hubkolbenmotoren. Handbuchreihe Energie. Gräfelfing: Technischer Verlag Resch, Köln: TÜV Rheinland 1983
47 Greiner, R.: Ein Beitrag zur Gemischbildung bei Ottomotoren unter Berücksichtigung von Abgasfragen. Dissertation, Universität Stuttgart 1976
48 Robert Bosch GmbH: Kraftfahrtechnisches Taschenbuch, 20. Aufl. Düsseldorf: VDI 1987
49 May, H.; Plassmann, E.: Abgasemissionen von Kraftfahrzeugen in Groß-Städten und industriellen Ballungsgebieten. Köln: TÜV Rheinland 1973
50 Hoenig, V.: „Untersuchung der Wirkungsmechanismen von Additiven für schweres Heizöl". Fortschr.-Ber. VDI Reihe 15 Nr. 84. Düsseldorf: VDI-Verlag 1991
51 Hoenig, V. und Baumbach, G.: Schadstoffminderung bei Schwerölfeuerungen durch Additive: Ruß, SO_3, NO_x. VGB-Band „Kraftwerk und Umwelt", Essen 1989
52 Baumbach, G.; Schnell, U.; Spliethoff, H.; Struschka, M.: Partikelemissionen bei verschiedenen Feuerungsanlagen für Holz, Kohle und Öl. Arbeitsgruppe Luftreinhaltung der Universität Stuttgart, Jahresbericht 1992
53 Kauppinen, E.I.; Pakkanen, T.A.: Coal Combustion Aerosols: A Field Study. Environ. Sci. Technol., 1990
54 Natusch, D.: „Size Distribution and Concentrations of Trace Elements in Particulate Emissions from Industrial Sources". VDI-Bericht Nr. 429, Düsseldorf: VDI-Verlag 1982
55 Fahlke, J.: Untersuchungen zum Verhalten von Spurenelementen an kohlebefeuerten Dampferzeugern unter Berücksichtigung der Rauchgasreinigungsanlagen. VGB Kraftwerkstechnik 73 (1973), Heft 3, S. 254/256
56 Struschka, M.: Holzverbrennung in Feuerungsanlagen: Grundlagen – Emissionen – Entwicklung schadstoffarmer Kachelöfen. Dissertation, Universität Stuttgart, 1993
57 Cooper, J. A.: „Environmental Impact of Residential Wood Combustion and Emissions and its Implications". Journal of the Air Pollution Control Association, 30 (1980) 8, pp. 855–861
58 Muhlbaier-Dasch, J.: „Particulate and Gaseous Emissions from Wood-Burning Fireplaces". Environ. Sci. Technol. 16 (1982) 10, pp. 639–645
59 Rau, J. A.; Huntzicker, J. J.: „Composition and Size Distribution of Residential Wood Smoke Aerosols". In: Proceedings 78th Annual Meeting, APCA, Paper N° 85-43,3, Detroit, June 1985
60 Bienek, D.: „Dioxin – Was ist das? Woher kommt es?" gsf mensch + umwelt, Magazin der Gesellschaft für Strahlen- und Umweltforschung München, 1985
61 Hagenmaier, H.: „PCDD und PCDF – Bestandsaufnahme und Handlungsbedarf." VDI-Bericht Nr. 745, Düsseldorf: VDI-Verlag 1989
62 Brunner, H.: Untersuchungen zu Herkunft und Vorkommen polychlorierter Dibenzodioxine und Dibenzofurane in der Umwelt. Dissertation, Universität Tübingen 1990
63 Abel, J.: „2,3,7,8-Intoxikation beim Menschen." VDI-Bericht Nr.634, Düsseldorf: VDI-Verlag 1987
64 Neubert, D.: „Mutagenic and Carcinogenic Potential and Potency of PCDD's and PCDF's." VDI-Bericht Nr.634, Düsseldorf: VDI-Verlag 1987
65 Schlatterrer, B.: „Beurteilung der Dioxinbelastung von Mensch und Umwelt." Vortrag bei der Informationstagung „Dioxin – Bewertung, Meßtechnik, Ausblick" des TÜV-Südwest, Stuttgart 27.11.1991
66 Beck, H.: „Isomerspezifische Bestimmung von PCDD, PCDFa in Human- und Lebensmittelproben." VDI-Bericht Nr. 634, Düsseldorf: VDI-Verlag 1987
67 NATO; CCMS: „Pilot Study on International Information Exchange on Dioxin and Related Compounds." Report 176, Aug. 1988: International Toxicity Equivalency Factors (I/TEF) Method of Risk Assessment for Complex Mixtures of Dioxins and Related Compounds. Report 178, Dez. 1988: Scientific Basis for the Development of the International Equivalency Factors (I/TEF) Method of Risk Assessment for Complex Mixtures of Dioxins and Related Compounds.
68 Poiger, H.; Pluess, N.; Schlatter, C.: „Subchronic Toxicity of some Polychlorinated Dibenzofurans." Dioxin '88 Umeå (veröffentlicht in Chemosphere)

69 Ahlborg, U.G.: „Nordic Risk Assessment of PCDDs and PCDFs." Dioxin '88 Umeå, (veröffentlicht in Chemosphere)
70 Umweltbundesamt: Sachstand „Dioxine". 262, Berlin 1984
71 Barnes, D.G.; Bellin, J.; Cleverly, D.: „Interim Procedure of Estimating Risk Associated with Exposures to Mixtures of Chlorinated Dibenzodioxins and Dibenzofurans (CDDs and CDFs)". Chemosphere 15, 1895 (1986)
72 17. Verordnung zur Durchführung des Bundes-Immissionsschutzgesetzes. Verordnung über Verbrennungsanlagen für Abfälle und ähnliche brennbare Stoffe – 17. BImSchV vom 23.11.1990, BGBl. I, S. 2545, 2832
73 Rotard, W.: „PCDD/PCDF in Wasser, Sediment und Boden." VDI-Bericht Nr.634, Düsseldorf: VDI-Verlag 1987
74 Gutachten des Bundesgesundheitsamtes zur Bodenbelastung mit PCDD/PCDF in Crailsheim-Maulach vom 12.05.1989
75 Stuttgarter Zeitung: „Äcker sind kaum mit Dioxin belastet." Ausgabe vom 23.01.1992
76 Rappe, C.: „Levels, Profiles and Patterns." 11th International Symposium on Chlorinated Dioxins and Related Compounds, Sept. 23rd–27th, 1991; Research Triangle Park, North Carolina USA; Introduction p.17
77 Valet, P.-M.: „Ergebnisse des Dioxinminimierungsprogramms in Baden-Württemberg." ECOPLAN/TÜV Südwest – Symposium „Dioxine – Belastung – Quellen – Verbleib", 6./7. 10. 1992, Tagungsbericht Stuttgart 1992
78 Hagenmaier, H.; Tichaczek K.-H.; Brunner H.; Mittenbach G.: „Application of $DeNO_x$-Catalysts for the Reduction of PCDD/PCDF and other PICs from Waste-Incineration Facilities by Catalytic Oxidation." Second annual international speciality conference „The Municipal Waste Combustion", April 15th–19th, 1991, Tampa, Florida
79 Richter, E.: „Die Abscheidung von polychlorierten Dioxinen und Furanen aus Abgasen mit Aktivkoksverfahren." Chem.-Ing.-Tech. 64 (1992) Nr.2, S.125–136
80 Stützle, R.; Hagenmaier H.; Hasenkopf O.; Schetter G.: „Erste Erfahrungen mit der Demonstrationsanlage in der MVA Stuttgart-Münster zum Abbau von PCDD und PCDF in Flugaschen", VGB Kraftwerkstechnik 71 (1991), Heft 11, S.1038–1042
81 Vehlow, J.; Vogg, H.: „Thermische Zerstörung organischer Schadstoffe". Müllverbrennung und Umwelt (1991), S. 447–467
82 Seifert, U.: „Aufgaben der Automobilforschung". ATZ (1983), 653
83 Umweltbundesamt: „Luftreinhaltung '81 – Entwicklung – Stand – Tendenzen". Materialien zum 2. Immissionsschutzbericht der Bundesregierung an den Deutschen Bundestag. Berlin: Erich Schmidt Verlag 1981
84 Bussien, R. (Begr.); Goldbeck, G. (Hrsg.): Automobiltechnisches Handbuch, Ergänzungsband zur 18. Auflage. Berlin: de Gruyter Verlag 1978
85 Eberius, H.: „Untersuchungen zur Restkohlenwasserstoff-Emission von Flammen in zylindrischen abgeschlossenen Feuerräumen". Bericht des 1. TECFLAM-Seminars „Schadstoffe bei Verbrennungsvorgängen", TECFLAM, DLR-Stuttgart, 1985
86 Berg, W.: „Verfahren zur Feldüberwachung in den USA und Kalifornien". VDI-Bericht 639, S.87–125, Düsseldorf: VDI-Verlag 1987
87 Nicht limitierte Automobil-Abgaskomponenten. Druckschrift, Volkswagen AG, Forschung und Entwicklung, Wolfsburg 1988
88 Hassel, D.; Brosthaus, J.; Dursbeck, F.; Jost, P.; Sonnborn, K.-S.: „Das Abgas-Emissionsverhalten von Nutzfahrzeugen in der Bundesrepublik Deutschland im Bezugsjahr 1980". Umweltbundesamt-Forschungsbericht 104 05 740/02, UBA-FB 83–030. Berlin: Erich Schmidt Verlag 1983
89 Hertkorn, W.: „Veränderungen des Kraftstoffverbrauchs und der Abgasbelastungen durch Geschwindigkeitsreduktion in untergeordneten städtischen Straßennetzen". Schlußbericht zum FE 77040, Institut für Straßen- und Verkehrswesen der Universität Stuttgart, 1989
90 Hassel, D.; Dursbeck, F.; Brosthaus, J.; Jost, P.; Hofmann, K.: „Das Abgas-Emissionsverhalten von Personenkraftwagen in der Bundesrepublik Deutschland im Bezugsjahr 1985". Forschungsbericht 104 05 143, UBA-FB 87-036. Umweltbundesamt, Berichte 7/87. Berlin: Erich Schmidt Verlag 1987

91 Hassel, D.; Weber, F.-J.: Ermittlung des Abgas-Emissionsverhaltens von PKW in der Bundesrepublik Deutschland im Bezugsjahr 1988. Zwischenbericht, UFOPLAN-Nr. 104 05 152, UBA-FB 91–042. Umweltbundesamt Texte 21/91, Berlin 1991
92 Meier, E.; Plaßmann, E.; Wolff, C.: Abgas-Großversuch – Untersuchung der Auswirkungen einer Geschwindigkeitsbegrenzung auf das Abgas-Emissionsverhalten von Personenkraftwagen auf Autobahnen. Abschlußbericht. Köln: Verlag TÜV Rheinland 1986
93 Staab, J.; Klingenberg, H.; Schürmann, D.: Strategy for the Development of a New Multicomponent Exhaust Emission Measurement Technique. SAE Technical Paper Series N° 830 437, Detroit, MI, March 1983
94 Grauer, A.; Baumbach, G.: Messung der Abgasemissionen am fahrenden Fahrzeug mit einem mobilen Meßsystem. VDI-Bericht Nr. 1059, S. 467/480, Düsseldorf: VDI-Verlag 1993
95 Dr.-Ing.h.c. F. Porsche AG; Institut für Chemische Technologie und Brennstofftechnik der TU Clausthal; TÜV Rheinland: Aromaten im Abgas von Ottomotoren – Materialienband. Abschlußbericht der drei Forschungsstellen im Auftrag der Deutschen Wissenschaftl. Gesellschaft für Erdöl, Erdgas und Kohle e.V., Hamburg, der Forschungsvereinigung Verbrennungskraftmaschinen e.V., Frankfurt, und des Umweltbundesamtes, Berlin 1988
96 Kraftfahrt-Bundesamt (Hrsg.): Schadstoff-Typprüfwerte von Personenkraftwagen, Kombinationskraftwagen und Kleinbussen mit Allgemeiner Betriebserlaubnis. 1. Ausgabe, Flensburg 1991
97 Pester, W.: Direkteinspritzung führt zu erheblich sparsameren Ottomotoren. VDI Nachrichten Nr. 15, 16.4.1993
98 VDI-Handbuch Reinhaltung der Luft. Berlin, Köln: Beuth, continuously supplemented
99 Theodore, L.; Buonicore, A. (Eds.): Air Pollutions Control Equipment – Selection, Design, Operation and Maintenance. Berlin, Heidelberg: Springer 1994
100 Umweltbundesamt: Luftreinhaltung '88 – Tendenzen – Probleme – Lösungen. Materialien zum Vierten Immissionsschutzbericht der Bundesregierung an den Deutschen Bundestag. Berlin: Erich Schmidt Verlag 1989
101 Laing, R.; Baumbach, G.; Lamberth, B.; Trukenmüller, A., Baumann, K.; Friedrich, R.; Pfeiffer, F.; Rott, U.; Voß, A.: Environmental Monitoring and Impact Assessment in Nigeria. Final Report, EU Project No. 6100.52.41.025/(6/ACP-UNI/025). Institute for Energy Economics and Rational Use of Energy, University of Stuttgart, 1994
102 Boden, T.A.; Kaiser, D.P.; Sepanski, R.J. and Stoss, F.W. (eds.): Carbon Dioxide Emissions – Introduction. pp. 497-500 in: Trends '93: A Compendium of Data on Global Change: ORNL/CDIAC-65. Carbon Dioxide Information Analysis Center, Oak Ridge National Laboratory, Oak Ridge, Tenn., USA, 1994
103 Umweltbundesamt und statistisches Bundesamt: Umweltdaten Deutschland 1995. Umweltbundesamt, Berlin und statistisches Bundesamt Wiesbaden, Germany, 1995
104 U.S. Environmental Protection Agency (EPA): National Air Quality and Emissions Trends Report, 1993. EPA Office of Air Quality Planning and Standards, EPA Report 454/R-94-026, Research Triangle Park, North Carolina 27711, October 1994
105 Kraftfahrtbundesamt: Bestand an Personenkraftwagen und Schadstoffgruppen. Statistische Mitteilungen, Reihe 1, Kraftfahrzeuge, Heft 6. Stuttgart: Metzler-Hoesch-Verlag, 1995
106 Boysen, B.; Friedrich, R.; Müller, T.; Scheirle, N.; Voß, A.: Feinmaschiges Kataster der SO_2- und NO_x-Emissionen in Baden-Württemberg und im Oberrheintal - Emissionsuntersuchungen im Rahmen des TULLA-Prjektes. Bericht über das 2. Status-Kolloquium des PEF, Band 2, S. 481/492, Kernforschungszentrum Karlsruhe 1986
107 Eigenberger, G.: Abluftreinigung - Schadgase und Gerüche. ALS Arbeitsgruppe Luftreinhaltung der Universität Stuttgart, Jahresbericht 1988
108 Obermaier, A.; Friedrich, R.; John, C.; Voß, A.: Zeitlicher Verlauf und räumliche Verteilung der Emissionen von flüchtigen organischen Verbindungen und Kohlenmonoxid in Baden-Württemberg. Forschungsbericht KfK-PEF 78, Kernforschungszentrum Karlsruhe 1991

3 Air Pollutants in the Atmosphere

Before becoming effective air pollutants emitted into the atmosphere can be subject to a variety of influences. These are, e.g., physical dilution, chemical transformations, enrichment or removal processes such as leaching. Life, deposition and impact of the air pollutants are considerably determined by these factors. These influences are again dependent on atmospheric processes which vary according to weather conditions.

3.1 Meteorological Influences on the Dispersion of Air Pollutants

The dispersion of air pollutants depends on wind direction, wind velocity and vertical turbulence. Wind and turbulence are part of a mass exchange dominating the entire weather system. This mass exchange compensates for the temporally and locally varying effects of radiation by turbulent air flows in the entire atmosphere (troposphere, s. section 3.2), in the boundary layer, in particular.

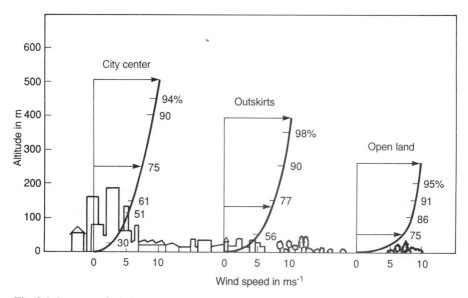

Fig. 3.1. Increase of wind speed with height relative to different ground unevenness [2, 3]

3.1.1 Wind

Emission dispersion occurs in the direction of the wind. The wind has a diluting effect on the flue gas concentrations behind a stack which is approx. proportional to its horizontal velocity. The exponential increase of the wind velocity with altitude favors the dispersion of flue gases from high stacks, s. Fig. 3.1. At the same altitudes, wind velocity over level ground is higher because of low ground roughness than over highly structurized areas with high ground roughness, e.g., over cities. The wind can also veer when changing altitudes [1] under the influence of friction. Wind velocity has a mean diurnal course reaching a maximum during daytime and a minimum during nighttime.

3.1.2 Turbulence

The constant air flows in the atmosphere are subject to irregular movements called turbulences. Turbulences have two components which cannot be considered separately despite their different meteorological causes, as they frequently interchange. Friction turbulence prevailing at ground level increases with higher wind velocities and roughness of the ground, e.g., over urban and industrial sites but also over orographic obstacles, and, like the wind, reaches a daily maximum. This friction turbulence manifests itself in rapid changes of wind direction and velocity; one speaks of gusty winds.

The second component, the so-called thermal turbulence, is of considerable significance for the vertical exchange. It penetrates into very high altitudes and is also known as "convection". It occurs only in unstable temperature layers.

The stability or instability induced by temperature layers in the atmosphere develops as follows.

The layer of air around the earth is not heated directly by solar radiation but by heat radiation or convective heating from the ground after the ground has been warmed up by high-energy solar radiation, thereby causing a reduction of temperature from the layers close to the ground to those higher up.

Because of the weight of the air, the density and subsequently the pressure increase down to the ground level, in other words the pressure decreases with height. As a result, when rising air reaches areas of lower external pressure, it can expand and consequently cools down. Sinking air, on the other hand, enters areas of higher external pressure where it is compressed and thus warmed up. These temperature changes caused by differences in pressure during vertical movements are called *adiabatic lapse rates*. This is based on the assumption that energy is neither supplied nor removed, representing an idealized assumption for a closed system which usually does not apply so absolutely in nature. The magnitude of the adiabatic lapse rate varies in terms of time and space depending on pressure and humidity conditions; the most frequent values of lapse rates are between 0.5 and 0.8 °C per 100 m [4-6]. Superadiabatic lapse rates, i.e., higher temperature drops with height lead to unstable layers, while subadiabatic gradients lead to stable ones. Superadiabatic layers are illustrated in Fig. 3.2. Let us suppose an air parcel at an altitude a with the tempera-

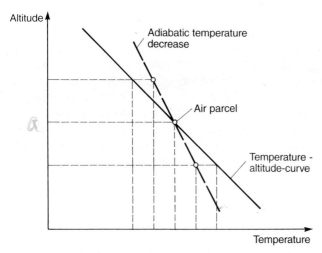

Fig. 3.2. Superadiabatic (unstable) layers [7]

ture ϑ. If this air parcel is lifted up, it cools down adiabatically (ideally speaking). According to the lapse rate representing the temperature gradient in the ambient air at a certain place and time, it may be observed that the air parcel is now warmer than its surroundings. Consequently, owing to its lower density, it experiences an uplift. Analogous to this, a dropping of the air parcel leads to an approximate adiabatic warming. As its temperature is now lower than that of its surroundings it will further sink. Thus, the slightest causes can trigger off radical changes in the atmosphere which do not lead back to the original state. These superadiabatic layers are therefore called instable. As these changes do not represent steadily moving flows (upward movements in one place, downward movements in another) one speaks of thermal turbulence. Examples of great instability result from dramatic temperature drops in higher altitudes due to rapid penetration of warm air into cold air masses or vice versa. Such situations manifest themselves violently as thunderstorms, rain, snow and hail showers [7].

Violent turbulences generally lead to noxious pollutants being well dispersed in the atmosphere, which is the reason why there are no local peak concentrations of long duration. The influences of wind and turbulence on the spreading of plumes are shown schematically in Fig. 3.3.

The formation of unstable temperature layers in the lower atmosphere (by convection) is, in the majority of cases, caused by the warming of the ground through solar radiation. Thus, the maximum of thermally caused turbulence occurs at noon or in the early afternoon hours, whereas the minimum can be observed at night. Just as there is a diurnal cycle, there is also an annual cycle with its maximum in summer. In the mean, these cycles dominate the turbulent structure of the atmosphere and its mixing qualities. The artificial warming of the lowest layers of air over large urban and industrial sites, often exceeding the winterly radiation of 160 J/(cm^2 and day), can also be the cause of additional turbulence in an unstable lower layer [1, 8, 9].

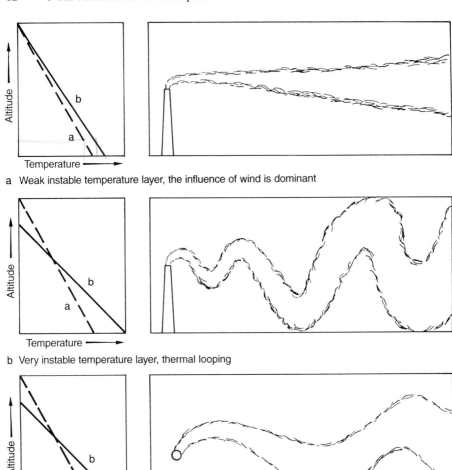

a Weak instable temperature layer, the influence of wind is dominant

b Very instable temperature layer, thermal looping

c Horizontal looping due to varying wind directions and instable layers

Fig. 3.3. Influences of wind and turbulence on the dispersion of smoke plumes.
Fields on left: temperature layers (*a.* adiabatic temperature-height curve; *b.* actual prevailing temperature-height curve; fields on right: behavior of smoke plume)

Topographic factors such as mountains and valleys as well as buildings and vegetation frequently have a substantial influence on air flows and thus on the dispersal of air pollutants [10]. E.g., so-called lee eddies can form behind obstacles, possibly leading to brief peak concentrations of pollutants [2]. Fig. 3.4 shows several specific meteorological and topographic influences on the dispersion of plumes.

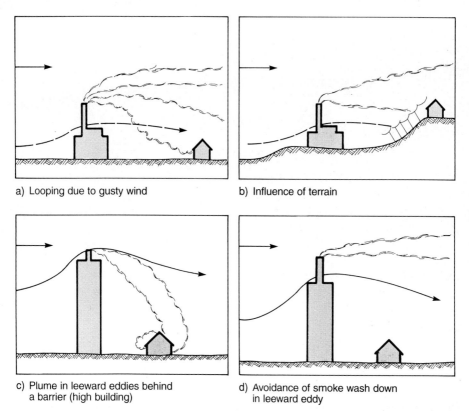

a) Looping due to gusty wind

b) Influence of terrain

c) Plume in leeward eddies behind a barrier (high building)

d) Avoidance of smoke wash down in leeward eddy

Fig. 3.4 a-d. Specific meteorologic and topographic influences on the dispersion of smoke plumes [2], → wind direction

3.1.3 Inversion

If the temperature decreases less with higher altitudes than corresponds with the adiabatic lapse rate, then one speaks of subadiabatic or stable layers. Fig. 3.5 illustrates this case.

If the air parcel at the altitude a and with the temperature δ is raised, it will then, ideally, cool down adiabatically. As the ambient temperature is now higher than the temperature of the air parcel, it will sink to its original altitude a due to its higher density. If it moves down from the altitude a, it warms up again approximately adiabatically, becoming warmer than its surroundings, and thus rises again to its original altitude a. Hence, any vertical movement is completely neutralized; the atmosphere has stable layers. One speaks of neutral or indifferent layers when the temperature decrease with higher altitudes is equal to the adiabatic gradient.

Air layers are particularly stable when the temperature not only decreases slightly with higher altitudes but actually increases. As the normal temperature-altitude curve is reversed here, this case of atmospheric layering is called inversion. In an inversion, being a pronounced case of subadiabatic layers, no up or down movements are pos-

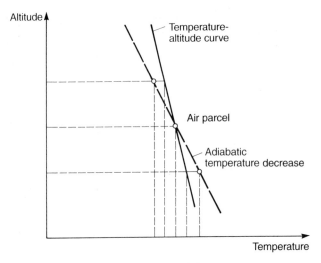

Fig. 3.5. Subadiabatic (stable) layers [7]

sible. The inversion functions as a barrier for convective processes and thus limits the dispersion of air pollutants.

3.1.3.1 Inversion Types

One distinguishes between two types of inversion – ground or surface inversions and elevated inversions. Fig. 3.6 illustrates the temperature behavior of these two types of inversion with the help of a diagram.

With the *surface inversion* a cold layer of air is resting on the surface of the ground; the temperature increases from the ground to the upper boundary of the inversion. Above this upper boundary it decreases again in accordance with natural principles.

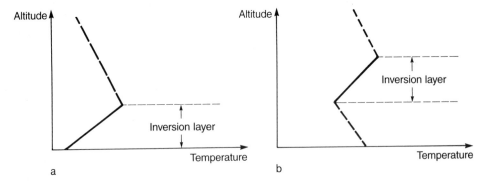

Fig. 3.6. Temperature curves in surface and elevated inversions. **a.** surface inversion; **b.** elevated inversion

3.1 Meteorogical Influences on the Dispersion of Air Pollutants   85

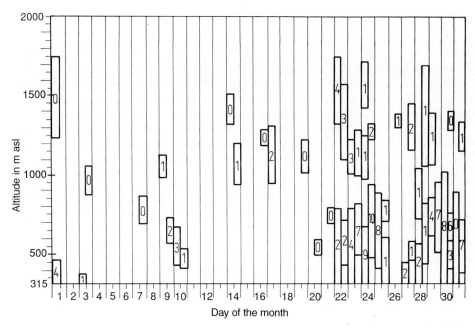

Fig. 3.7. Diagram of inversion layers, using one January as an example [11]

With the *elevated inversion* the temperature first decreases beginning at the ground upward, and from a certain altitude onwards (at the lower boundary of inversion) it starts to increase, to ultimately decrease again at the upper boundary of the inversion in accordance with natural principles.

Regular investigations of the vertical temperature layers are carried out by worldwide meteorological services via tethersonde measuring systems with telemetric transmitters. The balloon ascents are usually carried out every day at 0:00 and 12:00 o'clock. Fig. 3.7 graphically illustrates the inversion altitudes of one month above the meteorological station in Stuttgart, Germany. The lower lines of the bars mark the altitude at which the inversion starts. The bar stops at the altitude starting from which the temperature decrease takes place in accordance with the natural principles. The figures inside the bars indicate in K the rise in temperature from the lower to the upper limit of the inversion respectively.

3.1.3.2 Inversion Formation

Radiational inversion

Surface inversions are formed by radiation cooling of the ground. With slight cloudiness and only light wind – characteristics of high pressure weather conditions – the ground releases its heat after sunset via radiation; i.e., the balance of radiation absorbed and released becomes negative (negative radiation balance). With an overcast sky radiation cooling either does not occur or does so in a much weaker form as the clouds reflect and absorb radiation.

Air cannot warm up convectively from the ground cooled by radiation; on the contrary, the air in contact with the ground convectively conducts its heat to the ground. The layers of air close to the ground thus cool down more than the upper layers. The formation of the layer of cold air on the ground is further reinforced by sinking processes. The cold air, being the stable layer, remains on the ground and increases in volume during the night. During the following day the ground warms up again due to solar radiation, dissolving the inversion from below owing to the convection starting on the ground. If, during the day, the inversion does not break up completely from below, one speaks of an elevated inversion [12]. In Fig. 3.7, e.g., one can see, that for the period from January 22nd to the end of the month the nocturnal surface inversions (determined at midnight) dissipated somewhat from ther ground during the day (12:00 noon).

The occurrence of inversions is not seasonal in nature but possible throughout the whole year if suitable conditions prevail. Long-term statistics, e.g. for Stuttgart, Germany, show that nocturnal inversions occur the most frequently in summer and in autumn; stable weather conditions are more frequent in winter and spring than in these two seasons [11]. However, inversions in winter have a higher extension than in summer as radiational cooling lasts longer during the long nights. As a result, inversions continuing into the day occur predominantly in the winter months. The huge inversion layers shown in Fig. 3.7 (e.g., from January 22 to 25) did not dissolve during the day, but merely rose.

As a result of the more intense warming of the ground in summer, the inversions dissolve much faster than in winter.

The formation of ground inversions also depends strongly on the shape of the terrain. Cold air masses tend to flow into valleys, so that the ground inversions there are more frequent and greater in mass than in the mountains or on their upper slopes. The cooling of close-to-ground air can advance to such a degree that the temperature falls below the dewpoint. In this way ground fog, also called radiational fog, is formed.

Subsidence inversion

One characteristic of high-pressure weather conditions is the descent (subsidence) of air over a large area [6], during which the sinking cold air warms up adiabatically. Part of this heat causes cloud droplets to vaporize, thus leading to a dissolution of clouds and haze. Clear air and good visibility at high altitudes are the result. The air masses only sink to a certain distance from the ground. The lowest layers of air are warmed up on the ground via convection during the day and rise as so-called thermals. They cool down while rising, thus forming a dynamic barrier at a certain height for the sinking air. Hence, both the sinking warming air as well as the rising, cooling air can only escape laterally. A barrier layer with a horizontal direction of movement in its middle is formed having a warm upper section and a cold lower one, viz. an inversion. Elevated inversions such as these are called dynamic or subsidence inversions. Fig. 3.8 schematically illustrates the formation of a subsidence inversion.

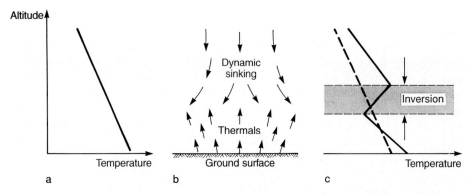

Fig. 3.8. Diagram of the flow and temperature conditions during the formation of a subsidence inversion [6]. **a** initial stage, temperature drop with increasing height; **b** flow conditions; **c** warming of sinking air, cooling of the heated near-ground air, inversion in between

As there are fewer and weaker thermals in autumn and winter, subsidence inversions can sink lower than in the summer; they can even settle on top of a ground inversion and thus form huge and extremely stable layers [13].

Frontal inversions

Elevated inversions can be formed when large air masses settle on top of each other. If, in winter or in autumn, cold, continental mainland air has flowed in over a longer period from the east, then, when there is a weather change, warm fronts coming in from southwest or west are frequently unable to displace this cold air immediately. The warm air flows on top of the cold one, forming an inversion. If the cold air reaches down to the ground a huge surface inversion has formed; if the surface inversion dissolves from below during the day then it is an elevated inversion. If the differences in air pressure are slight, such conditions – cold air at the bottom, warm air on top – can sometimes be sustained for a longer period of time, e.g., in January 1982 over Stuttgart, Germany [14].

If the warm air in the higher altitudes is very humid, condensations ensue at the boundary to the cold air. The clouds thereby formed at high altitudes can inhibit solar radiation, so that the cold air just above the ground cannot warm up. In this case, the inversion remains intact down to the ground the whole day.

3.1.4 Mixing Layer and Barrier Layers

The lowest section of the troposphere in which the air pollutants emitted mix with the surrounding air is the so-called *mixing layer*. As described above, good mixing results from wind and turbulence. There is a reduced incidence, or even total supression of vertical exchange due to the stable layers in the atmosphere. Stable layers of air with merely a slight drop in temperature with altitude or inversions act as barrier layers. Inversions prevent a vertical distribution of the pollutants released in these

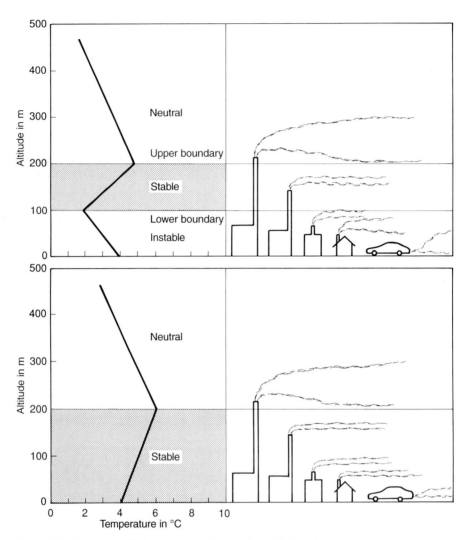

Fig. 3.9. Behavior of pollutant emissions being released below, in or above inversion layers

layers, unless these pollutants can penetrate the layer with the help of the buoyancy of exhaust plumes, s. Fig. 3.9. In surface inversions the mixing layer in which the dispersion of pollutants takes place is extremely shallow. With the warming effect of solar radiation in the morning the air just above the ground is also warmed up, dissolving the surface inversion from below. Thus the altitude of the mixing layer rises. Apart from the influence of the surface inversion, the mixing layer is generally bound by an elevated inversion or a stable layer of air on its upper side. Giebel's measurements [15] led to the discovery that the mixing layer is usually limited by a main barrier layer on its upper side. He defined this main barrier layer, as compared to the several weak inversion layers usually prevalent, as the air layer with the most sudden decrease of the aerosol concentration. A sudden decrease in the relative hu-

Fig. 3.10. Clearly visible effect of the main barrier layer during an autumnal high pressure weather situation: below the barrier layer there is haze with an increased aerosol concentration and humidity of the air, above there is excellent visibility; Roßberg, Swabian Jura, Oct. 27, 1985 (Photo: Günter Baumbach)

midity of the air can also frequently be observed here. From within the mixing layer up into the stable inversion layer, the aerosol concentration combined with the high humidity can lead to a marked decrease in visibility due to haze, whereas above the main barrier layer there is completely clear air with good visibility. Situations such as the one described above occur predominantly in autumn in stable high pressure weather situations, s. Fig. 3.10. The barrier effect prevails both above and below the stable inversion layer: air masses and exhaust plumes spreading above the inversion are prevented from sinking to the ground, and air pollutants released below the stable inversion layer are trapped in the air layer close to the ground, s. Fig. 3.9.

With the beginning of solar radiation the surface inversion is dissolved by convective heating. The upper limit of the mixing layer rises in the course of the day until the afternoon due to the increase of warm air masses close to the ground. Particularly in winter, surface inversions can drastically impede the rising of the main barrier layer, namely by preventing the formation of warm air in the lowest layers. Rise time and absolute height of the mixing layer's upper limit depend, in particular, on the time of year, the intensity of solar radiation, on air temperature and on wind velocity. Fig. 3.11 shows the mean diurnal course of the mixing layer altitude by means of a schematic diagram.

The behavior of atmospheric stable layers in the Ruhr District of Germany has been exhaustively investigated and described by Giebel [15, 16]. According to his findings and the inversion statistics of the German Meteorological Service, the mixing layer altitude shows values of between 0 m (at night and during the cold season)

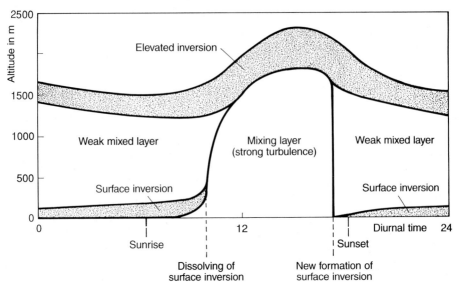

Fig. 3.11. Diurnal profile of barrier and mixing layer levels during summer high-pressure weather conditions over open terrain

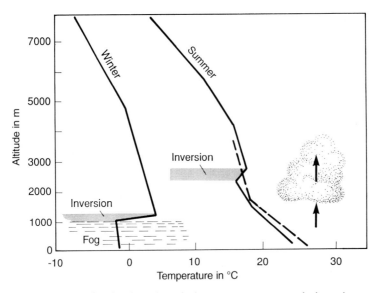

Fig. 3.12. Inversion levels and vertical temperature courses during winter and summer high-pressure weather conditions [13]. ——— vertical temperature course of air layers, --- vertical temperature course of rising air

and in excess of 3000 m (during high pressure weather conditions in the summer). In addition, due to the great radiation heat of the ground there is a much higher lapse rate in the mixing layer in summer during the day than in winter, resulting in more violent turbulences. Fig. 3.12 presents the different lapse rates and inversion layer altitudes for winter and summer in a schematic diagram.

3.1.5 Inversion Layers and Air Pollutants – Examples of the Dispersion of Pollutants

The behavior of exhaust gas plumes below, in and above barrier layers was illustrated with the help of Fig. 3.9. Along with emissions, barrier layers are of great significance in the occurrence of high pollutant concentrations. The well-known disastrous incidences of smog with numerous fatalities have happened exclusively during low exchange weather situations with pronounced and persistent inversions [17, 18]. The most recent smog alarm situations in Germany, too, occurred during weather situations with special inversion layers [14, 19, 20].

How drastically air pollution can increase during inversions as compared to days with good air circulation is shown in Fig. 6.9 by the example of the surroundings of an arterial road with heavy traffic.

The following two sections contain one example each of the long-range air pollutant transport (during elevated inversion situations) and of the short-range dispersion in a highland valley.

3.1.5.1 Widespread Dispersion of Air Pollutants: SO_2 Long-Range Transport

In large areas of the Federal Republic of Germany high SO_2 concentrations could occur as a result of transregional long-range transport from eastern directions. Repeated measurements carried out from aircraft during eastern winds along the borders of the former GDR and CSSR disclosed that transboundary SO_2 can be found along a wide front in high concentrations [21, 22]. During southeasterly winds North Germany received the brunt of it [23], during direct easterly winds Hesse and even North Rhine-Westphalia are at the receiving end, e.g., during the smog of January 1985 [24], and during northeasterly winds, South Germany. In February 1986, e.g., recurring conditions favoring long-range transport in South Germany during northeasterly winds led to the situation that the "Clean Air" measuring stations of the Federal Environmental Agency in the Bavarian Forest (Brotjacklriegel) and in the Black Forest (Schauinsland) registered the highest SO_2 diurnal and monthly mean values since the series of measurements began in the sixties [25].

On the basis of the situation of February 1986 during which very high SO_2 concentrations were also measured in the greater Stuttgart area and in the northern part of the Black Forest, long-range transport will be explained in greater detail [26]. As can be seen from the wind vector diagram in Fig. 3.13a Baden-Wuerttemberg and Bavaria had winds from northeasterly directions almost the entire month of February. Almost the entire month pronounced elevated inversions with their lower limits between 800 and 1200 m (above NSL) also prevailed. The behavior of these inversion layers is illustrated in Fig. 3.13b.

The elevated inversion layers prevented upward dispersion of the pollutants. On the other hand, few low altitude and surface inversions occurred, so that the "clouds

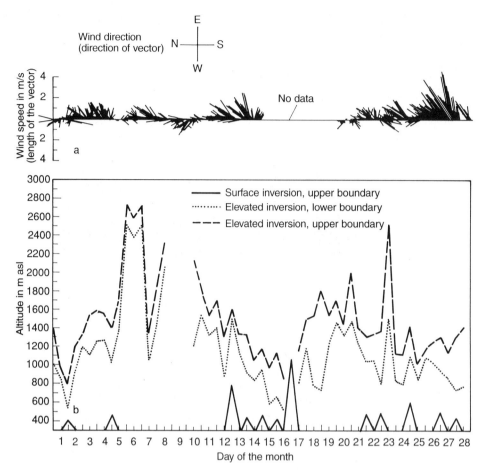

Fig. 3.13. February 1986 – wind conditions and behavior of the inversion layers in the Stuttgart area. **a** wind directions and speeds measured at the air pollution measuring station of the Institute for Process Engineering and Power Plant Technology in the forests of Schönbuch (between Stuttgart and Tübingen) [27]; **b** level of the inversion layers above Stuttgart (according to radio sonde ascents of the German Weather Service, Aerological Station Stuttgart, 315 m above mean sea level [28])

of pollutants" were not restricted to the lower air layers and hence were not stopped by the highlands. Besides this, low temperatures and a dry snow cover throughout the country effected a reduced deposition of SO_2. Thus, all the preconditions for long-range transport of air pollutants particularly for SO_2 existed in February 1986. When and where high concentrations of this kind occurred depended essentially on the prevailing wind direction and wind velocity in the different levels.

Fig. 3.14 presents an overview of the half-hourly SO_2 values measured in February 1986 by different measuring stations in South Germany. Each half-hourly mean value is represented by a vertical line in this diagram.

High concentrations can be observed from February 6-13, 6-20 and 23-26, 1986. The transport of SO_2 could be observed particularly well by the increase of SO_2 on

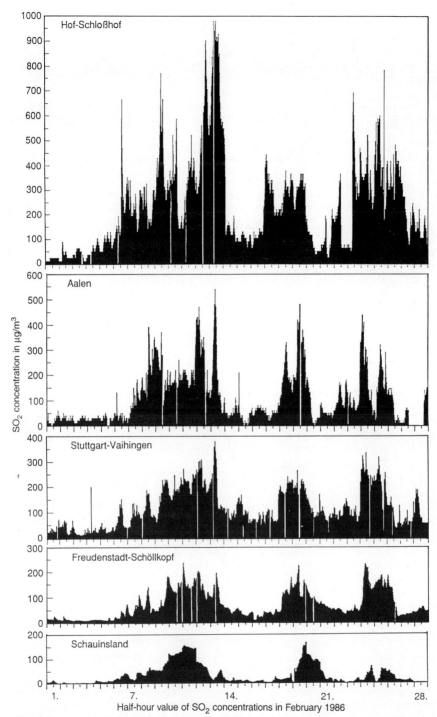

Fig. 3.14. SO$_2$ profiles in February 1986 at different measuring stations in South Germany [25, 27, 29-31]

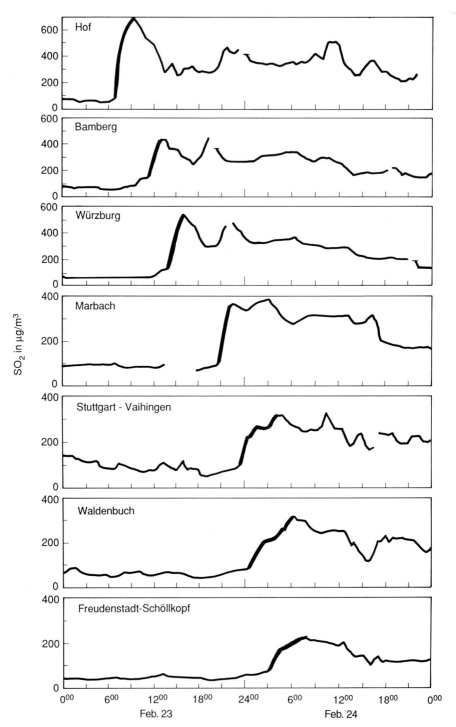

Fig. 3.15. SO_2 profiles on February 23 and 24, 1986, at different measuring stations in South Germany; the increases propagate from northeast to southwest

3.1 Meteorological Influences on the Dispersion of Air Pollutants   95

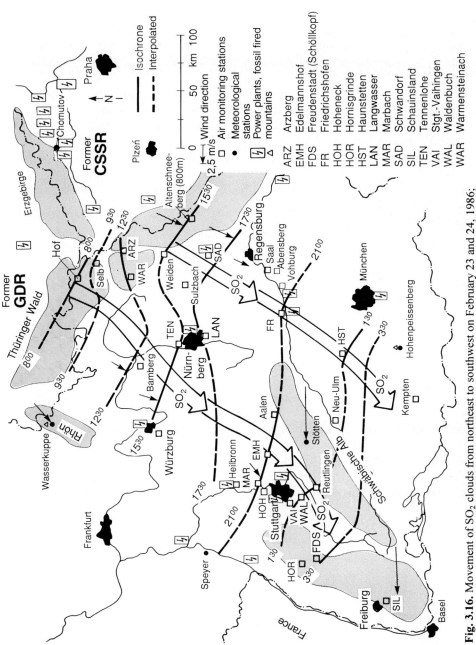

Fig. 3.16. Movement of SO$_2$ clouds from northeast to southwest on February 23 and 24, 1986; the propagation of the steep increase is shown by the isochrones according to Fig. 3.15

February 23/24. As an example, Fig. 3.15 shows that these increases of SO_2 were carried from the measuring stations in Hof in North Franconia, Germany, over the greater Stuttgart area to the Black Forest. The time lag at the individual measuring stations corresponds with the prevailing wind velocities of approx. 3 to more than 5 m/s [26].

From the SO_2 increases measured in numerous measuring stations in Northeast Bavaria and Baden-Wuerttemberg, the path of the "SO_2 clouds" during the conditions on February 23 and 24, 1986, (shown in Fig. 3.16) could be reconstructed [26]. Besides the western "SO_2 cloud" originating in the former GDR and spreading from Hof via Bamberg, Würzburg and Stuttgart into the Black Forest, there was an eastern cloud which was additionally fed from the CSSR and which made its way via Arzberg, Weiden to the area of Neu-Ulm. The SO_2 clouds observed in South Germany were transported over distances of more than 400 km from the industrial areas of the former GDR and CSSR.

The transport directions during the conditions on February 23/24, 1986, could also be reproduced with trajectories of the German Meteorological Service [26].

During the night from February 23 to February 24 SO_2 concentrations increased from 80 to approx. 320 μm/m³ in the Schönbuch (Waldenbuch), Germany, and from 50 to 220 μm/m³ at the Schöllkopf (near Freudenstadt). As the wind had been blowing from the same direction for hours before the increase, the influences of long-range transport and local sources on SO_2 pollution in the Schönbuch and Black Forest could be estimated for this situation. Approx. 22-25 % of the SO_2 can be assumed to have originated from local sources, e.g., from the Middle Neckar Region, and approx. 75-78 % were carried in via long-range transport.

Comparative investigations between *gaseous and particle-bound sulfur pollution* in the forested area near Freudenstadt revealed that with the particles transported from remote areas, increased sulfur loads were spread over the country [34]. The origin of the air pollutants could also be proved on the basis of the particle composition. However, it became apparent that in conditions favoring long-range transport such as the one in February 1986, the largest part of the sulfur occured in the form of SO_2 during the gaseous phase. In the long-term mean, however, particle-bound sulfur pollution (e.g., in the Black Forest) was of the same magnitude as the one caused via the gaseous phase [34].

The conditions in February 1986 cannot be regarded as isolated cases. In South Germany, e.g., similar situations occurred in January 1985 and in January 1987, leading to comparable SO_2 pollution [32, 33].

3.1.5.2 Short-Range Dispersion of Air Pollutants in Highland Valleys

As part of the research into the investigation of the causes of the most recent forest damage, the air pollution in a Black Forest valley caused by a factory was investigated intensively by measurements [27, 34]. Fig. 3.17 shows a sectional view of the valley with the factory and the locations of the measuring stations. The factory produces cardboard. The process heat required was generated by a boiler plant with heavy fuel oil firing (28 MW thermal output). The oil used had sulfur content of

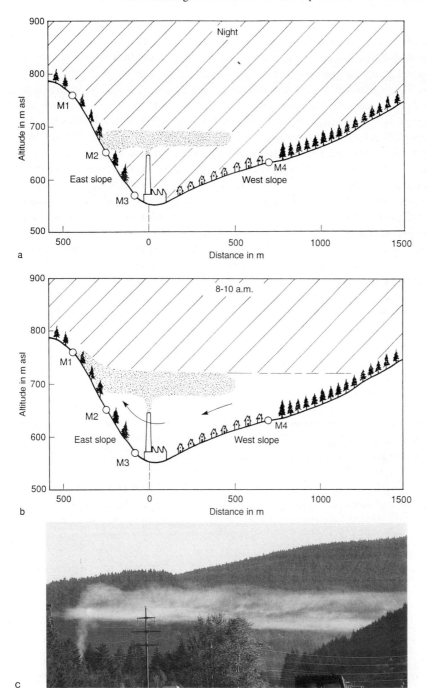

Fig. 3.17. Dispersion of a smoke plume in the Black Forest during inversion conditions, diagrammatic; M1-M4: continuously working SO_2 and temperature measuring stations; altitudes indicated in m above mean sea level (m.s.l.). **a** during the night until 8 a.m.; **b** 8-10 a.m.; **c** photographic record of the clearly visible dispersion of a smoke plume in a (different) Black Forest valley during an inversion situation (Photo: Günter Baumbach)

between 1 and 2 %. The plant was operated day and night at almost constant load, so that SO_2 emissions were nearly constant, i.e., approx. 50 kg SO_2/h. In this case, therefore, SO_2 concentrations in the ambient air were hardly influenced by varying emissions but were primarily dependent on the dispersion conditions of the flue gases.

For over a period of six months the measuring stations 1-4 shown in Fig. 3.17 carried out continuous measurements of both SO_2 concentrations and of temperatures which are a necessary factor for the determination of possible inversions. Besides this, during phases of intensive measurement, components of suspended dusts and dust depositions were also investigated at the measuring stations.

It became apparent that during windy and rainy weather only very low SO_2 concentrations were to be observed at the measuring stations. However, during stable high-pressure situations with inversion layers in the valley, SO_2 values which were very high in part, were measured on the wooded eastern slope, particularly at a certain altitude, usually at the stations halfway up the slope. Fig. 3.18 shows a diagram of the half-hourly values (vertical lines) measured in October 1985 at this station halfway up the slope. The peaking of the SO_2 concentrations is merely a result of the dispersion conditions.

From October 2-7, 1985, huge inversions settled into the valley. At the same time, the operation of the factory was closed down for inspection. As, during this

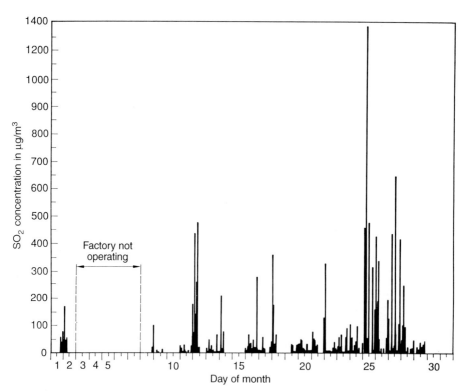

Fig. 3.18. Half-hourly values of the SO_2 concentrations in October 1985, medium-level station M2. 655 , above m.s.l..

period, no SO$_2$ at all could be measured in the forest air, the SO$_2$ concentrations measured at other times in this month stated to have originated in the factory. A further indicator was the brief, temporary, peaking of high concentrations. If these had been transregional transport processes, the concentration fluctuations would have been more balanced (as is to be seen in Fig. 3.14, Freudenstadt-Schöllkopf).

The results of the measurements carried out are presented in greater detail in [35]. Here the situational variations of the SO$_2$ concentrations are shown with the example of a day with an inversion (Fig. 3.19). Practically no SO$_2$ was detected at the measuring station with the lower location throughout the entire day, with the exception of a slight concentration increase from 0-1 o'clock. Increased SO$_2$ concentrations were also extremely rare at the measuring station M4 on the western slope. The highest concentrations occurred until 8 o'clock at the mid-way station M2. This station was located just at the level of the inversion layer, where the smoke plume spread out almost only horizontally after it had cooled down. The dispersion of the smoke plume in this situation is to be seen schematically in Fig. 3.17a. When the inversion layer lifted – as can be seen in Fig. 3.19a by the temperature profile from 8 o'clock

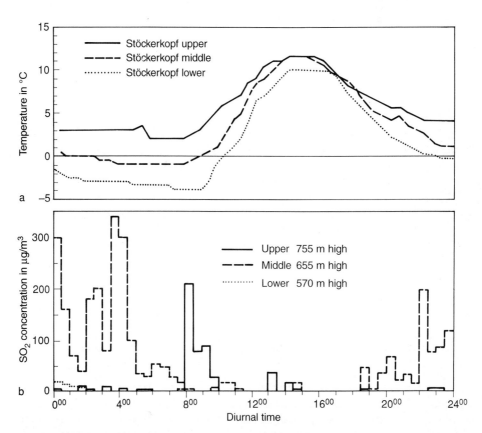

Fig. 3.19. Diurnal profiles of temperatures (**a**) and SO$_2$ concentrations (**b**) on October 26, 1985, in the Black Forest valley

on – the upper measuring station M1 was in the range of the high SO_2 concentrations from 8-10 o'clock. This case is shown as a diagram in Fig. 3.17b. As the slope was an eastern one, it was warmed by the sun in the forenoon. In addition to breaking up the inversion from below, this effected an upward flow on the slope [36] which assisted the smoke plume to relocate from the middle station to the upper station. From noon to the evening hours the inversion on the slope had broken up to such a degree that only a little SO_2 could be traced. In the evening hours when cold air again filled the valley the SO_2 concentration started to rise again drastically at the mid-way station from 18:00 hours onwards, and the process described above started anew.

Depending on the extent of the inversion, the SO_2 concentrations varied on the slope [34].

To further ensure that the SO_2 concentrations occurring in the valley and on the slope were to be attributed to the exhaust gases of the factory, measurements for suspended dust and dust deposition were carried out at several points and the amounts of nickel, vanadium and sulfur, the dust components specific for heavy fuel oil furnace exhaust gases, were determined [37]. The highest nickel content in *dust deposition* was measured at the valley measuring station M3 located near the factory. The further the distance to the factory the more the nickel deposition decreased. These facts obviously show that the largest part of the sedimenting particles of the exhaust gases settled on the ground in the immediate vicinity of the factory.

The highest concentrations of vanadium, nickel and sulfur in *suspended dust* also occured at the location of the slope where the highest SO_2 concentrations were measured (near M2).

The smoke plume so distinctly visible on the photograph in Fig. 3.17c due its high particle content very clearly shows that the dispersion in valleys during inversion situations indeed follows the pattern illustrated in Fig. 3.17a and 3.17b. With a weak flow from left to right (looking at the photograph) the exhaust gases remain exactly in the vertical layer where they were transported due to their buoyancy, without expanding vertically. To the right, the whole valley is covered with smoke at this altitude. In those places where the smoke plume meets the slope, air pollution concentrations rise sharply. In many cases the exhaust gas plumes are not visible, but with stable exchange conditions their dispersion follows the pattern described above.

These examples show that in valleys of low and high mountain regions, relatively slight emissions can lead to very high pollutant concentrations due to the effect of the atmospheric barrier layers. This was also observed at various other locations, e.g., [38-41].

3.2 Chemical Transformations of Pollutants in the Atmosphere

Most emitted pollutants are unstable and are transformed in the atmosphere by chemical reaction processes, partly by passing through a variety of intermediate products. On the one hand, the application of dispersion models, which are to take not only processes of transport and diffusion but also chemical reactions into consideration, require knowledge on the reaction kinetics of the chemical transformations in the atmosphere. On the other hand, for the determination of pollutant concentrations and the correct selection of measuring equipment it is important to know which changes the originally emitted pollutants can undergo on their way through the atmosphere. Thus, e.g., it will not suffice to judge the effects of SO_2 emissions exclusively by the SO_2 concentrations measured, as SO_2 can be transformed into sulfuric acid or sulfates. These can occur in precipitation or in fog as aerosols, adsorbed to dusts, and become active there.

There are numerous and complicated reactions taking place in the atmosphere. Special literature contains a wide range of information on individual reactions and transformation velocities of air-polluting substances based on detailed laboratory tests. In the atmosphere itself, however, concentrations of atoms and radicals can either not be measured at all or not yet routinely. Thus, their concentrations can often only be determined with the help of model calculations. Assuming that the model calculations provided accurate results, concentrations of atoms and radicals could only be indicated for "mean" conditions due to variations in radiation conditions and trace gas composition. In specific individual cases, e.g., in the air plume of an urban sprawl or the exhaust gas plume of a stack, considerable deviations from these "mean" concentrations can occur, s. Schurath in [42].

Possible transformations of the pollutants emitted will be demonstrated in the following with the help of some selected reaction equations, however limiting them to the final products which are of particular relevance for ambient air measurements. Intermediate products and their reactions amongst each other shall not be dealt with. Special literature is available for more information on this, e.g., [42-47].

3.2.1 General Considerations

3.2.1.1 Atmosphere and Air Pollution

Most of the chemical reactions taking place in the atmosphere are triggered off by sunlight or are accelerated by it, whereby different wave length ranges are effective for the individual reactions. The spectral distribution of sunlight is illustrated in Fig. 3.20. This diagram also illustrates the layers and the temperature profile of the atmosphere (from [43]).

The temperature peaks are to be attributed to light absorption. The troposphere is heated by the earth's surface. By absorbing UV light the stratospheric ozone layer causes a dramatic increase in temperature at the range of the stratopause, whereas in the thermosphere, temperatures of up to 700 °C are generated by the light absorption of ionized atoms.

What is popularly understood as weather takes place in the lowest layers of the atmosphere, the troposphere. The dispersion and chemical transformations or decomposition reactions of most of the atmospheric pollutants occur in this range. The water's cycle of vaporization, cloud formation and precipitation represents the most important "cleansing mechanism" here. Particles and water-soluble gases assist cloud formation and are washed out with the precipitation. Thus they remain in the atmosphere only for a short period of time, on an average from some days to some weeks.

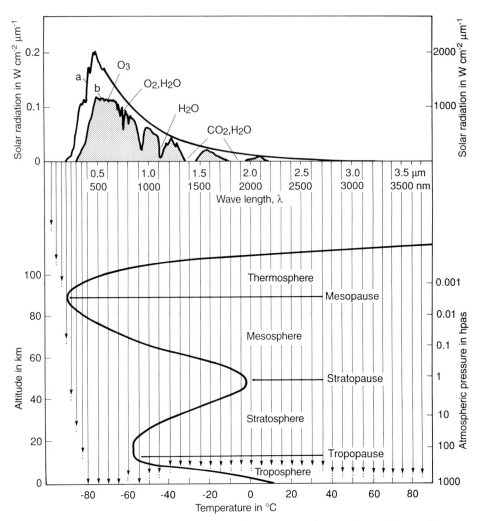

Fig. 3.20. Spectral distribution of solar radiation (upper section): **a** beyond the earth's atmosphere; **b** on the ground. Temperature distribution and levels division of the atmosphere (lower section). The warmer layer with a temperature maximum in the stratopause level is a result of the radiation absorption caused by ozone. The vertical arrows indicate schematically the depth to which individual wavelengths of solar radiation penetrate into the atmosphere [43]

Generally, the mean life span of an atmospheric trace substance is determined by the velocity of the decomposition mechanisms (decomposition rate). Spatial distribution is closely linked to the life span. Substances with slow decomposition rates with a life span of many years, can be dispersed evenly around the globe by the winds and can even reach regions above the troposphere. Substances whose life spans last several months can be well mixed within one hemisphere (northern or southern half of the globe), but concentration differences between the hemispheres do occur. The appearance of short-lived components such as OH and HO_2 radicals whose life span is less than an hour is mainly determined by local composition and decomposition mechanisms. SO_2, e.g., normally remains in the atmosphere for just a few days. Hence, its distribution is mainly restricted to industrial nations, whereas regions far from the sources, i.e., across the seas and in developing countries, only show extremely slight concentrations. Fig. 3.21 provides some basic facts on the life spans of trace gases.

Many trace gases released into the atmosphere are at low degrees of oxidation (e.g., CH_4, NO, CO). In contrast to this, the substances returning to the earth via rainfall are completely oxidized (e.g., HNO_3, CO_2, H_2SO_4).

An overview of the atmospheric chemistry of the man-made (anthropogenic) gases SO_2, NO_x and hydrocarbons is to be seen in Table 3.1. It is evident that nitrogen oxides are much more actively involved in atmospheric chemistry reactions than, e.g., SO_2. Hydrocarbons have both active and inactive components.

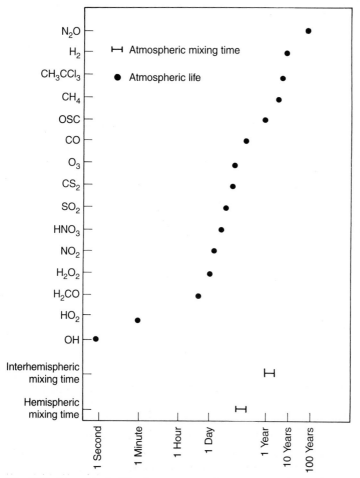

Fig. 3.21. Atmospheric life of trace gases – from one second to one century [43]

Table 3.1. Air-chemical behavior of SO_2, NO_x and hydrocarbons (acc. to Schurath in [42])

	SO_2	NO, NO_2	Hydrocarbons (HC) a.o.
Gas phase			
Photochemical behavior	Inactive (absorption of light without chemical consequences)	NO_2 very active (anthropogenic ozone formation)	HC inactive; aldehydes a.o. active (photochemical HO_2 source)
Radical-chemical behavior	Passive (weak OH radical interceptor)	Very active (e.g., transformation $HO_2 \rightarrow OH$ by NO; radical buffer, radical source, radical sink depending on the situation)	Very active (source of peroxide radicals; cause chain branchings by reacting with OH, to a lesser degree with O atoms)
Behavior towards ozone	Non-reactive	NO very reactive; NO_2 moderately reactive (night reaction)	Only unsaturated hydrocarbons moderately reactive
Phase transition			
Formation of primary aerosols	Important, condensation nucleus formation very important	Inactive	Insignificant
Absorption and transformation in cloud droplets, in humid aerosol		Important, but not so profoundly researched	Insignificant (HC barely soluble, oxidation products more soluble)
Final products	Sulfuric acid and sulfates, acid rain	Nitric acid and nitrates, acid rain	CO_2, CO, organic oxide compounds in rain, in the aerosol

3.2.1.2 Calculation of Reaction Rates

Reaction rates are determined for investigating the decomposition of air pollutants in the atmosphere. Generally, reaction velocity is defined as the time-related variation of concentration of the reactant of interest. It is then divided by the corresponding stoichiometric factor [44]:

$$aA + bB \rightarrow cC + dD \tag{3.1}$$

$$v = -\frac{1}{a}\frac{dc(A)}{dt} = -\frac{1}{b}\frac{dc(B)}{dt} = \frac{1}{c}\frac{dc(C)}{dt} = \frac{1}{d}\frac{dc(D)}{dt}. \tag{3.2}$$

A negative reaction velocity signifies a decreasing concentration, a positive one an increasing concentration.

The general formula for rate law for a reaction between two substances is:

$$v = kc(A)^m c(B)^n, \tag{3.3}$$

whereby v stands for reaction velocity; A-D for reactants; c (A)- c (D) for concentrations of the reactants; $a - d$ for stoichiometric factors of the reactants; k for velocity constant (temperature-related), m and n for exponents: mostly whole numbers, sometimes also fractional numbers. The conformity of the exponents with the stoichiometrical factors is purely incidental; $m+n$ indicate the reaction order.

To determine the concentration after a certain time t during a reaction, the reaction velocity must be integrated. If, e.g., during a first order reaction ($m=1$, $c(B)$=const), the concentration of the co-reactant A with the initial concentration c (A)$_0$ (at the time $t=0$) after time t is required, then – e.g., also [44] – applies :

$$\ln\frac{c(A)}{c(A)_0} = -kac(B)t \tag{3.4}$$

or formulated differently:

$$c(A) = c(A_0)e^{-kac(B)t}. \tag{3.5}$$

$c(B)$ = const.

Therefore, the concentration decreases exponentially with time.

3.2.2 Oxidation of SO_2

Before SO_2 is deposited on vegetation or on soil it is partially oxidized to sulfuric acid or sulfate. For this, gaseous phase reactions (homogenous), liquid phase reactions or those involving particulate matter (heterogenous reactions) are possible [42, 45, 46]. Research has shown that the particulate sulfate formed has a longer residence time in the atmosphere than SO_2. For SO_2 a medium residence time of one day is listed, for sulfate 3-5 days [46].

3.2.2.1 SO_2 Oxidation in the Gas Phase

The homogenous gas phase oxidation of SO_2 can, shown in a simplified way, follow the two mechanisms below [46]:
- direct photooxidation,
- oxidation through photochemically formed components (e.g., radicals).

However, direct photooxidation is insignificant for the conversion of SO_2 in the atmosphere [42, 46].

Of the reactions with photochemically formed radicals, the following three reactions with OH or RO_2 radicals play the major role in the gas phase reactions of the SO_2 [42]:

$$SO_2 + OH + M \rightarrow HOSO_2 + M \tag{3.6}$$

$$SO_2 + HO_2 \rightarrow OH + SO_3 \tag{3.7}$$

$$SO_2 + RO_2 \rightarrow RO + SO_3. \tag{3.8}$$

where R is organic group, e.g., alkyl groups, and M free, non-reactive collision partner in a triple collision, e.g., N_2. The collision partner absorbs part of the energy released during the reactions which would otherwise lead to the product's decomposition.

Of these, the most important reaction and thus the most important gas phase reaction of the SO_2 is the one with OH radicals, reaction (3.6). In literature the conversion velocity of SO_2 according to this reaction is given as 0.4 up to a maximum of 4.0 %/h [42]. Accordingly, the life span of SO_2 in European latitudes in summer is 3-5 days. In winter photochemical activity drops drastically, thus decreasing the contribution of the homogenous gas phase oxidation to the conversion of SO_2 [46] during this period. The $HOSO_2$ radical created in this reaction (3.6) obviously leads to the rapid formation of an aerosol which consists mainly of sulfuric acid, s. Beilke in [46]. The sulfuric acid is either adsorbed by particles or dissolved in water drops and is then finally washed out with the rain.

3.2.2.2 SO_2 Conversion in Liquid Phase and on Solid Particles

Conversions within liquid droplets such as cloud, rain, fog droplets, snow and dew as well as conversions on solid particles are of special significance for the oxidation of SO_2. As several phases are involved here (at least two), one speaks of heterogenous reactions.

Two processes are of significance for conversion within liquid droplets :
1. the physical solution of the SO_2 in water droplets,
2. the chemical conversion in the water droplets.

3.2 Chemical Transformations of Pollutants in the Atmosphere

SO₂ solution in water droplets

SO$_2$ dissolves in water droplets as hydrate and maintains a balance between the gaseous and liquid phase:

$$(SO_2)_{gas} + (H_2O)_{liquid} \rightleftarrows (SO_2 \cdot H_2O). \tag{3.9}$$

How much SO$_2$ is dissolved in water depends on the partial pressure or the concentration of the SO$_2$ in the air on one hand and on the solubility of the SO$_2$ in water on the other (Henry coefficients, s. [48] and [49, pg. 662]). The solubility is temperature dependent.

Depending on the pH value of the water the SO$_2$ hydrate dissociates to hydrogen sulfite (HSO$_3^-$) in an acidic environment or (when there are high pH values) to sulfite (SO$_3^{2-}$) in an alkaline environment, whereby protons (H$^+$) are released which are then present in the water as hydronium ions (H$_3$O$^+$):

$$SO_2 \cdot H_2O + H_2O \leftrightarrow H_3O^+ + HSO_3^-$$
in a slightly acidic range $\tag{3.10}$

$$SO_2 \cdot H_2O + 2H_2O \leftrightarrow 2H_3O+6 + SO_3^{2-}$$
in an alkaline range $\tag{3.11}$

The dependency of the dissociation and with it of the solubility on the pH value is illustrated in Fig. 3.22.

For small droplets and pH values below 6 the equilibrium between SO$_2$ in the gaseous phase and hydrogen sulfite or hydrate in the liquid phase is quickly reached. With decreasing pH values it shifts towards the gaseous phase. The factor determining the velocity of the SO$_2$ oxidation in cloud, fog and dew water are the oxidation

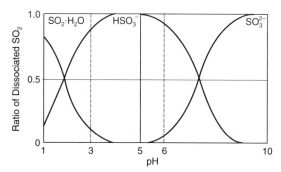

Fig. 3.22. Amount of dissolved SO$_2$ species relative to the pH value at 25 °C (acc. to Barrie in [46])

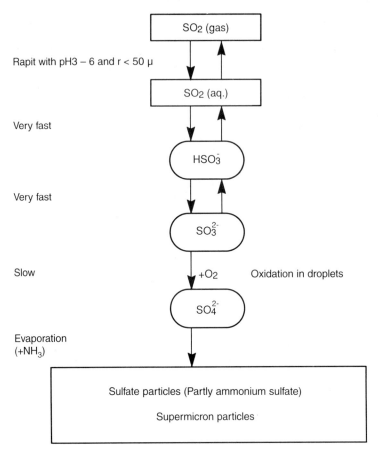

Fig. 3.23. Diagrammatic illustration of SO_2 absorption and sulfate formation in water droplets (acc. to Beilke in [46])

stages and not the SO_2 absorption into the drop or the transportation processes within the drop [50]. The path of the SO_2 absorption up to the oxidation and neutralisation to sulfate particles is illustrated by means of a diagram in Fig. 3.23.

If, before the SO_2 oxidizes in the cloud and fog droplets, these droplets vaporize, then the SO_2 passes back into the gas phase.

SO_2 oxidation in water droplets

The processes of SO_2 oxidation in water droplets depend on many different factors and are not yet comprehensively known [46].

The oxidation of SO_2 by O_2 can be accelerated in polluted air by catalysts, in particular by manganese and ferrous ions [50].

This type of oxidation plays a particular role in large exhaust gas plumes (e.g., from power plants) in connection with emitted dusts. In less polluted air SO_2 oxidation is accelerated by oxidizing agents such as O_3 and H_2O_2 dissolved in water droplets.

In clouds with a liquid water content of 0.1 g/m³, pH values of between 4 and 5 and ozone concentrations between 20 and 100 µg/m³ SO_2 oxidation, e.g., takes place between 0.02 and 2.8 %/h; with easily water-soluble H_2O_2, SO_2 oxidation rates as high as between 2.3 and 9.9 %/min have been determined in the laboratory. In contrast to this, oxidation with O_2 proceeds much more slowly [46].

Findings so far have revealed that in summer at higher temperatures and in more intense sunlight, oxidation conditions are better than in winter. Due to lower conversion rates at low temperatures, SO_2 transport over longer distances is rather more likely in winter in combination with certain meteorological situations than in summer.

SO_2 and acid rain

If, as per the above explanations, sulfuric acid is formed in rain drops and if there are only few neutralization agents such as NH_3 or lime particles, then the rain water acidifies due to the strong dissociation of the sulfuric acid:

$$H_2SO_4 + 2H_2O \rightarrow 2H_3O^+ + SO_4^{2-} . \tag{3.12}$$

Rain water contains sulfuric acid in an extremely dilute form and is therefore 100% dissociated, i.e., the dissociation constant $K_{H_2SO_4}$ takes on very high values:

$$K_{H_2SO_4} = \frac{c(H_3O^+)^2 \cdot c(SO_4^{2-})}{c(H_2SO_4)} , \tag{3.13}$$

$c(H_2SO_4)$ approaches 0,

$c(H_3O^+), c(SO_4^{2-}), c(H_2SO_4)$ concentrations in water.

It is assumed today that approx. 2/3 of the acid rain is caused by the acidification with sulfuric acid formed from SO_2, and 1/3 with nitric acid coming from nitrogen oxides.

SO_2 oxidation on solid particles

The reaction of SO_2 with O_2 on aerosol particles depends on the following factors:

- SO_2 concentration in the gas phase,
- catalytic properties of the particles,
- specific surface,
- acidity,
- relative humudity.

SO_2 oxidation on solid particles in smoke plumes is of great significance. SO_2 conversions were studied [45, 46] in smog chambers (under lab conditions) and in the exhaust gas plumes from power plants. These tests established that the conversion rate is proportional to the SO_2 concentration, i.e., high conversion rates during the initial phase rapidly decline. The capacity of aerosol for heterogenous SO_2 conversion is obviously of a limited nature, probably due to the degree of saturation of the aerosol surface. It may generally be stated, that the mean velocity of the SO_2 conversion in smoke plumes is relatively low with homogenous gas phase reactions probably predominating over heterogenous reactions as far as particles are concerned [46]. Oxidation on solid particles in heavy fuel oil furnace exhaust gases have obviously not been investigated so far. It is possible that higher conversion rates occur here, as particles particularly active catalytically, such as vanadium pentoxide (V_2O_5) and soot particles are emitted. Investigations by Novakov [46] have shown, that in critically polluted areas with a high soot content in the ambient aerosol, large amounts of SO_2 can be converted to sulfate.

3.2.3 Reactions of Nitrogen Oxides in the Atmosphere

The oxides of nitrogen (NO, NO_2, NO_3) can become active in atmospheric chemistry in a variety of ways, e.g., as a reaction promoter or reaction inhibitor, as they have an

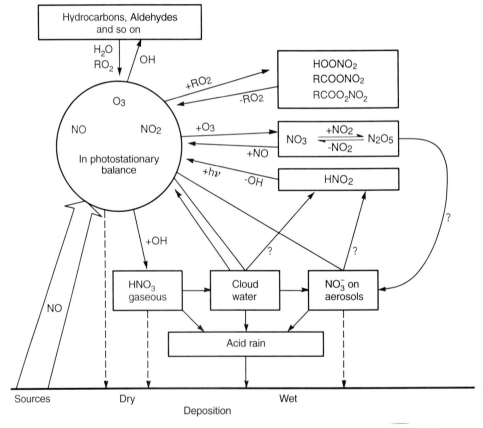

Fig. 3.24. Diagram of nitrogen oxide reactions in the atmosphere (acc. to Schurath in [42])

unpaired electron which gives them the character of a radical. The most significant ones are NO, which is released largely during combustion processes and NO_2, which is formed in the air from NO. Fig. 3.24 shows the reactions of the nitrogen oxides in the atmosphere in a simplified form. From the great variety of possible reactions the most important ones will be shown in the following pages.

3.2.3.1 NO Oxidation and the Formation of Ozone

The NO emitted is oxidized to NO_2 in the atmosphere according to, among others, the following two reactions, s. Schurath in [42] and [51]:

$$2NO + O_2 \xrightarrow{k_1} 2NO_2 \tag{3.14}$$

$$NO + O_3 \xrightarrow{k_2} NO_2 + O_2, \tag{3.15}$$

with k being velocity constants.

NO oxidation with atmospheric oxygen (3.14) is a trimolecular reaction (third order), i.e., the NO concentration enters the reaction velocity quadratically. Thus, when NO concentrations in the atmosphere are low, the reaction proceeds very slowly.

The most important conversion process for NO in the atmosphere is oxidation by ozone (3.15). This reaction takes place relatively fast. Without the action of light, e.g., at nighttime, the deficient reactant is used up entirely in a short time, s. Schurath in [42]. Thus, almost no NO is found in areas far from pollution sources, e.g., in wooded areas of the Black Forest; instead large amounts of ozone are present. In the vicinity of pollution sources, however, e.g., on the edge of highways, almost all the ozone is used up due to the constant supply of NO from car exhaust gases. The comparison of the diurnal courses of the components NO, NO_2 and O_3 measured at a forest and at a highway measuring station is illustrative of this situation, s. Fig. 3.25.

There are numerous other reactions with NO which, however, will not be discussed further here for reasons of clarity, s., e.g., [3, 42, 43, 49].

In the presence of sunlight NO_2 is split up by photolyses during which process NO, O and finally O_3 are again formed [42, 51]:

$$NO_2 + h\nu \ (290-430 \text{ nm}) \xrightarrow{k_3} NO + O, \tag{3.16}$$

where k_3 is dependent on intensity of solar radiation $h\cdot\nu$, e.g., $k_3 = 0.5$ min^{-1} (summer, midday sun),

$$O + O_2 + M \rightarrow O_3 + M^1 \text{ (fast secondary reaction, not speed limiting)} \tag{3.17}$$

As the oxygen concentration c (O_2) is constant and the fast secondary reaction (3.17) does not limit speed, $c(O_2)$ in k_3 is not taken into account.

---

[1] M = third (body) molecule, e.g., N_2 which absorbs the excess energy

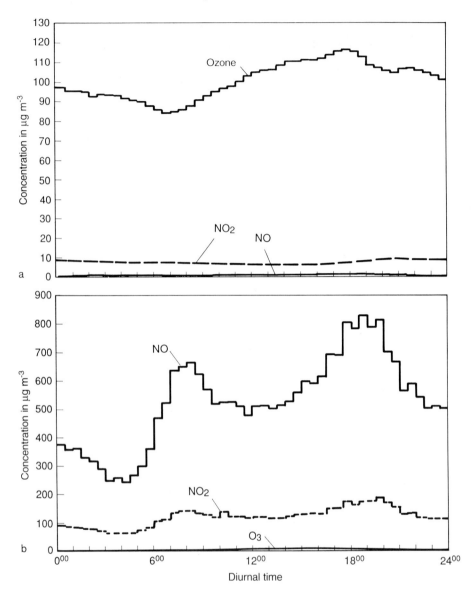

Fig. 3.25. Comparison of measured mean diurnal courses of NO, NO$_2$ and O$_3$. **a** in a forest area near Freudenstadt, Germany, on sunny summer days in 1987; **b** on the edge of highway A8 near Wendlingen, Germany, in October 1985 (from [52])

Degradation and formation of O$_3$ and NO compete with the speed constants k_2 and k_3 with k_3 depending on light intensity. Between O$_3$, NO$_2$ and NO the so-called photostationary equilibrium which is described by means of the following equations results:

$$NO_2 + O_2 \underset{k_2}{\overset{k_3(h\nu)}{\rightleftarrows}} NO + O_3 \tag{3.18}$$

$$c(O_3) = \frac{c(NO_2)}{c(NO)} \cdot \frac{k_3(hv)}{k_2}. \tag{3.19}$$

Hence, the ozone concentration depends on the NO_2/NO ratio and on the effective light intensity (via k_3). With a steady supply of NO, e.g., near high traffic roads or in cities in general, the NO_2/NO concentration ratio remains small, i.e., even in intense sunlight only a little ozone is formed, unless other trace gases cause an oxidation of the NO, thus increasing the NO_2/NO ratio in this way. This will be dealt with more thoroughly in the next section.

Reaction rates will be calculated here as examples of (3.14)-(3.18) with the calculation principles shown in section 3.2.1.2 .

Example 1: NO oxidation with atmospheric oxygen, reaction (3.14)

$$v = -\frac{1}{2}\frac{dc(NO)}{dt} = -\frac{dc(O_2)}{dt} = \frac{1}{2}\frac{dc(NO_2)}{dt}, \tag{3.20}$$

(Third order reaction; concerning NO secondary order: m=2, n=1),

$$\frac{dc(NO)}{dt} = 2k_1 c(NO)^2 \cdot c(O_2), \tag{3.21}$$

Assumptions:
- NO concentration: $\quad c(NO)=100$ ppb$=0.1$ ppm,
- O_2 concentration in the air: $\quad c(O_2)=0.21 \cdot 10^6$ ppm,
- velocity constant: $\quad k_1 = 7.5 \cdot 10^{-10}$ ppm^{-2}min^{-1} [51].

$$\frac{dc(NO)}{dt} = -2 \cdot 7{,}5 \cdot 10^{-10} \text{ ppm}^{-2} \cdot 0{,}1^2 \text{ ppm}^2 \cdot 0{,}21 \cdot 10^6 \text{ ppm}$$

$$\frac{dc(NO)}{dt} = -3{,}15 \cdot 10^{-6} \text{ ppm min}^{-1}.$$

With an NO concentration in the air of 100 ppb, $3.15 \cdot 10^{-6}$ ppm per minute are oxidized to NO_2 with the atmospheric oxygen, that means 0.00315 % per minute or 0.2 % per hour (reaction velocity divided by the concentration). So the reaction takes place extremely slowly. With decreasing concentration, reaction velocity decreases exponentially even further.

Example 2: NO oxidation with ozone, reaction (3.15)

$$\frac{dc(NO)}{dt} = -k_2 \cdot c(NO) \cdot c(O_3), \tag{3.22}$$

(bimolecular reaction of secondary order),

Assumptions:
- NO concentration: $\quad c(NO)=100$ ppb$=0.1$ ppm,
- O_3 concentration: $\quad c(O_3)=30$ ppb$=0.03$ ppm,

- velocity constant: $k_2 = 25\,\text{ppm}^{-1}\,\text{min}^{-1}$ [51], according to Schurath in [42, 51], but values differ in literature),

Assuming that there were a constant supply of NO (e.g., near a busy street), the NO concentration would then be constant (e.g., at 100 ppb) and the O_3 concentration would also remain constant (e.g., due to constant turbulent advections of more ozone from the surrounding air), then 75 ppb NO per minute will be oxidized with ozone to NO_2. This would be equivalent to an oxidation rate of 75 % per minute.

NO oxidation with ozone in the concentration range at around 100 ppb NO and at 30 ppb O_3 is therefore approx. $2.4 \cdot 10^4$ faster than oxidation with atmospheric oxygen. The latter reaction, therefore, is of little consequence for the NO oxidation in the ambient air as some ozone is always naturally present. NO oxidation by atmospheric oxygen is only significant with high NO concentrations, e.g., in exhaust gas plumes [50].

In this example in which a constant O_3 concentration of 30 ppb is assumed and provided there is only a limited supply of NO, the half-life period of NO (i.e., the drop to 50 ppb) is calculated according to equation (3.4) and equation (3.15) as follows:

$$\ln c(A) - \ln c(A_0) = -k \cdot c(B) \cdot t$$

$t = 55\,\text{s}$,

where $c(A)$ is the final NO concentration = 50 ppb, $6c(A_0)$ the initial NO concentration = 100 ppb, $c(B)$ the O_3 concentration = 30 ppb and $k = k_2 = 25\,\text{ppm}^{-1}\,\text{min}^{-1}$.

Example 3: photolytic NO_2 degradation and O_2 formation, reaction (3.16)

$$-\frac{dc(NO_2)}{dt} = \frac{dc(NO)}{dt} = \frac{dc(O)}{dt}. \tag{3.23}$$

As the consequent reaction (3.17) proceeds so fast that it does not limit speed,

$$\frac{dc(O)}{dt} = \frac{dc(O_3)}{dt}$$

can be set/replaced

$$-\frac{dc(NO_2)}{dt} = \frac{dc(O_3)}{dt} = k_3 \cdot c(NO_2),$$

assumptions:
- NO_2 concentration: $c(NO_2) = 20\,\text{ppb} = 0.02\,\text{ppm}$,
- velocity constant: $k_3 = 0.5\,\text{min}^{-1}$ for summer and midday sun,

$$\frac{dc(NO_2)}{dt} = -0{,}5\,\text{min}^{-1} \cdot 20\,\text{ppb},$$

$$\frac{dc(NO_2)}{dt} = -10 \text{ ppb min}^{-1}.$$

Hence, 10 ppb min^{-1} NO$_2$ are degraded photolytically, with 10 ppb min^{-1} O$_3$ being formed. However, this is limited by reaction (3.15), s. photostationary equilibrium, reaction (3.18). This is shown in example 4:

Example 4: Ozone in photostationary equilibrium, reaction (3.18)

$$\frac{dc(O_3)}{dt} = k_3 \cdot c(NO_2) - k_2 \cdot c(NO) \cdot c(O_3), \qquad (3.24)$$

Assumptions:
$c(NO_2)$ and k_3 as in example 3,
$c(NO) = 5$ ppb $= 0.005$ ppm,
$c(O_3)$ and k_2 as in example 2,

$$\frac{dc(O_3)}{dt} = 0.5 \text{ min}^{-1} \cdot 0.02 \text{ ppm} - 25 \text{ ppm}^{-1} \text{ min}^{-1} \cdot 0.005 \text{ ppm} \cdot 0.03 \text{ ppm}$$

$$\frac{dc(O_3)}{dt} = 0.00625 \text{ ppm min}^{-1} = 6 \text{ ppb min}^{-1}.$$

In relation to the initial O$_3$ concentration 20 % min^{-1} of ozone is formed.

With the given ratios (NO$_2$/NO ratio=4) and in the summer midday sun, ozone formation would exceed ozone degradation. With the above values (NO$_2$/NO ratio=4) in photostationary equilibrium and at constant conditions (which can be the case for a short period of time) the following O$_3$ concentration would ensue, (3.19):

$$c(O_3) = \frac{0.02 \text{ ppm}}{0.005 \text{ ppm}} \cdot \frac{0.5 \text{ min}^{-1}}{25 \text{ ppm}^{-1} \text{ min}^{-1}},$$

$$c(O_3) = 0.08 \text{ ppm} = 80 \text{ ppb}.$$

3.2.3.2 Participation of Hydrocarbons in NO Oxidation

When hydrocarbons of anthropogenic or natural origin react with hydroxyl (OH) radicals, peroxy radicals can be formed which, by re-forming OH radicals, tend to cause an oxidation of the NO to NO$_2$.

In the clean troposphere OH radicals react mainly with CO and CH$_4$ [42], in the polluted one they do so also with other hydrocarbons:

$$OH + CO \rightarrow CO_2 + H \qquad (3.25)$$

$$OH + CH_4 \rightarrow H_2O + CH_3. \qquad (3.26)$$

The H atoms and hydrocarbon radicals (CH_3 or generally RCH_3) attach themselves to oxygen molecules to form peroxy radicals HO_2 or RO_2 (R=alkyl groups), e.g.:

$$H + O_2 \rightarrow HO_2 \tag{3.27}$$

$$CH_3 + O_2 \rightarrow CH_3O_2 . \tag{3.28}$$

The peroxy radicals react with NO, whereby NO_2 is formed and OH radicals are reformed:

$$NO + HO_2 \rightarrow NO_2 + OH \tag{3.29}$$

$$NO + CH_3O_2 \rightarrow NO_2 + CH_3O \tag{3.30}$$

or generally

$$NO + RO_2 \rightarrow NO_2 + RO. \tag{3.31}$$

OH radicals are very reactive and enter into many different reactions in the atmosphere [42]. Particularly under the influence of solar radiation, OH radicals are present in significant concentrations.

Thus with the OH radicals, oxidizing agents (peroxide radicals) are formed from hydrocarbons, without the OH radicals being consumed in the process. Different hydrocarbons tend in varying degrees to the formation of oxidizing agents or to the reformation of OH radicals, and thus to the formation of ozone.

Schurath has investigated the "ozone formation potential" of various hydrocarbons [53]. It has long since been known about alkenes (olefins) that they are particularly effective in ozone formation. However, aromatic hydrocarbons can also have a high ozone formation potential [53].

So owing to the participation of these oxidizing agents the NO_2/NO ratio is increased, leading to a higher ozone concentration in intensive solar radiation. This process is illustrated in Fig. 3.26. It causes the "anthropogenic ozone formation", well-known particularly in the *photochemical smog* of Los Angeles [54]. Nitric oxides in themselves, then, do not cause very high ozone concentrations; hydrocarbons must also be involved.

If the NO_2/NO ratio is sufficiently large, the probability of RO_2 radicals adding on to NO_2 increases, leading to the formation of peroxide compounds (Fig. 3.24, upper right). Of these compounds the irritant peroxyacetyl nitrate (PAN, $CH_3CO_3NO_2$) is the most stable and has become well-known from the Los Angeles smog.

Fig. 3.27 shows a typical diurnal variation of the concentrations of hydrocarbons (C_nH_m), nitrogen oxides (NO, NO_2) and ozone (O_3) in Los Angeles. Due to rush hour traffic from 6 to 8 a.m. the pollutants C_nH_m and NO from automobiles increase sharply. The formation of the consequent products NO_2 and O_3 increases with sunlight. An increase in NO_2 and O_3 is characteristic of photochemical smog. Even in European cities this type of photosmog with high O_3 and NO_2 concentrations has been known to occur on warm summer days. For more information on the occurrence of photo-oxidizing agents, the significance of ozone and further underlying principles of photochemical reactions s. Becker et al. in [47].

3.2 Chemical Transformations of Pollutants in the Atmosphere   117

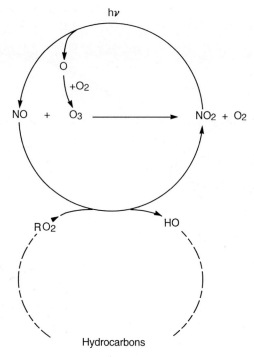

Fig. 3.26. Diagram illustrating ozone formation from NO_2 with hydrocarbons contributing

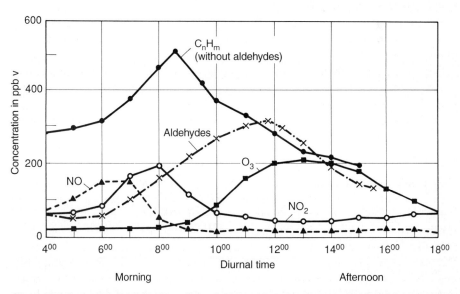

Fig. 3.27. Characteristic diurnal profiles of C_nH_m, NO, NO_2, O_3 and aldehyde concentrations in Los Angeles (acc. to [54])

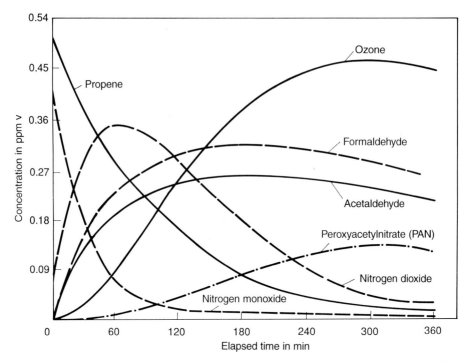

Fig. 3.28. Results of a typical smog chamber experiment; initial concentrations of reaction partners: 0.45 ppm NO, 0.05 ppm NO_2, 0.5 ppm C_3H_6, (acc. to [49])

Many experiments have been carried out to reproduce the formation of photochemical smog in reaction chambers under simulated sunlight. Fig. 3.28 shows the results of such an experiment. NO and, as representative of hydrocarbons, C_3H_6(propene) were used as initial substances. The formation of strong irritants, which are typical of photochemical smog, could be clearly proven: NO_2 (nitrogen oxide), O_3 (ozone), HCHO (formaldehyde), CH_3CHO (acetaldehyde) and PAN (peroxyacetyl nitrate).

3.2.3.3 NO_2 Oxidation

Due to the reaction of NO_2 with O_3, measurable concentrations of NO_3 and also N_2O_5 can occur [55] at night; these concentrations, however, are several magnitudes below those of NO, NO_2 and O_3:

$$NO_2 + O_3 \rightarrow NO_3 + O_2. \tag{3.32}$$

During the day NO_3 formation is reversed by the photolysis of the NO_3 and by a very quick reaction with NO.

The most important consequent reaction of the NO_2 takes place with OH radicals:

$$NO_2 + OH + M \rightarrow HNO_3 + M^2 \tag{3.33}$$

whereby M = triple collision partner.

In this low atmospheric layer the gaseous nitric acid thus formed is very stable and highly water-soluble and is therefore removed from the atmosphere by leaching or dry deposition. The NO_2 reaction (3.33) is approximately 11 times faster than the corresponding SO_2 reaction (3.6). Accordingly, the rate of reaction of the NO_2 is also 11 times faster than that of SO_2 and its life-span accordingly shorter, a mean value being one day, sometimes a mere few hours, s. Schurath in [42]. For this reason, NO_2 cannot be transported over the same long distances as SO_2. Due to the high reaction rates (3.33) other reactions of the nitric oxides are relatively insignificant.

3.2.3.4 NO_x and Acid Rain

Direct absorption and conversion of NO into NO_2 in liquid droplets is hardly possible because of the poor solubility of NO. Even NO_2 does not dissolve well enough in water for an absorption and conversion of this gas in the liquid phase to make significant contributions to the decomposition of nitric oxides [46]. What is decisive is the oxidation of the NO_2 to HNO_3 in the gaseous phase (3.33). Provided that the nitric acid is not directly absorbed on surfaces or forms particle-shaped ammonium nitrate (NH_4NO_3) with ammonia (NH_3), nitric acid is absorbed in the form of drops and contributes to the acidification of rain water due to its strong dissociation:

$$HNO_3 + H_2O \rightarrow H_3O^+ + NO_3^- . \tag{3.34}$$

Measurements of the sulfate (SO_4^{2-})-nitrate (NO_3^-) ratio in rain water show that nitrogen oxides contribute to the acidity (acidification) of the rain water by approximately 30 %. This corresponds to a molar ratio of $SO_4^{2-}:NO_3^-$ of 1.1:1 (1 mol H_2SO_4 provides 2 protons, 1 mol HNO_3 only 1 proton). As the molar emission ratio of $SO_2:NO_x$ is at least 2:1 in Germany, a larger share of the SO_2 must be removed from the atmosphere by some other means than by rain, as Schurath in [42] suggests possibly via direct absorption by vegetation.

3.2.4 Ozone in the Atmosphere

3.2.4.1 Ozone in the Stratosphere

Ozone is formed by exposing O_2 molecules to short-wave UV light:

$$O_2 + h\nu\,(<242\,\text{nm}) \rightarrow O + O \tag{3.35}$$

whereby M is a third (body) molecule that absorbs the excess energy.

$$O + O_2 + M \rightarrow O_3 + M, \tag{3.36}$$

---

[2] M = third (body) molecule, e.g., N_2, which absorbs the excess energy

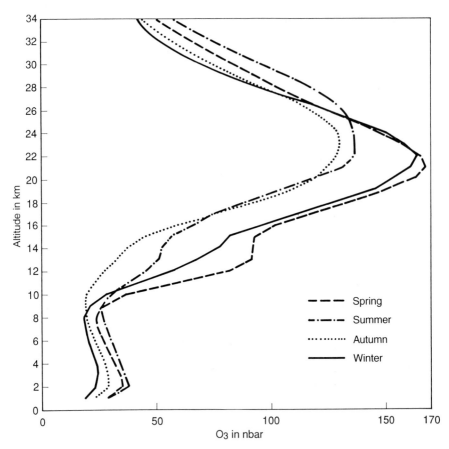

Fig. 3.29. Vertical profiles of the annual mean (1967-1982) ozone values above the weather station Hohenpeißenberg/Bavaria [56]. The volume concentration of ozone can be calculated from the O_3 partial pressure (in nbar) related to the air pressure prevalent at the relevant altitude (in mbar); e.g., at the ozone maximum at an altitude of 20 km the air pressure amounts to approx. 50 mbar, so the ozone concentration there reaches approx. 3.4 ppm in winter and spring

These reactions take place at high altitudes and lead to the so-called ozone layer in the upper stratosphere (s. also Fig. 3.20). An overview of the expansion of the ozone layer in the stratosphere at different times of the year is shown in Fig. 3.29.

The processes of ozone formation and ozone decomposition in the higher atmospheric layers are manifold. They have been described by Fabian in [43] with great clarity. The earth is shielded from short-wave UV radiation by the ozone layer.

It is feared today, that the protective ozone layer in the stratosphere is being depleted by man-made air pollutants. These problems will be dealt with in chap. 4 "Effects of Air Pollution".

3.2.4.2 Ozone in the Troposphere

As shown above, one part of the ozone is produced in the stratosphere from where it reaches the troposphere by mixing processes. Vertical profile measurements of the Meteorological Obeservatory of Hohenpeißenberg, Germany, have shown that the ozone concentration takes a dramatic plunge from the stratosphere to the tropopause, and below the tropopause, in the free troposphere, e.g., above Central Europe at an altitude of 12 km, has almost constant concentrations of 50-70 ppb (100-140 µg/m^3).

As the short-lived air-polluting substances emitted are restricted to few km above ground level due to atmospheric stable inversion layers (s. sect. 3.1), 50-70 ppb can be regarded as the natural level created by mixing processes with stratospheric ozone at an altitude of 12 km. Due to sinking processes from higher altitudes, increased O_3 concentrations could occasionally be detected at mountain measuring stations, e.g., at the Zugspitze, by comparing with radioactive tracers. Such phenomena, however, could so far only be observed in stations located at high altitudes [47, 57].

In non-polluted air layers of the lower troposphere close to the ground, mean O_3 concentrations are between 20 and 40 ppb, with maximum values of 40-60 ppb in the summer, s. Becker et al. in [47]. If higher O_3 concentrations occur in the air close to the ground, then one must distinguish between the "transregional type" formed during certain weather conditions with intense sunlight and relatively high temperatures [57, 58], where photochemical processes with air polluting substances can be involved and, on the other hand, ozone occuring regionally with a typical diurnal variation in the lowest air layer. In this case it was observed that high ozone concentrations (in excess of 70 ppb) occur much more frequently here than in elevated layers, measured at mountain stations at 1800 m and 3000 m altitude (NSL) [57], even if the mean O_3 concentrations are higher at high altitudes, s. Fig. 3.30.

The diurnal variation at the valley station of Garmisch shown in Fig. 3.30 is clearly influenced by photochemical reactions with air pollutants (nitrogen oxides and hydrocarbons, s. sect. 3.2.3). In the night hours the transregional ozone in the valley is consumed by reacting with nitrogen oxides; during the day, however, the sunlight causes so much ozone to form from the air pollutants, that even the mean concentrations are clearly higher than the ones measured at the mountain stations (Zugspitze and Wank). On certain days the concentration profiles are even more pronounced, with concentrations higher than 100 ppb occurring in the valley during the afternoon, whereas on the Zugspitze an almost constant level of approximately 50 ppb prevails [57]. O_3 concentrations on the Zugspitze are almost completely unaffected by direct reactions with anthropogenic, regionally occurring pollutants; at the Wank station, however, at an altitude of 1600 m a slight influence can still be observed. During the morning hours (approx. 8-12 a.m.) when slight concentrations of nitrogen oxides probably reach this station by a rising mixing layer, the O_3 concentration drops slightly. From 12 o'clock onwards it starts rising again and exceeds the values of the Zugspitze station, all of which is suggestive of photochemical processes.

In large cities on warm summer afternoons O_3 concentrations of 250-300 ppb caused by photochemical processes have already been measured [47]. During the night concentrations there drop back to extremely low values as a rule. During pre-

122   3 Air Pollutants in the Atmosphere

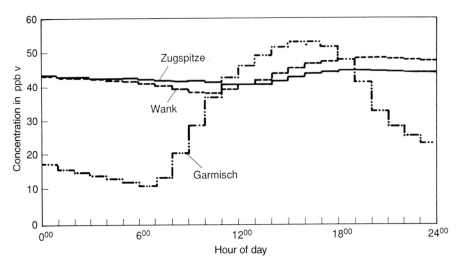

Fig. 3.30. Mean diurnal profile of the O_3 concentrations during the summer months June 1 to Sept. 30 on sunny days without precipitation in the years 1977-1980 at a valley station in Garmisch (740 m above m.s.l.), the mountain station Wank (1 600 m above m.s.l.) and the mountain station Zugspitze (2964 m above m.s.l.) (from [57])

cipitation and on very windy days O_3 concentrations in the valley (Garmisch) and at the medium altitude mountain station Wank are generally lower, and the diurnal variations are not as pronounced (lower O_3 production).

Ozone conditions similar to the ones in high mountain valleys or the surrounding mountains also occur in and above towns located in narrow or wide valleys. This, e.g., is presented graphically in the O_3 and NO_2 vertical profiles shown in Fig. 3.31 which were obtained on clear days over the town of Tübingen, Germany, by a tethersonde measuring system [59]. Below the nocturnal surface inversion (Fig. 3.31c) the ozone is almost completely decomposed by oxidation of primary exhaust gas components and by deposition up to an altitude of 200 m above the ground. Instead NO_2 occurs in a higher concentration up to this altitude. Above the inversion the air masses are largely detached from the ones below. The ozone concentrations here remain at a relatively high level during the night, whereas there are only very low values for NO_2. Accordingly, elevations bordering on the valley are exposed to higher ozone concentrations at night than the ground of the valley itself.

Induced by the warming of the ground by the sun there is a high vertical mixing and photochemical activity during the day which results in relatively constant NO_2 profiles (on a lower level) and O_3 profiles (on a higher level), s. Fig. 3.31b.

In the busy streets of the city the primary components NO and HC, as also NO_2, predominate during the day, whereas ozone remains at very low concentrations, s. the street canyon "Mühlstrasse" in Fig. 3.31a. In reduced traffic zones such as near the "Rathaus" (town hall), mean concentrations of primary substances sink obviously while ozone rises. In the neighboring elevations, again, ozone has a relatively high level and the other substances a relatively low ambient level.

3.2 Chemical Transformations of Pollutants in the Atmosphere

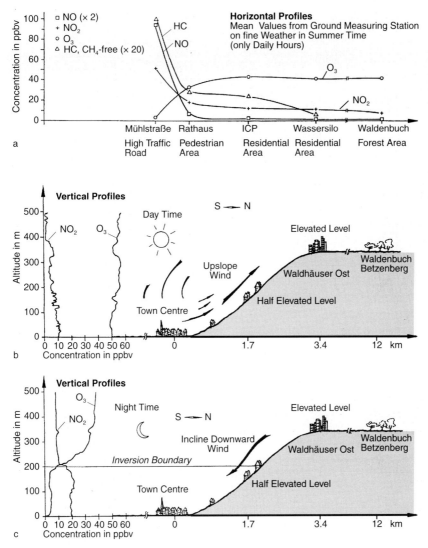

Fig. 3.31. NO_2 and O_3 vertical profiles during the day (**b**) and at night (**c**) over the town of Tübingen, Germany, (during fine weather) as well as diagram of horizontal distribution (**a**)

Conditions such as those were measured in summer in the town of Tübingen also occur in the same way in towns located in shallower valleys, e.g., in Heilbronn, where investigations were carried out for a special ozone-reduction model experiment [60]. Even over relatively flat terrain, e.g., the Swiss Middleland, which is, however, widely spread out between the Jura Mountains and the Alps, the same ozone decomposition at night and near the ground and the good mixing in the afternoons together with even ozone distributions on a high level can be observed. These dynamics become obvious in the isopleth diagrams of Fig. 3.32: e.g., lines of the same potential temperature and the same ozone concentration for time and altitude

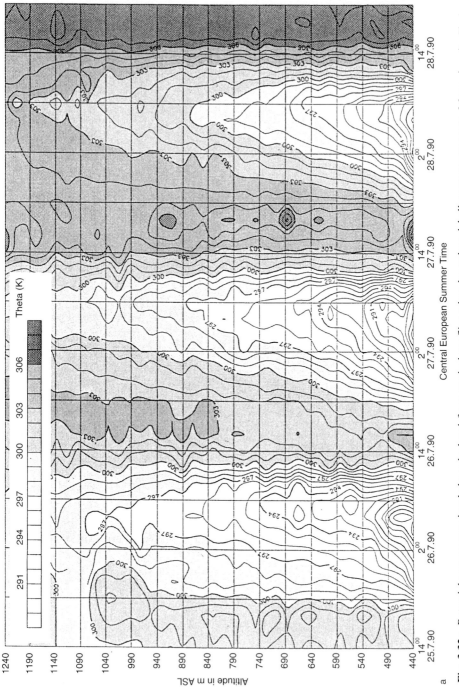

Fig. 3.32a. Potential temperature isopleths calculated from vertical profiles taken by tethered balloon measurements. Measuring site Siselen in the Swiss Middleland [61]

3.2 Chemical Transformations of Pollutants in the Atmosphere 125

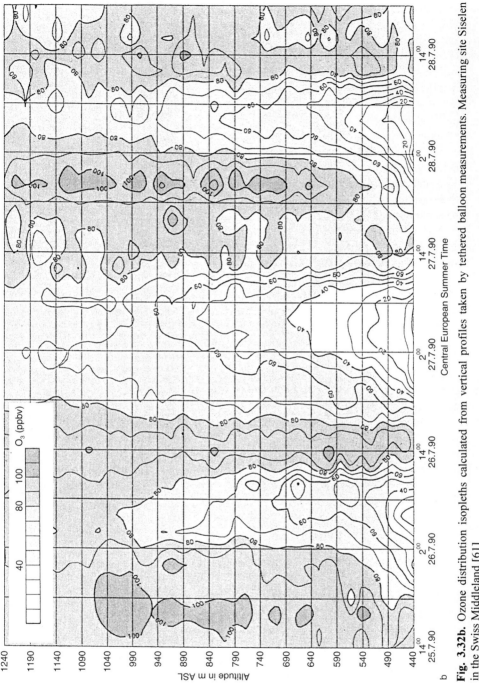

Fig. 3.32b. Ozone distribution isopleths calculated from vertical profiles taken by tethered balloon measurements. Measuring site Siselen in the Swiss Middleland [61]

were determined by interpolation from vertical profiles recorded by a tethersonde system every two hours from the ground up to an altitude of 800 m and more [61]. As for the potential temperature, an increase with higher altitudes, depending on the gradient, means neutral to stable layering. One can see in Fig. 3.32a that in the afternoon hours of the individual days relatively even temperature distributions prevail over the whole heights. From evening onwards through the night and into the morning hours one clearly finds lapse rates with proper "nests" in the second half of the night. These "nests" develop due to nocturnal surface inversions which reach their maximum expansion in the early morning hours.

The ozone isopleths determined by the tethersondes act in conjunction with the temperature layers, as can be seen in Fig. 3.32b: Every day starting at noon the ozone concentrations are distributed evenly over the entire altitude. The ozone sinkings are located in the nocturnal surface inversion "nests" (with higher NO_2 concentrations at the same time, s. [61]).

The fact that the influence of atmospheric chemistry on O_3 formation and consumption depends less on geographical altitude than on the distance to the pollution source has already been established by comparing ozone behavior in the Black Forest and near highways, s. Fig. 3.25. The O_3 is consumed by oxidation of the nitrogen oxides emitted by the automobiles. Nitrogen oxides caused by traffic were recorded only in reduced concentrations and without pronounced diurnal variation at a measuring station in the forest. Accordingly, there is only a slight drop in O_3 in the morning hours. During the afternoon hours the O_3 concentration rises owing to photochemical processes, but it is not as pronounced as in the vicinity of anthropogenic sources, e.g., in the valley town of Garmisch-Partenkirchen (Fig. 3.30). With an even greater distance to inhabited areas, the O_3 diurnal variation becomes more even, similar to the situation on the Zugspitze (s., e.g., [33, 62]). At these "clean air stations", however, it cannot be determined whether the ozone present is of natural origin or whether photochemically formed ozone close to the source has mixed so well with the ambient air while being transported to the so-called clean air areas that the pronounced diurnal variation has become more constant. Under certain meteorological circumstances the ozone can be transported to these stations located far from the source. Such transports, however, are not subject to a daily rhythm.

To sum up, it can be said that the occurrence of ozone in the lower troposphere above the atmospheric main stable inversion layer is determined by widespread concentration dispersal. In the air layers close to the ground below the stable inversion layers, ozone behavior is strongly influenced by reactions of atmospheric chemistry with anthropogenic air pollutants. Measurements have not yet shown with absolute certainty whether ozone occurrence in the higher troposphere is mainly due to ozone being transported there from the stratosphere or due to ozone production by photochemical processes [63, 64].

3.2.5 Carbon Compounds

3.2.5.1 Organic Carbon Compounds

Organic trace gases can be classified into pure hydrocarbons – alkanes, alkenes, alkines and aromatic hydrocarbons – which are released, e.g., during the production, distribution and incomplete combustion of fossil fuels, aldehydes and ketones which are formed due to partial oxidation during combustion processes, and other volatile compounds (e.g., alcohols, halogenated hydrocarbons, solvents etc.). The majority of organic gases are not water-soluble and are therefore decomposed mainly by homogenous gas reactions, the reaction with OH radicals being the most important one. Some also react with ozone [42].

Based on known reaction velocities Schurath calculated the lifespans of some of the organic gases in [42], s. Table 3.2. The reaction processes are complex and only partially proved.

Table 3.2. Calculated mean life-spans of some organic trace gases (Schurath in [42] and [43]); assumptions: OH concentration of $10^6 cm^{-3}$, O_3 concentration of 60 ppb

Trace Component	Mean Life-span	
Methane	8 years	
n-butane	4.5 days	
Ethene	1.13 days	
Propene	6.4 hours	
Ethine	45.4 days	
Benzene	9.7 days	
Toluene	1.8 days	
Acetaldehyde	17.4 hours	
Ethyl alcohol	3.9 days	water –soluble,
Formic acid	33 days	thus wet deposition
Carbon monoxide	2 months	

3.2.5.2 Inorganic Carbon Compounds

The decomposition of methane in the troposphere and stratosphere leads to the formation of carbon monoxide. It is also formed by various other processes, e.g., by the oxidation of hydrocarbons such as terpenes from coniferous woods or it is released from the soil in warm regions [65]. Large amounts of man-made CO are released by incomplete combustion processes in furnaces and particularly by motor vehicles. Vertical profile measurements of CO in different locations of the earth have shown that CO concentrations drop with increasing altitudes. This indicates that the majority of the CO is released or formed in the boundary layer close to the ground and that the higher troposphere is more a sink for CO [63].

The main cause for a CO sink in the atmosphere is, just as is the case with hydrocarbons, the reaction with OH radicals to CO_2 and H radicals which in turn enter into further reactions. Another important sink for CO is the soil into which CO is absorbed and converted into CO_2 by microorganisms at temperatures between 15 and 35 °C [66].

3.2.6 Particles in the Atmosphere

Dust particles in the atmosphere differ not only in size but also in their chemical composition, the reason for this being different mechanisms of formation. A summary of the aerosol processes in the atmosphere is shown in Fig. 3.33 [67]. It also shows the particle size range into which the individual components are classified according to their typical size [68].

Typical components of coarse particles, which reach the atmosphere mainly by wind erosion of minerals, are calcium, titanium and iron. Plant material, e.g., pollen, belongs to this size class as also sea salt particles which reach the South German region with maritime air masses.

Coarse industrial particle emissions, e.g., from the cement industry or from power plants have been greatly reduced owing to progress in dust removal technology. The release of fine particles from industrial processes is largely unimpeded, in particular during processes without or with low-efficiency dust precipitators. Key substances for the particle emissions from oil furnaces, heavy oil furnaces in particular, are the elements nickel and vanadium. Particles containing lead have chiefly been formed by motor vehicle traffic so far.

In the atmosphere ultrafine aerosol can also be formed when gases react with each other. A known example is the "gas-to-particle-conversion" of SO_2 to sulfate particles which are frequently neutralized by ammonia. The chemical composition of aerosols in the atmosphere is influenced by surface reactions and hydrate envelopes at higher atmospheric humidities.

There are no routine measurements for particulate components either. Table 3.3. contains an overview of the magnitudes of some anthropogenic heavy metals and organic compounds in aerosols occurring in rural and urban areas. Sources of heavy metals are metallurgical and combustion processes. Lead is caused by automobile traffic (this is greatly reduced when unleaded gasoline is used). Nickel and vanadium are typical of the combustion of fuel oils. Polycyclic aromatic hydrocarbons are byproducts of domestic heaters, diesel exhaust gas and of coking plants.

3.2.7 Precipitation Components

The chemical composition of rain or snow is determined by the absorption of gases and the incorporation of particles into the droplets. Due to the long residence time in the cloud, balances can develop between the droplets and the ambient air. Here, the degree of absorption of particles is determined by atmospheric turbulence, and hardly at all by the atmospheric residence time of the precipitation elements. Precipitation from stratus clouds, e.g., generally have lower component concentrations. These concentrations reflect the transportation paths of the air mass. After passing through industrial centers, concentration values increase.

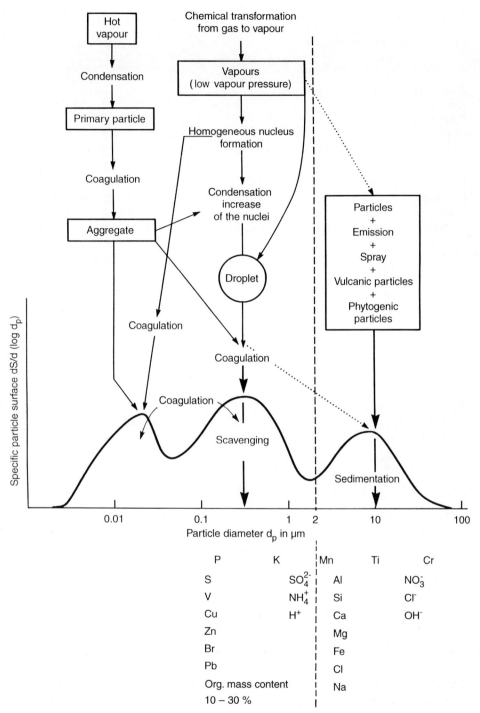

Fig. 3.33. Compilation of atmospheric aerosol processes and typical aerosol size distribution (specific surface distribution) acc. to Whitby [67], supplemented by mean aerosol sizes for individual components acc. to Klockow [68]

Table 3.3. Typical concentration ranges of non-routine/extra-routine measurements of aerosol components – heavy metals and polycyclic aromatic hydrocarbons [70]

	Substance	Aerosol Components/ng/m^3/concentr.range	
		Rural Area	Urban Area
Arsenic	As	1 ... 5	3 ... 30
Beryllium	Be		0.01 ... 2
Lead	Pb	20 ... 60	200 ... 1000
Cadmium	Cd	0.2 ... 2	2 ... 20
Chrome	Cr	1 ... 5	5 ... 30
Cobalt	Co	0.1 ... 1	0.5 ... 5
Copper	Cu	1 ... 10	20 ... 150
Manganese	Mn	10 ... 50	20 ... 100
Nickel	Ni	1 ... 10	5 ... 20
Mercury	Hg (part.)	0.05 ... 3	0.2 ... 2
Antimony	Sb	0.5 ... 2	2 ... 30
Selenium	Se	0.5 ... 3	1 ... 10
Vanadium	V	1 ... 10	10 ... 50
Zinc	Zn	50 ... 100	100 ... 1000
Benzo[a]pyrene	BaP	0.5 ... 3	2 ... 20
Dibenzanthracene	Db (ah) A		1 ... 15
Benzopaphtothiopene	BNT	0.5 ... 3	1 ... 15
Benzo[a]anthradene	BaA	0.5 ... 3	2 ... 40
Chrysene	CHR	1 ... 10	5 ... 50
Indenopyrene	IND	1 ... 5	2 ... 30

Thundershowers in the course of which cloud droplets are often flung violently through the vertical air column for several kilometers reflect the local conditions of the atmosphere. Great variations in space and time have been observed. Component concentrations can increase dramatically [69].

At the beginning of a precipitation, wash-out processes determine the precipitation composition and concentration, i.e, absorption and incorporation of gases or particles in the lower air layers. Drastically increased precipitation concentrations have been observed here.

3.3. Distribution and Temporal Development of Air Pollutants in Unpolluted and Polluted Areas

The air pollution concentrations show both strong spatial differences as well as temporal variations and developments. There will be some examples for both cases in the following sections.

3.3.1. Spatial Distribution

On a global scale there are differences in air quality between largescale areas, but differences in pollution can also occur within a smaller area. Basically, different climatic and vegetation zones also have different conditions of air quality. In the arid zones of deserts or areas bordering these deserts it is primarily the dust raised by wind which causes the air to be polluted by particulate matter. E.g., in the African countries bordering the Sahara the particle concentration increases steeply when the wind blows from the direction of the Sahara. In the West African country Nigeria particle concentrations during Harmatten wind are up to ten times higher when compared to conditions without this Saharan wind [71], s. Fig. 3.34.

Another source of air pollution occurring over large areas are forest, savanna and bushland fires which are sometimes kindled by natural causes like lightning. Most of the time, however, it is man that causes them. For instance, as a result of population growth fires are intentionally lit for deforestation in order to gain land for settling and agriculture . These fires occur globally but mainly in the tropical regions of the earth during the dry season, with occasional occurrences in the boreal forests of Siberia or Canada [72]. Biomass burning can contribute extensively to the budgets of several gases which are important in atmospheric chemistry [73]. The emitted pollutants are: particulate matter, large amounts of carbon monoxide (CO), hydrogen (H_2), methane and other reactive hydrocarbons, nitrogen oxides, especially N_2O and NO_x (depending on the nitrogen content of the burned biomass), cabonyl sulfide (COS) and methyl chloride (CH_3Cl). These air pollutants are widespread in tropical countries because of the distribution of the sources and the atmospheric mixing capacity in the tropics. Figure 3.35 shows a photograph of a savanna fire. The air pollution caused by the fire is to be plainly seen. Other sources of globally distributed air pollution are natural decomposition processes, e.g., in swamps, moors and marshy lands, soil exhalations, volcanic eruptions, transformations and vented gases or oceanic spraying. Besides, substances caused by human activities such as combustion and industrial proceses and intensive animal breeding reach the atmosphere and

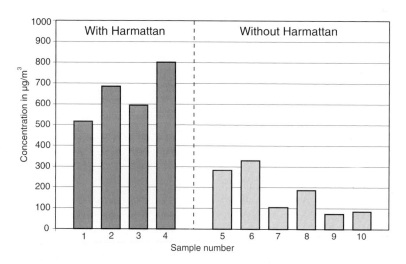

Fig. 3.34. Suspended particulate matter (SPM) with and without Harmattan (Sahara wind) in Lagos/Nigeria [71], 24 h mean values

Table 3.4. Typical concentrations of several trace compounds in different polluted areas (acc. to [74,75])

Compound	Major sources	Background, in general	Rural areas	Polluted areas (urban, industrial)	Lifetime	Remarks
Hydrogen Peroxide H_2O_2 in ppb	Photochemical Reactions	< 0.1	0.1 – 1	0.03 – 3		also in liquid phase
Carbon Monoxide CO in ppb	Burning, combustion processes, oxidation of natural hydrocarbons	100 – 200 (northern hemisphere) 30 – 90 (southern hemisphere)		1000, up to 50 000		increasing trend
Carbonyl Sulfide COS in ppt	Soils, coastal marshes, volcanic eruptions, biomass burning	400 – 600			44 years	slow decomposition
Carbon Disulfide CS_2 in ppt	Oceans, soils, biological processes	15 – 30	– 300	100 – 200	12 days	oxidation to SO_2
Dimethyl Sulfide (DMS), $(CH_3)_2S$ in ppt	Ocean algal decomposition	< 10 1-800 (oceanic air)	1 – 20	2 – 500	0.6 days	oxidation to SO_2
Hydrogen Sulfide H_2S in ppt	Bacterial reduction, soils and wetlands	30 – 150	– 400	30 – 800, up to 2000	4 – 5 days	photochemical reactions act as sink
Sulfur Dioxide SO_2 in ppb	Anthropogenic, volcanic, oxidation of other sulfur comp.	0.005 – 0.09	2 – 50	15 – > 500	2 – 4 days	decreasing trend in polluted areas

3.3 Distribution and Temporal Development of Air Pollutants 133

Compound	Major sources	Background, in general	Rural areas	Polluted areas (urban, industrial)	Lifetime	Remarks
Sulfate SO_4^{2-} in µgS m^{-3}	Sea spray, oxidation of SO_2	0.1 – 0.3	5	-10	1 week	wet and dry deposition
Nitrous Oxide N_2O in ppb	Soil	310			170 years	increasing by 0.2 % p.a.
Ammonia NH_3 in ppb	Animals, soil, biomass burning	0.1 – 5	6 – 20	> 100	6 days	conversion to NH_4^+
Ammonium NH_4^+ in µg m^{-3}	Conversion from NH_3	0.05	1	10 – > 30	5 days	wet and dry deposition
Nitrogen Monoxide NO in ppb	fossil fuels, biomass burning, soils	< 0.1	0.5 – <10	50 – 800	< 2 days	oxidation to NO_2 by O_3
Nitrogen Dioxide NO_2 in ppb	Oxidation of NO, Lightning	0.01	1 – 20	2 – 100	< 2 days	Oxidation to NO_3, Photodissociation
Peroxyacetylnitrate PAN in ppb	Photochemical reactions	0.01 – 0.5		5 – 50 (smog)		
Nitrate NO_3^- in µg m^{-3}	Oxidation of NO_2	0.2 – 0.5		10 – 90 (smog)	5 days	wet and dry deposition
Hydrochloric Acid HCl in ppb		1 (maritime air) 0.3 (continental air)	- 0.3	~ 0.6	0.6 - 21 days	

Fig. 3.35. Photograph of a bush burning (Photo: Günter Helas, Max Planck Institute for Chemistry, Mainz)

change its natural composition. Table 3.4. shows as examples the concentrations of some atmospheric trace compounds to be found in different polluted areas of the world. To complete the picture Table 3.5. will provide an insight into the concentrations of organic substances commonly found in urban and rural areas. Graedel describes more than 500 compounds, mainly organic substances, which can occur in the ambient air [76]. In Table 3.5 only a selection of typical representatives of some of the important substance classes is shown [70]. As measurements were not carried out on a regular basis it is not possible to indicate statistical distributions; only concentration ranges are given. The number of measured values upon which these data are based varies greatly for the different substances, in part only a few individual results make up the data base.

Hydrocarbons such as alkanes, alkenes, alkines and aromatics originate mainly in automobile traffic, other organic and inorganic compounds originate in industrial processes (chemical industry, degreasing of metals etc.). Fluorochlorinated hydrocarbons stem from spray can gases, from coolants and the foaming of plastics.

One of the air pollutant components to be found dispersed over large areas and in part also over small ones is tropospheric ozone. When UV radiation is high it is formed from anthropogenic exhausts, and perhaps also from products of other combustion processes like the above-mentioned biomass burning. Starting in the areas of large cities it spreads over vast stretches of land in the summer. In the USA mainly the area around Los Angeles, California, is known for its high ozone concentrations, but high ozone concentrations also occur in some areas of the densely populated eastern states of the USA [77], s. Fig. 3.36. Even in relatively small countries like Germany there are regional differences in ozone pollution due to different formation preconditions such as strong solar radiation and source capacity of precursor substances and due to varying decomposition conditions in urban areas. As an example Fig. 3.37 shows the ozone concentration distribution in Germany on a warm summer day [78]. For assessing and comparing the ozone concentrations of different cities maximum pollution values are used. Table 3.6 shows representative peak one-hour concentrations in various cities in the seventies and between 1980 and 1992.

Table 3.5. Typical concentration ranges of non-routinely detected volatile organic compounds in rural and urban areas [70]

	Compound	Concentration ranges in $\mu g\,m^{-3}$	
		Rural	Urban
Methane	CH_4	1200	1200...3000
Ethane	C_2H_6	1 ...5	3...15
n-pentane	$n\text{-}C_5H_{12}$	1 ...3	5...50
n-octane	$n\text{-}C_8H_{18}$	0.2...1	2...10
Cyclohexane	$cyclo\text{-}C_6H_{12}$	0.1...1	1...10
Ethene	C_2H_4	0.5...5	5...30
Butene	C_4H_6	1 ...2	1...10
Ethine	C_2H_2	0.2...3	5...30
Isoprene		1 ...10	
Benzene	C_6H_6	1 ...5	5 ...30
Toluene	$C_6H_5CH_3$	0.5...2	5 ...50
Xylenes	$C_6H_4(CH_3)_2$	0.1...1	5 ...50
Methyl alcohol	CH_3OH		10 ...20
Ethyl alcohol	C_2H_5OH		10 ...50
Formaldehyde	$HCHO$	0.5...2	10 ...20
Acetaldehyde	CH_3CHO	1 ...2	0.5...15
Acetone	$CH_3CO\,CH_3$	0,1...1	10 ...50
Methyl chloride	CH_3Cl	1 ...2	1 ...2
Dichloromethane	CH_2Cl_2	0.2...0.5	1 ...5
Chloroform	$CHCl_3$	0.2.....0.5	0.5...3
Tetrachloromethane	CCl_4	0.5...1	1 ...3
1,1,1-trichloroethane	CH_3CCl_3	1 ...3	5 ...10
Chloroethene	$CH_2CH\,Cl$	0.1	0.1...1
Trichloroethylene	C_2HCl_3	0.2...1	2 ...15
Perchloroethylene	C_2Cl_4	0.5...2	2 ...15
Dichlorobenzene	$C_3H_4Cl_2$		1 ...10
Dichlorodifluoromethane (F 12)	CF_2Cl_2	1 ...2	1...5
Trichlorofluoromethane (F 11)	$CF\,Cl_3$	1 ...2	1...4

Globally, the main load of anthopogenic air pollutants is concentrated in the urban areas, in particular the metropolitan areas. Whereas in some metropolitan areas industrial exhaust gases and, e.g., wild waste burning processes are sources of considerable air pollution, in other places it is automobile traffic which is the main source of air pollution. On the one hand the emission intensity of automobiles has a strong influence on air quality: in many of the world's large cities many automobiles

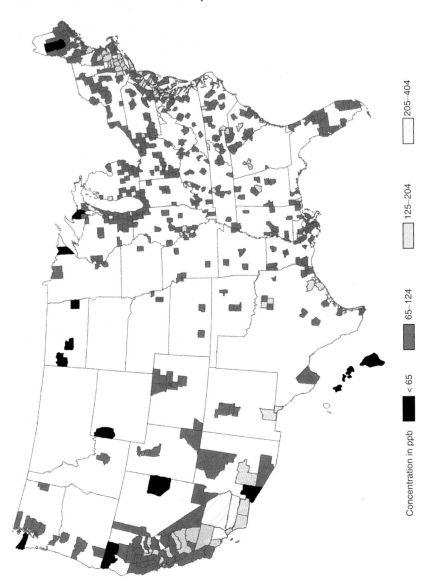

Fig. 3.36. Ozone air quality concentrations in the United States, 1993 – highest second daily 1-hour maximum (acc. to [77])

in poor condition and with high exhaust emissions are in use. On the other hand atmospheric dispersal and chemical transformation conditions (e.g., ozone formation) as well as topographic location have an influence on air quality in metropolitan areas. The most unfavorable locations are basin-type areas with nightly inversions and strong solar radiation during the day. The cities Athens, Los Angeles and Mexico City are examples of this type of adverse topography. The World Health Organization undertook a classification of the air quality situation in different large cities of

Table 3.6. Maximum one-hour ozone concentrations of various cities in the 1970s and between 1980 and 1992 [47, 53, 77, 79, 80, 81, 82, 83]

City	Country	O_3 peak concentration in ppb
Mexico City	Mexico	470
Los Angeles	United States	360
Vlaardingen	Netherlands	270
Biddinghuizen	Netherlands	162
Harwell (rural site near London)	United Kingdom	260
Stevenage	United Kingdom	164
Athens	Greece	>250
Cologne	Germany	255[1)]
Mannheim	Germany	190[2)]
Schauinsland (mountain station)	Germany	179
Martiques	France	> 200
Nice	France	170
Paris	France	164
Ilmitz	Austria	>200
Tokyo	Japan	> 00
Oslo	Norway	200
Sydney	Australia	> 160
Payerne	Switzerland	165
Rome	Italy	140
Zagreb	Yugoslavia/Croatia	140

[1)] 1/2 h value,   [2)] 3 h average

the world. Pollution was classified by the different pollutants SO_2, suspended particulate matter (SPM), lead (Pb), CO, NO_2 and O_3. This type of classification is represented in Fig. 3.38 [84]. While industrial sources are chiefly to be blamed in cities with SO_2 problems, particulate matter problems are caused jointly by industry, traffic and natural sources, e.g., in the city of Cairo they are caused by the nearby desert. Lead, NO_2 and most CO problems are typical traffic-specific substances. Apart from strong solar radiation, cities polluted by ozone are also exposed to high air pollution by precursor substances.

Air pollution over large cities can usually be recognized by a clearly visible veil of haze (smog). An example of this is shown in Fig. 3.39, a photograph of the smog

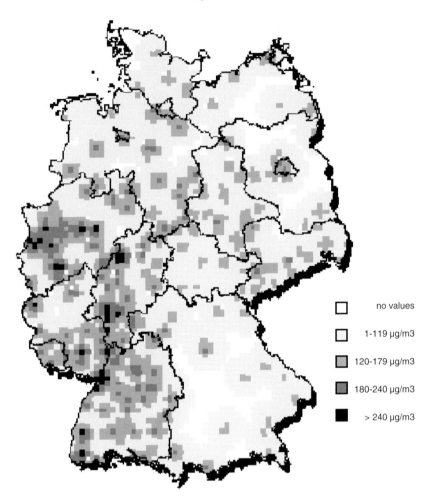

Fig. 3.37. Ozone air quality concentrations in Germany during a warm summer week (3rd to 7th of July, 1995) – highest half-hour values [78]

over Los Angeles. This low visibility is caused by aerosols [85], e.g., nitrates and sulfates, which are in part formed by gaseous pollutants when sun radiation is high. There are, of course, other large cities not mentioned in Fig. 3.38 which are also affected by high air pollution. More examples of this are the cities of Ankara, Athens [86, 87], Lagos [71], Milan [88] and Santiago de Chile [89].

In Central European cities air quality has improved greatly over the past few years (see next section). However, traffic-related air pollution problems in the cities are still a way from being solved. Pollution reaches a peak particulary in the downtown areas with their heavy traffic and high building density, whereas the areas on the outskirts show lower concentrations. Figure 3.40 shows an example of local distribution of air pollutants in a city. Spatial distributions such as this one are obtained by random sampling at the grid corner points of a 1 x 1 km grid. In Germany 26 measurements are executed on every corner point over the period of one year for this

3.3 Distribution and Temporal Development of Air Pollutants 139

City \ Substance	SO$_2$	SPM	Pb	CO	NO$_2$	O$_3$
Bangkok	▫	■	▫	▫	▫	▫
Beijing	■	■	▫	▫	▫	▫
Bombay	▫	■	▫	▫	▫	▫
Buenos Aires	▫	▫	▫	▫	▫	▫
Cairo	▫	■	■	▫	▫	▫
Calcutta	▫	■	▫	▫	▫	▫
Dehli	▫	■	▫	▫	▫	▫
Jakarta	▫	■	▫	▫	▫	▫
Karachi	▫	■	■	▫	▫	▫
London	▫	▫	▫	▫	▫	▫
Los Angeles	▫	▫	▫	▫	▫	■
Manila	▫	■	▫	▫	▫	▫
Mexico City	■	■	▫	■	▫	■
Moscow	▫	▫	▫	▫	▫	▫
New York	▫	▫	▫	▫	▫	▫
Rio de Janeiro	▫	▫	▫	▫	▫	▫
Sao Paulo	▫	▫	▫	▫	▫	■
Seoul	■	■	▫	▫	▫	▫
Shanghai	▫	■	▫	▫	▫	▫
Tokyo	▫	▫	▫	▫	▫	■

■ Serious poblem, WHO guidelines exceeded by more than a factor of two

▫ Moderate to heavy pollution, WHO guidelines exceed by up to a factor of two (short-term guidelines exceeded on a regular basis at certain locations)

▫ Low pollution, WHO guidelines are normally met (short-term guidelines may be exceeded occasionally)

Fig. 3.38. Classification of the air quality situation in different large cities of the world (from World Health Organization [84])

Fig. 3.39. Smog over Los Angeles (Photo: Ulrich Greul, University of Stuttgart)

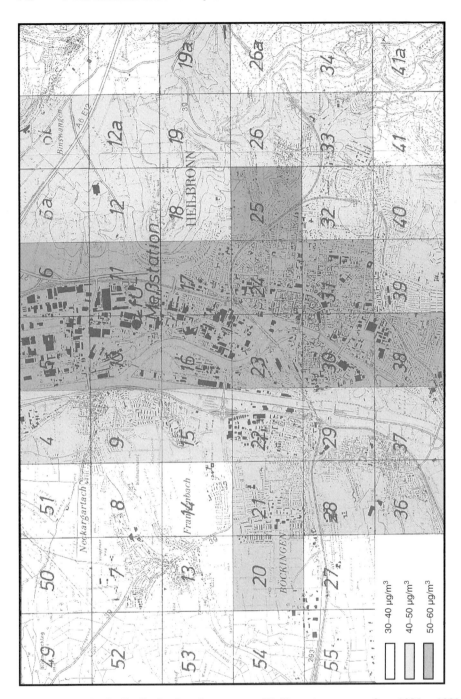

Fig. 3.40. Local NO$_2$ distribution in a German town (Heilbronn) – averages from 1985 to 1989 [90]

purpose. The 1 km² surface value is the mean value of the measurements taken at the corner points. Thus, the mean pollution distribution of a city is obtained and one can see which areas of a town are in particular need of clean air measures. Most frequently, the downtown areas are polluted the most due to the high traffic concentration. Within the 1 x 1 km² grid squares there can, however, be great differences in air pollution. E.g., measurements directly on dense traffic streets show high values of nitrogen oxides, carbon monoxide, hydrocarbons and particulate matter, particularly when traffic runs through so-called street canyons. Table 3.7 shows examples of peak concentrations of nitrogen oxides in the street canyons of German cities. With increasing distance from the street concentrations decrease exponentially, more or less strongly depending on wind direction and velocity, s. Fig. 3.41a. In this way, 50 m away from the street only 40 % of the original pollution of the side of the street remains (s. also Chapter 6.2.5.4). Directly at the side of the street there are high NO concentrations, s. Table 3.7, but with increasing distance to the street the NO_x/NO_2 ratio decreases or the NO_2/NO_x ratio increases, s. Fig. 3.41b. Apart from the concentrations near the side of streets in cities it is not NO but NO_2 which is the main pollutant characterizing air pollution by nitrogen oxides.

Table 3.7. Examples of NO and NO_2 values measured in German street canyons or directly at high traffic roads [91]

Place	Year	NO µg/m³	NO_2 µg/m³
Cologne, Neumarkt	1980 – 1985	533 – 634	215 – 356 98 Percentiles
Stuttgart, Marienplatz	1985/1986	200 – 1100	100 – 300 second highest half-hour values
Wendlingen, Highway	1985	750 – 2000	100 – 450 highest values
		580 – 1300	86 – 300 98 Percentiles
Tübingen, Mühlstraße	1987/1988	750 – 1250	150 – 200 highest values
Stuttgart-Wangen, Highway B 10	1989	–	180 – 200 highest values
Stuttgart-Vaihingen, Highway, 40 m distance	1990/1991	478 – 856	125 – 212 highest values
		262 – 552	75 – 157 98 Percentiles

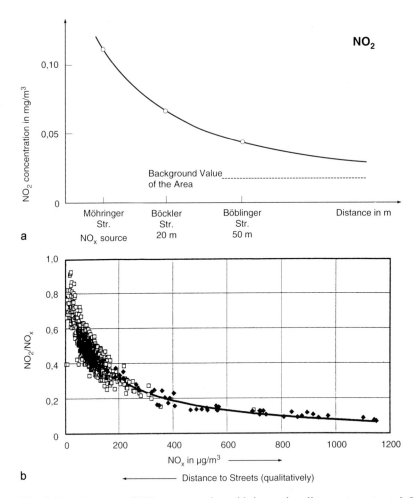

Fig. 3.41. a Decrease of NO_2 concentration with increasing distance to a street, **b** Increase of NO_2/NO_x ratio with decreasing NO_x concentrations (that means increasing distance to streets; measuring values from 210 different stations, 1982-1993 [92]

3.3.2. Temporal Trends in Air Quality

3.3.2.1. Carbon Compounds

Carbon Dioxide (CO_2)

Atmospheric carbon dioxide concentrations have been measured on many background sites of the world since 1968 [93]. At the Mauna Loa Station on Hawaii and at the South Pole measurements have even been carried out since 1957/58 [94]. The CO_2 values for before this period have been reconstructed from samples of air occlusions in ice cores drilled at the Antarctic Siple Station [95].

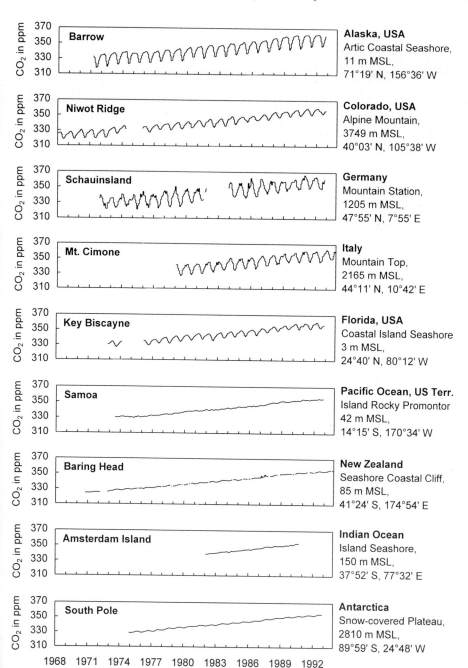

Fig. 3.42. Monthly atmospheric CO_2 concentrations at several measuring sites of the world (acc. to [96, 97, 98, 99])

144   3 Air Pollutants in the Atmosphere

At all stations an increasing CO_2 trend is to be observed, giving rise to discussions about the greenhouse effect (s. also Chap. 4.2.1.2). Fig. 3.42 shows a selection of CO_2 courses of several of the world's measuring stations. The annual increase within the last ten years on these sites is calculated to be between 1.40 to 1.45 ppm per year [93].

In the northern hemisphere a distinct seasonal pattern of CO_2 is to be found with a drawdown occurring in August and September and the annual peak in March to May. The seasonal variation is caused by the growth season of plants in the middle to the high latitudes in the northern hemisphere. Over the southern Pacific Ocean nearly no pattern is to be observed; at the South Pole the amplitude of the seasonal pattern is very low. The reason for the smaller amplitudes in the southern versus the northern hemisphere is the lack of major land vegetation there. However, the CO_2 concentration differences between the hemispheres are low because the atmosphere is generally well mixed. But the CO_2 difference between the northern and the southern hemisphere increased from 1 ppm in 1960 to 3 ppm in 1985, a development parallel to the (northern hemisphere) increase in fossil fuel combustion sources [100].

Methane (CH_4)

Like carbon dioxide concentrations historical atmospheric methane concentrations have also been reconstructed from ice core drillings (e.g., [101]). Regular atmospheric records have been taken from sites in the NOAA/CMDL air sampling network since 1983 [102] (NOAA/CMDL = US National Oceanic and Atmospheric Administration, Climate Monitoring and Diagnostic Laboratory).

Like carbon dioxide methane shows an increasing trend. In Fig. 3.43, which contains values of six sites of the world from Alaska in the North to the South Pole, the measured courses of methane concentrations are shown. The globally averaged atmospheric CH_4 annual means derived from this NOAA/CMDL air sampling network increased from a value of 1625.6 ppm in 1984 to 1714.1 ppb in 1992. This means an average increase rate of 11.1 ppb/year. In the southern Oceanic and the Antarctic sites the CH_4 concentrations occur at lower levels than at the northern sites of the measuring network (s. Fig. 3.43).

Although the quantitative importance of the different factors contributing to the observed increase is not yet well known, it can be stated that human activities are the main reasons for the increase [100].

Halocarbons

The most important halocarbons in respect to radiative forcing (greenhouse effect) are fully halogenated chlorofluorocarbons CFC-11 (CCl_3F) and CFC-12 (CCl_2F_2) [100]. Trend data for these gases have been gathered by the Atmospheric Lifetime Experiment (ALE) and the Global Atmospheric Gases Experiment (GAGE) [103] as well as by the NOAA/CMDL flask sampling program [104]. With the last program the global and hemispheric trends were calculated by averaging the monthly values

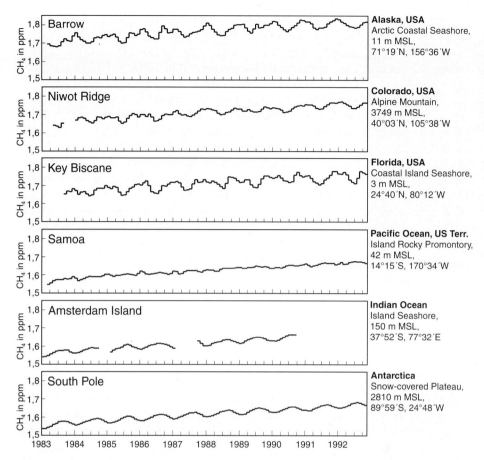

Fig. 3.43. Monthly atmospheric CH_4 concentrations at several measuring sites of the world [102]

of seven measuring stations. These trends are shown in Fig. 3.44. It can be seen that the atmospheric concentrations of these exclusively man-made gases have been steadily increasing. Although the concentrations are much lower than those of CO_2 and CH_4 these halocarbons have a great importance because of their high global warming potential [100].

3.3.2.2. Sulfur Dioxide (SO_2)

Apart from particulate matter sulfur dioxide emitted during the combustion of sulfurous fuels used to be the main component in anthropogenic problems of air pollution in the past. In the former German Democratic Republic and Czechoslovakia SO_2 concentrations, for instance caused by the combustion of highly sulfurous lignite, were the reason for severe air pollution until recently. Long-range transport across

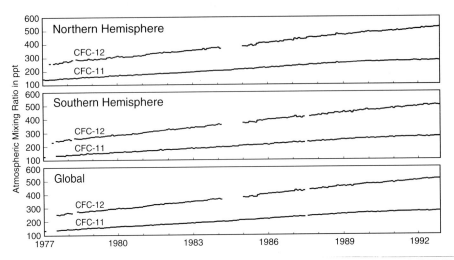

Fig. 3.44. Global and hemispheric trends of the halocarbons CFC-11 (CCl$_3$F) and CFC-12 (CCL$_2$F$_2$), monthly averages from seven measuring sites of the NOAA/CMDL sampling program [104]

hundreds of kilometers brought SO$_2$ pollution to remote areas far from the sources (s. Chap. 3.1.5.1). Owing to the use of fuels containing less sulfur, the installment of flue gas desulfurization units and shut-downs of out-of-date firing plants, e.g., in power stations (s. also Chap. 2.2), SO$_2$ concentrations have slackened off in most industrial nations in the past few years. This development becomes obvious when looking at the profiles of the SO$_2$ annual mean values of the past 20 to 30 years as shown for West and East Germany in Figs. 3.45a and b and for some other Western European countries in Fig. 3.45c.

In U.S. and Canadian cities th2e trend is also a decreasing one, as can be seen from Fig. 3.46. The second highest 24-hour concentrations have been used in this diagram and not the annual mean values. This type of representation is more realistic as the effect of SO$_2$ as an irritant gas depends on repeated short-term exposure to peak concentrations. However, when the annual averages decrease the frequency of peak concentrations also decreases, so that the trends shown in Fig. 3.45 give a representative picture nonetheless. These SO$_2$ downward trends cannot be observed globally, however. In many of the world's cities the population continues to be exposed to high SO$_2$ concentrations, as can be seen from the examples shown in Fig. 3.47.

3.3.2.3 Suspended Particulate Matter (SPM)

Owing to the effects of exhaust gas purification measures concentrations of suspended particulate matter, just as with the component SO$_2$, have been distinctly reduced in the industrial nations in the past years. This trend is shown in Fig. 3.48a for Germany, in Fig. 3.48b for selected European cities, in Fig. 3.48c for selected U.S.

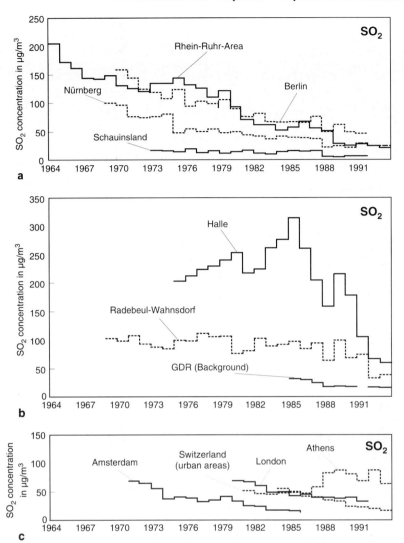

Fig. 3.45. Trends of SO$_2$ concentrations (annual averages). **a** West Germany: in the industrial area Rhine-Ruhr, in the cities Berlin, Nuremberg and at the background station Schauinsland (Black Forest) [105, 106, 107, 108], **b** East Germany: In the cities of Halle and Radebeul-Wahnsdorf and at background sites [105, 106, 109], **c** in different areas of western Europe [77, 83, 86, 110]

and Canadian cities and in Fig. 3.48d for selected cities of the world. Just as is the case with the component SO$_2$ there are still considerable air pollution problems in many of the world's cities where suspended particulate matter is concerned. Examples of this are Mexico City, Beijing, Delhi and Bangkok, s. Fig. 3.38.

At four background stations operated by the National Oceanic and Atmospheric Administration (NOAA) aerosol measurements have been carried out since 1977 by measuring of direct solar irridiance??? and calculated as aerosol optical depth (AOD)

148   3 Air Pollutants in the Atmosphere

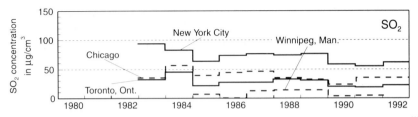

Fig. 3.46. Trend in second highest 24-hour SO_2 concentrations in selected U.S. and Canadian cities, 1983-1991 [77]

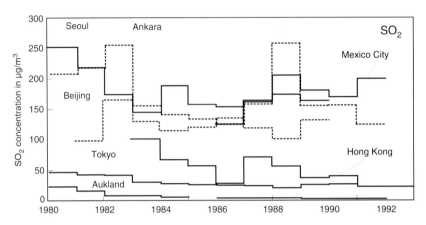

Fig. 3.47. Trends in annual average SO_2 concentrations in selected cities of the world [77, 81, 111, 112]

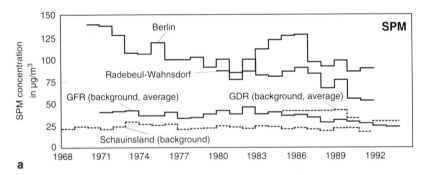

a

Fig. 3.48. Trends in annual average suspended particulate matter (SPM) concentrations. **a** in Germany: in the city of Berlin, in Radebeul-Wahnsdorf and at background stations of the GFR (average), the GDR (average) and Schauinsland [105, 107, 109], **b** in selected European cities [77, 83, 86], **c** in selected U.S. and Canadian cities [77], **d** in selected cities of the world [77, 81, 112]

3.3 Distribution and Temporal Development of Air Pollutants 149

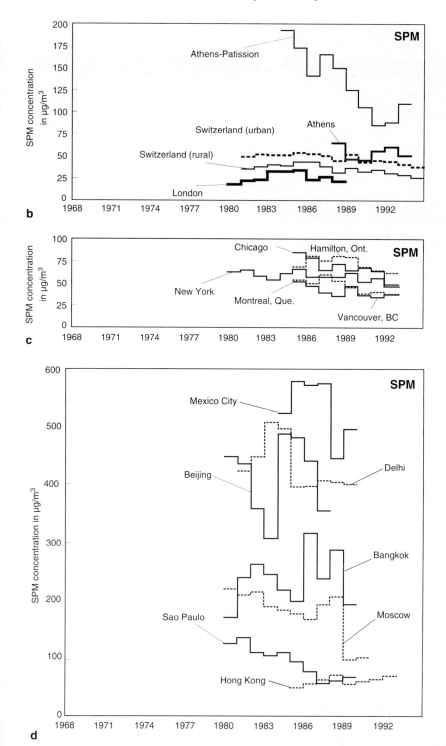

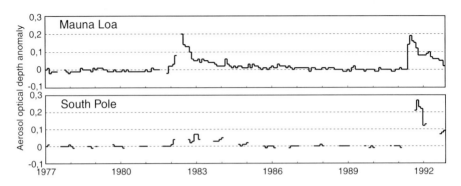

Fig. 3.49. Aerosol optical depth anomalies at the NOAA background stations on the Mauna Loa, Hawaii, and at the South Pole [113]

[113]. From two stations, Mauna Loa, Hawaii, and the South Pole the courses of aerosol optical depth are shown in Fig. 3.49. In Hawaii, the predominant features are the large increases in AOD that followed the volcanic eruptions of El Chichón (Mexico, 1982) and the Pinatubo (Philippines, 1991). At the South Pole, the effects of the El Chichón eruption were much smaller at this site than in the northern hemisphere. In contrast/on the contrary??, the maximum AOD anomaly at the South Pole after the Pinatubo eruption was greater than at the other three NOAA measurement sites (Barrow, Alaska; Mauna Loa, Hawaii; Samoa Pacific Island) [113].

3.3.2.4 Nitrogen Oxides

By improving combustion processes in, e.g., automobile engines, it has been possible to reduce both fuel consumption and carbon monoxide and hydrocarbon emissions. The resulting higher combustion temperatures, however, have caused nitrogen oxide emissions to rise. In the seventies, e.g., measurements in the Netherlands show a drastic NO_2 increase. When, in the eighties, catalytic converter technology for automobiles was introduced in many countries the effects of this exhaust gas purification technique were for the time being completely neutralized by the constantly rising number of automobiles. Finally, in 1992, readings in measuring stations showed a slight downward trend in NO_2 concentrations in West Germany, s. Fig. 3.50a. In European cities, too, slightly lower NO_2 concentrations are to be seen only in the past few years, s. Fig. 3.50b. In the USA a slight drop in the NO_2 concentrations since 1990 in a mean of 201 measuring sites can be recognized, s. Fig. 3.50c.

3.3.2.5 Tropospheric Ozone (O_3)

Formation and decomposition processes of tropospheric (surface) ozone have been dealt with in Chap. 3.2.4.2. Due to a variety of many influences it is not easy to compare one year's ozone concentrations with those of another year. E.g., sunny

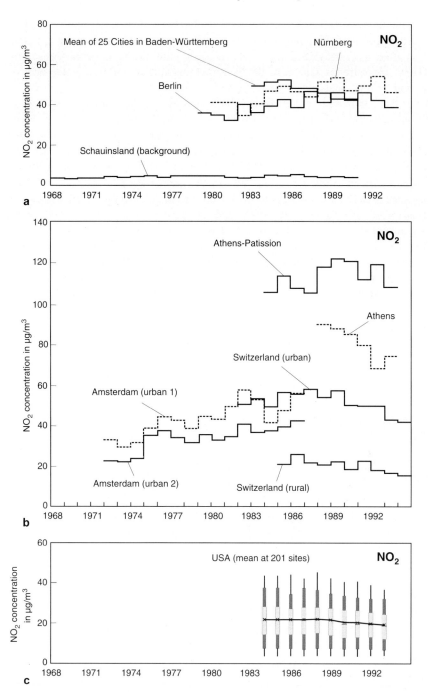

Fig. 3.50. Trends in annual average nitrogen dioxide concentrations, **a** in Germany [105, 107, 108, 114], **b** in European cities [83, 86, 110], **c** in U.S. cities (boxplot comparisons of NO_2 concentrations at 201 sites) [77]

warm summers have much higher ozone concentrations than rainy ones. On sites influenced by primary components, e.g., near streets with a lot of traffic, the nitrogen oxides emitted there may shift the photostationary equilibrium to ozone consumption. Thus, clear trends can only be measured on sites uninfluenced by nearby sources and only in measuring series over many years.

In the late nineteenth and early twentieth centuries ozone measurements were carried out in Montsouris, south of Paris [115, 116], in Zagreb, Croatia, Yugoslavia [117]. To be able to compare the measuring results then with today's values the measuring methods used then had to be duplicated and the values corrected according to today's standards. Kley et al. [115] and Bojkov [116] found out that in Montsouris, uninfluenced by the wind directions of the greater Paris area, ozone concentrations were between 15 and 20 ppb. In Zagreb the annual average volume fractions of the surface ozone ranged between 35 and 25 ppb between 1889 and 1900 with differentiation between day and night-time hours, s. Fig. 3.51. The curve for Montsouris is also drawn in this diagram according to Bojkov's data [116].

In this century ozone measurements have been carried out at different locations in the world since the early fifties. Fig. 3.52 gives an overview of the measuring programs. In Germany, the longest measuring series are from the measuring stations in Arkona on the island of Rügen, Baltic Sea, the city of Dresden-Radebeul, Mount Fichtelberg in the Ore Mountains and the city of Kaltennordheim in the Thuringian Forest. As examples, Fig. 3.53 shows the ozone profiles of the stations in Arkona and Dresden-Radebeul. Included in the chart for comparison are the profiles of the south German mountain station Hohenpeissenberg where the measuring series started in

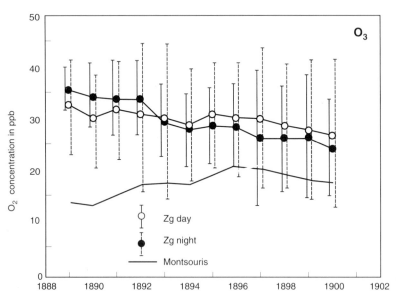

Fig. 3.51. Annual average volume fractions of surface ozone (with annual amplitude inside bars), Zagreb 1889-1900 [117]; the curve of Montsouris (near Paris) is drawn according to Bojkov's data [116]

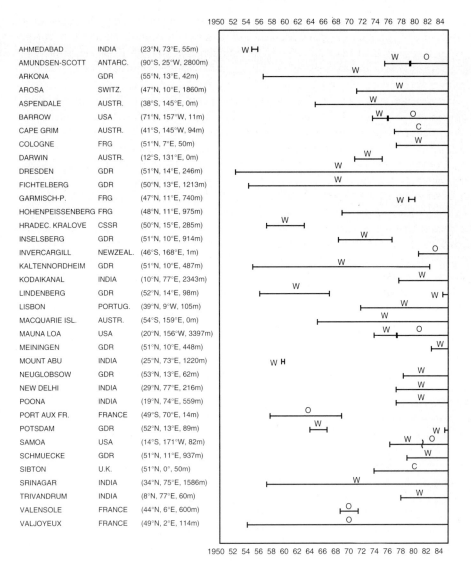

Fig. 3.52. Periods of surface ozone records across the world from the 1950s until 1984 [118, 119], W = wet-chemical, O = optical, C = chemiluminescent

the early seventies. The wet-chemical KI measuring method used then shows an interference with SO_2, leading to an underestimation of the ozone values. As a result, the O_3 values recorded in Dresden-Radebeul from 1952 to 1972 were too low. From 1972 onwards SO_2 filters were installed for measurement [119]. The leap in the measured values is undeniable. Due to very low SO_2 concentrations the O_3 values measured in Arkona were never influenced. Hohenpeissenberg's measuring values of the years 1971 to 1975 were submitted to an SO_2 corrrection, so that the originally low values were raised. In this way drastically increasing trends could no longer be

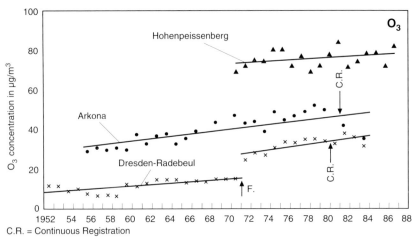

Fig. 3.53. Annual mean values of surface ozone at the German stations Arkona (Rügen Island in the Baltic Sea), Dresden-Radebeul (town in Saxony) and Hohenpeissenberg (Bavarian mountain, 17 km north of the European Alps, 975 m above m.s.l.), [119, 120]

observed [120]. Arkona, however, shows an increase in O_3 from 30 µg/m³ in 1950 to over 50 µg/m³ in 1980. From 1972 to 1984 a rising trend in the ozone values is also to be observed in Dresden-Radebeul.

As ozone concentrations have a strong annual course with low concentrations in winter and high values in summer, it is necessary to take into consideration not only the annual averages but also the peak values of one month or, where long-term measuring series are concerned, the maximum values for a year. Using monthly averages and highest half-hour values Fig. 3.54 shows the development of surface ozone concentrations of a German city (Heilbronn) and of a forest area in the northern Black Forest (Freudenstadt-Schöllkopf) for the years 1984 to 1991. The ozone maxima in the summer months and the minima in the winter are quite distinct. As a result of the oxidation of primary pollutants most of the ozone in the cities is consumed at night, especially when the sky is clear and there are surface inversions. Thus, ozone mean values are not as high in the cities when compared to, e.g., the monthly mean values of Freudenstadt-Schöllkopf or the annual mean values on the Hohenpeissenberg. But the cities' maximum values can exceed the peaks in forest and mountain regions.

In the USA the EPA (Environmental Protection Agency) has tried to develop techniques for adjusting O_3 trends for meteorological influences [122]. Their report features a statistical model in which the frequency distribution of O_3 concentrations is described as a function of meteorological parameters. The model has been used to calculate "meteorologically adjusted" estimates of the upper percentiles of daily maximum concentrations for each year. Fig. 3.55 displays ambient air quality trends and meteorologically adjusted trends for 43 metropolitan areas. The "adjusted" trend

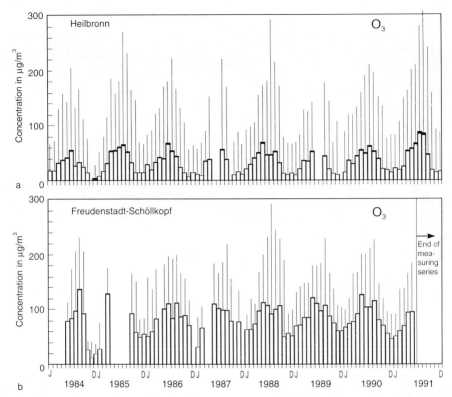

Fig. 3.54. Recent ozone concentrations in Germany. Monthly averages (bars) and monthly highest half-hour values (vertical lines), **a** in an urban area (Heilbronn) [121] and **b** a forest site in the northern Black Forest (Freudenstadt-Schöllkopf) [33]

indicator shown in Fig. 3.55 is the composite mean of the meteorologically adjusted 99th percentile daily maximum one-hour concentrations across each of the 43 individual metropolitan areas. The smoothing introduced by meteorological adjustment is especially evident in the peak O_3 year 1988 which was followed by years less conducive to O_3 formation. A steady downward trend is clear. The composite average of the 99th percentile daily maximum one-hour concentrations in 1993 is 12 % lower than the 1984 level. Coincidentally, the 10-year percent change in both the adjusted and unadjusted composite average of the 99th percentile concentration for these 43 cities is exactly the same percentage change as the national second daily maximum one-hour trends statistic for 509 sites [77]. The total volatile organic compounds (VOC) emissions are estimated to have decreased 9 % between 1984 and 1993. During this same period NO_x emissions, the other precursor of O_3 formation, increased 1%. Whether this drop in ozone is a direct result of the VOC reduction has yet to be confirmed with the help of future statistics taking into special account emissions on warm "ozone days".

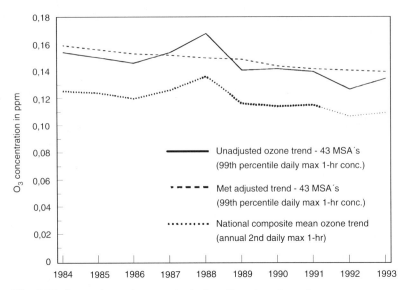

Fig. 3.55. Comparison of meteorologically adjusted, and unadjusted, trends in the composite average of the second highest maximum one-hour ozone concentration of 43 metropolitan statistical areas (MSAs), 1984-1993 [77]

3.4 Models of Pollutant Dispersion (B. Steisslinger)

The air quality in the lower troposphere is, on one hand, determined by the pollutant amounts emitted, and by the individual conditions for dispersion on the other. Apart from the topographical circumstances, meteorological influences, above all atmospheric layering and wind conditions, are particularly important here.

The correlation between emissions, atmospheric distribution and pollutant concentrations has often been simulated with mathematical meteorological models.

3.4.1 Objective and Application of Mathematical Meteorological Simulation Models

The central purpose of dispersion modelling is to describe the relationship, shown in the diagram in Fig. 1.4 between pollutant emission, transmission and ambient air concentrations of one or several air pollutants as a function of space and time in a mathematically exact way. This is done with the calculation depending on emission volume, individual meteorological conditions and, if necessary, a number of parameters which take into account transformation and deposition processes in the atmosphere.

Simulation calculations can be used to establish criteria for planning the location of industrial plants and complexes requiring official approval, for determining minimum stack heights, for developing and assessing air quality control strategies (to

3.4 Models of Pollutant Dispersion

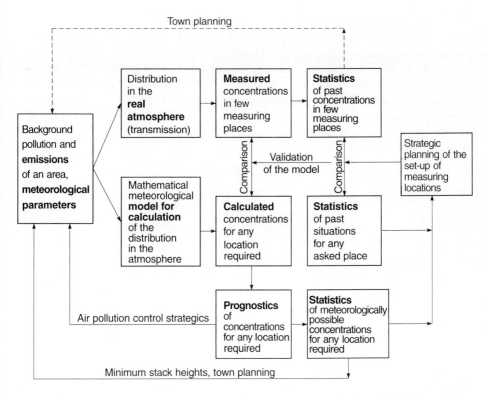

Fig. 3.56. Main tasks and applications of mathematical-meteorological simulation models (acc. to [123])

investigate the effects of emission restrictions on air quality) and thus for maintaining or restoring air quality (short-term and long-term prognostics). Fig. 3.56 shows an overview of tasks and possibilities, input quantities and data findings of mathematical meteorological models:

Table 3.8. Classification of application fields for dispersion models in relation to space and time scales [124]

Field of Application	Space Scale	Time Scale
Regional to supraregional (mesoscale)	50 km to 2 000 km	1/2 day to 1 week
Urban to regional	1 km to 100 km	1 hr. to 1 day
Point source (power plant)	500 m to several km	1/2 hr. to several hrs
Line sources (motor vehicle expressway without buildings nearby)	100 m to several km	1/2 hr. to several hrs
Street canyon (motor vehicle traffic – in cities)	several meters (1 m to 100 m)	several min. to 1 hr.

The spatial scales of simulation calculations range from the investigation of individual emitters to urban climatological questions (investigation of the effects of buildings, street blocks or industrial settlements on air quality) to the investigation of entire urban agglomeration areas and supraregional areas. Due to these highly differentiated requirements which vary very much in scale, it is understandable that an individual calculation model can never do justice to all questions, i.e., from short-range to country-to-country pollution transport. For this reason different groups have developed varying model concepts which are tailored to the individual requirements of space and time, s. Table 3.8.

3.4.2 Model Concepts

The dispersion of air-borne substances in the atmosphere depends on two different transport mechanisms:

For one, air pollutants are carried along with existing wind flows (advection), for another they are distributed by diffusion movements due to atmospheric turbulence, these movements being superimposed on the average wind field. Based on the knowledge of the flow field, with its temporal and spatial variations, the course of pollutant dispersion can be calculated by also taking turbulence effects into account. The extent of turbulence influence is determined by terrain features (orography), the structure of the ground (type and extent of building, vegetation, roughness), weather conditions, wind velocity and diurnal variations in radiation intensity (atmospheric layers).

The quality of dispersion calculations, above all in the case of short-term prognoses, depends definitively on precise prognostication of the time-related development of the dispersion conditions and with it on the flow field. For these reasons, in a simulation model for the computation of atmospheric dispersion both information on the wind field to be investigated and information on the individual turbulence condition of the atmosphere must be taken into account.

The basic precondition for the computation of dispersion processes is that flow and turbulence models must be dealt with first.

3.4.2.1 Flow and Turbulence Models

Flow models can be carried out with two basically different model concepts. Diagnostic and prognostic simulation models will be used depending on the task and objective concerned .

Diagnostic models

One of the main applications of diagnostic models is for simulating the actual situation of a wind field based on a statistically proven series of measurements and a known, experimentally investigated framework of conditions. This type of model is used mainly for long-term investigations and case studies during which a wide variety of dispersal situations is simulated.

Another important application of diagnostic simulation models is for the calculation of undiverging, mass-consistent flow fields by inter- and extrapolating wind data which were registered within a certain area. Before starting on dispersion calculations it must be generally guaranteed that the individual volume elements of the area to be investigated are neither subject to a mass increase nor a mass loss, if flow processes are being exclusively investigated with emission, transformation and deposition processes, however, being excluded.

Prognostic models

Unlike diagnostic model concepts, prognostic simulation models can predict the development of a flow field in a spatial scale of 2.5 to 2500 km and time-wise from just a few hours to several days.

Flows described by a model of this type are, on the one hand, imbedded in a larger scale floating flow field (geostrophic wind), but they must, on the other hand, also be able to take into account smaller scale phenomena (topographically or thermically induced turbulences) [125].

The mathematical treatment of flow processes with prognostic models is based on a set of partial non-linear differential equations. A three dimensional grid, for which

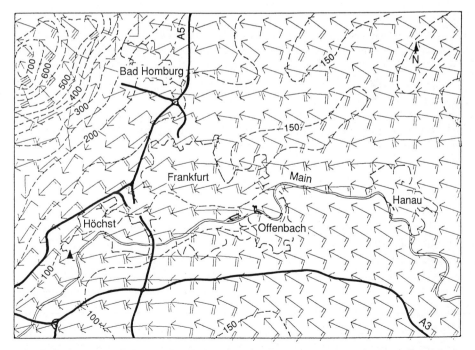

Fig. 3.57. Example of simulation calculations with flow models: wind field calculated with REWIMET in the Main-Taunus region of Germany [129]

this equation system must be solved with numerical methods, is superimposed on the model area where a flow field is to be determined [125]. By determining the suitable coordinate system and choosing the grid points carefully, the model's resolving power can be adapted to the requirements given by the orography of the area under investigation. The most important prognostic concepts developed in the Federal Republic of Germany are listed in [126].

When using models of this type one characteristic feature needs special attention – compared to diagnostic flow models, the modelling work is distinctly more work – intensive and generally also requires very much computing time.

One possibility of shortening computing times is to restrict modeling to a few vertical layers. For example, the three-layer model "REWIMET" designed by Heimann [127] represents a compromise between computing time and degree of detail of the flow field.

According to VDI guideline 3783, sheet 6 [128], this model for the calculation of wind fields is to be used in Germany within the framework of investigations on the regional distribution of air pollution over complex terrain. As an example of flow models Fig. 3.57 shows a wind field recorded by REWIMET in the Main-Taunus region (range: 60 km x 40 km).

3.4.2.2 Modeling of Pollution Dispersion

To create a predictable relation between emission and air quality one may employ different processes [130, 131]:

For long-term planning, concentration calculations are, as a rule, carried out with the help of models with a physical background (deterministic models). They are based on an emission register dependent on meteorological variables. For short-term concentration forecasts involving as little effort as possible, statistical and empirical models are used. Depending on meteorological factors and other parameters pollutant concentrations can be determined based on statistically guaranteed relationships which have been set up empirically from a number of measured data. To investigate topographic influences and effects of buildings in the immediate surroundings of emission sources physical models which reproduce natural conditions on a laboratory scale (wind tunnel experiments) are frequently used.

The deterministic model concept based on physical principles with the most varied applications will be explained in the following with the help of examples:

All flow and distribution models of this type are based on mathematical meteorological calculation methods originating from the conservation theorems of mass, energy and impulse. The statistical theories of Taylor are the basis for the mathematical description of turbulence effects [132]. With these concepts, the physical interrelationships of turbulent wind velocity variations which overlay the medium altitude flow can be determined.

Lagrange models

From the statistic diffusion theory which is based on the fact that air pollutants are always carried along with the average wind flow, although diffusion is statistically random, a class of distribution models can be directly derived – the so-called Lagrange models (Monte-Carlo-simulation models, particle simulation, trajectory models).

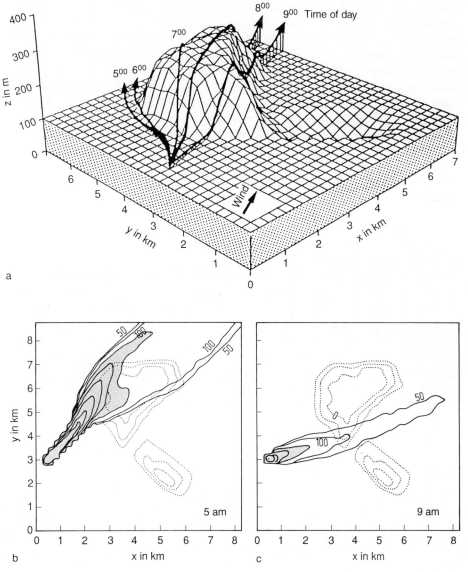

Fig. 3.58. Trajectories (**a**) and concentration fields at 5 a.m. (**b**) and 9 a.m. (**c**) calculated with FITNAH [133]

With these models the path of an air parcel containing pollutants (particles) is calculated in a flow field for a longer period of time, and the temporal and spatial movements of the air element are traced. In the course of this movement an air element absorbs pollutants on its path across sources and releases them over sinks. If this process is simulated for particles which are released by certain emission sources, pollutant concentration can be determined from the spatial distribution of these particles according to a random given dispersion time at a certain location.

Fig. 3.58a shows as an example paths of air parcels (trajectories) which were calculated with the simulation model "FITNAH" developed by Gross. Fig. 3.58b shows a ground concentration field determined with this model.

To make the change in flow direction visible on a hill in the transitional period from night to day air parcels were "started" between 5 and 9 o'clock in the morning above the ground and their trajectories were calculated. The influence of the thermal layers on the trajectory is clearly to be seen. Whereas, until 6 o'clock, the air parcels flow around the hill due to the stable layers close to the ground and cause relatively high pollutant concentrations there, after sunrise one can see the increasingly obvious relationship of the trajectories to the direction of the geostrophic wind. From 9 o'clock onwards the effects of the topography on the trajectory is barely noticeable. Once the air exchange has noticeably improved, pollution concentrations in the atmospheric layers close to the ground drop dramatically.

K-models

When using Lagrange models the volume of the air parcel being examined changes on its trajectory depending on the ambient pressure that it is exposed to. When using K- or gradient models, however, the geometric space where atmospheric pollution transport is to be investigated, remains unchanged.

K-models are based on the fact that turbulent diffusion can be treated analogous to molecular (Brownian) diffusion. When applying models of this type, turbulence-dependent substance transport is treated as being proportional to the concentration gradient of a pollutant gas between inlet and outlet of a controlled volume. The proportionality factor is, in this case, the turbulent exchange coefficient K, for which this type of model is named. As turbulence-induced movements of air parcels differ in vertical and horizontal directions, they must be assigned different diffusion coefficients.

The central equation for K-models is the advection-diffusion equation (3.27) which can be derived from the continuity equation in a general formulation [134]:

$$\frac{\partial c(\vec{x}, t)}{\partial t} = -\vec{u}(\vec{x}, t) \cdot \nabla c(\vec{x}, t) + \nabla (K(\vec{x}, t) \cdot \nabla c(\vec{x}, t))$$
$$+ Q(\vec{x}, t) + S(\vec{x}, t), \tag{3.37}$$

$\vec{x}$ location vector,
t time,
$c(\vec{x}, t)$ pollutant concentration at location x at time t,
$\vec{u}(\vec{x}, t)$ time and location dependent (medium altitude) wind field,

∇	Nabla operator,
$K(\vec{x}, t)$	diffusion tensor containing as diagonal elements diffusion coefficients in all three coordinate directions,
$Q(\vec{x}, t)$	mass flow of time and location dependent pollutant sources,
$S(\vec{x}, t)$	mass flow of time and location dependent pollutant sinks.

This differential equation for the description of time-related change in pollutant concentrations at a defined location at a certain time is composed of a transport term (advection term) caused by the driving average wind field, a diffusion term taking into account turbulence-induced movements and spatially and temporally varying source and sink terms.

Generally formulated, the advection-diffusion equation cannot be solved completely (mathematically definitely). For this reason numerical processes must be employed to solve this equation system. The differential quotients of the advection-diffusion equation are then replaced by differential quotients between spatial and temporal points, whereby a comprehensive system of equations is available for the calculation of concentrations closely linked in space and time. By following a certain framework of conditions, this system can be solved step by step [135].

Depending on the type of concept used to solve the advection-diffusion equation one differentiates between different types of K-models:

When working with Euler's grid models a suitable, fixed spatial grid is imposed on the distribution area and the space to be investigated is partitioned into a multitude of individual volume elements. At the individual grid points the advection-diffusion equation is solved with the help of finite difference equations for discrete sections of time. As these computing operations are highly time-intensive and require large memory capacities, the use of Euler's models is largely restricted to scientific questions [136].

In particle-in-cell models diffusion velocity is determined via the concentration gradient and imposed on the advection movement determined by the wind field. With the thus derived pseudotransport velocity the continuity equation follows on from the advection equation via a formal simplification. This continuity equation can then be solved in a fixed spatial grid with numerical processes [125].

Box models

The simplest concept for the calculation of an air volume's composition is a Box model. Limits of the area to be investigated are the ground as lower limit and, as a rule, an inversion layer as upper limit. The lateral limit of the box is determined by the horizontal expansion of the test area, s. Fig. 3.59.

As a rule it is considered a precondition that the air inside the volume investigated be ideally mixed. Air movements induced by turbulent diffusion and wind flows inside the box are not taken into consideration. Changes in concentration result merely from deposition and transformation processes.

Box models take into account the horizontal transport (advection) and removal of air pollution in the form of a constant flow through the lateral limits of the volume

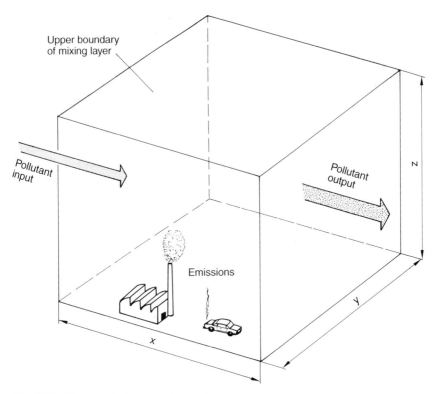

Fig. 3.59. Diagram of a box model investigation area

element being investigated. They are particularly well-suited to simulating transformation reactions concerning atmospheric chemistry as pollutant concentrations must be basically calculated at only one single point. This type of model, however, can neither be employed to investigate areas with inhomogenous sources at different locations nor for sinks [136].

Gaussian models

To calculate the dispersion of non-sedimenting substances from point sources which are not subject to physical and chemical transformations during transport, a calculation model based on Gaussian diffusion is used for approval procedures (calculations of stack heights, air quality prognoses).

This dispersion model is based on the analytical solution of a highly simplified form of the advection-diffusion equation. Preconditions for this are constant emission, constant wind field in terms of location and time with unchanging wind direction, spatially and temporally unvarying diffusion parameters, a flat, vacant dispersion area and a complete reflection of pollutants impacting the ground [137].

The coordinate system is positioned into the space investigated in such a way that the origin is set at the base of the emission source in question and the x axis points in the direction of the average wind, s. Fig. 3.60.

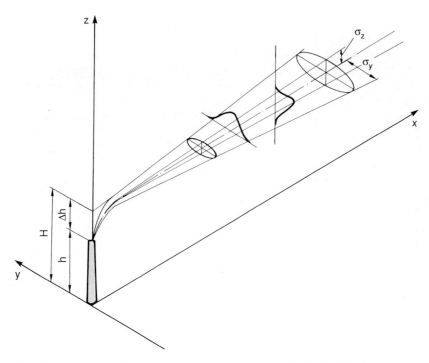

Fig. 3.60. Dispersion of smoke plumes acc to a Gaussian distribution [137]

In addition to the above conditions, advection in x direction must prevail over turbulence-induced processes, so that the effects of diffusion in this direction may be disregarded. Furthermore, it is assumed in a simplified form that the diffusion movements in y and z direction proceed according to a Gaussian distribution [136].

For emissions from point sources the ambient air concentration can then be described by the two-dimensional so-called Gaussian plume equation (3.28) at a certain location leewards of the source [137, 138]:

$$c(x,y,z) = \frac{Q}{2\pi\sigma_y\sigma_z u_h} \exp\left[-\frac{y^2}{2\sigma_z^2}\right] \cdot \left(\exp\left[-\frac{(z-H)^2}{2\sigma_z^2}\right] + \exp\left[-\frac{(z+H)^2}{2\sigma_z^2}\right]\right) \quad (3.38)$$

x, y, z	cartesian coordinates of the point in the distribution direction (x), vertical to the distribution direction horizontal (y) and vertical (z),
c (x, Y, z)	mass concentration of air pollution at the point of impact with the coordinates (x, y, z) for every single distribution situation,
z	altitude of the point of impact above ground,
Q	emission mass flow of the emitted air polluting substance from the emission source.

H effective source altitude,
σ_y, σ_z horizontal and vertical dispersion parameters,
u_h wind velocity as a function of altitude.

The above-mentioned assumptions which, for the moment, distinctly restrict the field of application of Gaussian models, are partially compensated for by the use of special input parameters, which were taken from dispersion experiments. Based on such experiments, stability classes were defined by different authors (Pasquill [139], Turner [140], Klug/Manier [141, 142]), taking into account exchange conditions during typical weather conditions. These stability classes characterize the turbulent condition of the atmosphere and are limited by wind velocity as well as the type of layering which provides the information on radiation conditions, the degree of cloudiness and cloud altitude.

In the version of the TA Luft (Technical Directive on Air Pollution Control) [138] currently valid the stability classification is laid down by Klug/Manier. To set up prognoses the pollution concentrations calculated for different wind directions and stability classes are taken into consideration according to the statistical frequency of these conditions.

3.4.3 Consideration of Chemical Transformations in Dispersion Models

An important area in the field of simulation models is the determination of the pollution from secondarily formed components: photochemical species (Los Angeles smog), acid air pollution (London smog). The reliability of such simulation calculations depends on whether the physical and chemical transformations taking place in the atmosphere can be simulated with sufficient accuracy.

By including transformation reactions a dispersion model is considerably enlarged and the required computing time correspondingly prolonged. Except when using simple box models, the mathematical concepts for the calculation of the transport of reactive air pollution, including dry and wet deposition, require the solving of the general advection-diffusion equation which takes into account the chemical transformations via an additional reaction term. With the help of a system of coupled differential equations, this term has the effect that the concentrations of individual pollutants are not determined independently of each other, but by taking their mutual influence into consideration.

Owing to the great variety of possible reaction processes in the atmosphere simplifications are necessary, but above all, the number of compounds to be examined must be restricted. The influence of organic compounds is either considered by combining several hydrocarbons into certain groups in which decomposition mechanisms follow the same pattern, or by tracing such components which can be regarded as being representative substances of entire reaction classes [132, 136]. In the case of inorganic pollutants only the most important main reactions are represented as models.

3.4.4 Summary and Overview of Model Concepts

Fig. 3.61 once again shows the correlation between dispersion theories and model concepts in brief.

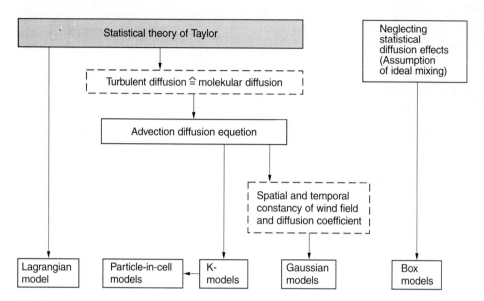

Fig. 3.61. Overview showing the correlation between distribution theories and model concepts [125]

3.5 Bibliography

1. Schirmer, H.: Stadtklima und Luftreinhaltung. VDI-Berichte Nr. 477, Düsseldorf 1983, S. 101–109
2. Schalz, J.: Das Stadtklima, Ein Faktor der Bauwerks- und Städteplanung. Karlsruhe: C. F. Müller 1974
3. VDI-Kommision Reinhaltung der Luft (Hrsg.): Stadtklima und Luftreinhaltung. Berlin: Springer 1988
4. Reuter, H.: Die Wissenschaft vom Wetter. Berlin: Springer 1978
5. Weischet, W.: Einführung in die allgemeine Klimatologie. Stuttgart: Teubner 1983
6. Meyers kleines Lexikon: Meteorologie. Mannheim: Meyers Lexikonverlag 1987
7. Häckel, H.: Meteorologie. Uni-Taschenbücher 1338, Stuttgart: Eugen Ulmer 1985
8. Robel, F.; Hoffmann, U.; Riekert, A.: Daten und Aussagen zum Stadtklima von Stuttgart auf der Grundlage der Infrarot-Thermographie. Beiträge zur Stadtentwicklung, Chemisches Untersuchungsamt der Landeshauptstadt Stuttgart 1978
9. Baltrusch, M.; Schütz, G.: Wärmeinsel. Beitrag in [3]
10. Beckröge, W.: Veränderungen des Klimas im Stadtbereich. Beitrag in [3]
11. Deutscher Wetterdienst, Wetteramt Stuttgart: Monatliche Darstellung der bodennahen Inversionsschichten. Nicht veröffentlichte Auswertungen, Stuttgart 1982
12. Malberg, H.: Meteorologie und Klimatologie, Berlin: Springer 1985

13 Deutscher Alpenverein: Alpin-Lehrplan 9, Wetter Lawinen. München: BLV Verlagsgesellschaft 1983
14 Baumüller, J.; Reuter, U.; Hoffmann, U.: Analyse der Smog-Situation in Stuttgart − Januar 1982. Chemisches Untersuchungsamt der Stadt Stuttgart, Klimatologische Abteilung, Mitteilung Nr. 4, 1982
15 Giebel, J.: Verhalten atmosphärischer Sperrschichten. LIS-Bericht Nr. 12, Landesanstalt für Immissionsschutz, Essen 1981
16 Giebel, J.: Untersuchungen über Zusammenhänge zwischen Sperrschichthöhen und Immissionsbelastung. LIS-Bericht Nr. 29, Landesanstalt für Immissionsschutz, Essen 1983
17 Stern, A.C.; Boubel, R.W.; Turner, D.B.; Fox, D.L.: Fundamentals of air pollution. 2nd ed. Orlando, Florida: Academic Press 1984
18 Leithe, W.: Die Analyse der Luft und ihrer Verunreinigungen. Stuttgart: Wissenschaftliche Verlagsgesellschaft 1974
19 Giebel, J.; Bach, R.-W.: Ursachenanalyse der Immissionsbelastung während der Smogsituation am 17.1.1979. Schriftenreihe der Landesanstalt für Immissionsschutz, Essen, Heft 47 (1979) 60−73
20 Külske, S.; Pfeffer, H.K.: Smoglage vom 16.−20. Januar 1985 an Rhein und Ruhr. Staub Reinhalt. Luft 45 (1985) Nr. 3, S. 136−141
21 Fa. Geosens, Rotterdam: Messungen des grenzüberschreitenden Transports von SO_2, NO_x und Ozon zwischen der Bundesrepublik Deutschland und der DDR/CSSR. Monatsberichte aus dem Meßnetz 6/85, Umweltbundesamt Berlin 1985
22 Paffrath, D.; Peters, W.; Rösler, F.; Baumbach, G.: Fallstudie über den Beitrag des Ferntransports von SO_2 zur lokalen Luftverschmutzung in der Bundesrepublik Deutschland. Staub Reinhalt. Luft 47 (1987) Nr. 7/8, S. 35−41
23 Bruckmann, P.; Reich, T.; Schrader, W.: Die Hamburger Smogepisode im Dezember 1983. Staub Reinhalt. Luft 45 (1985), Nr. 6, S. 307−312
24 Bruckmann, P.; Borchert, H.; Külske, S. et al.: Die Smog-Periode im Januar 1985. Staub Reinhalt. Luft 46 (1986) Nr. 7/8, S. 334−342
25 Umweltbundesamt: Monatsberichte aus dem Meßnetz Nr. 2/86, Berlin 1986
26 Heinze, J.; Baumbach, G.: Untersuchung von SO_2-Ferntransporten bei Nordostwinden am Beispiel von Situationen im Februar 1986. Institut für Verfahrenstechnik und Dampfkesselwesen der Universität Stuttgart, Abt. Reinhaltung der Luft, Bericht Nr. 15 − 1989
27 Baumbach, G.; Baumann, K.; Dröscher, F.: Luftverunreinigungen in Wäldern − Ergebnisse von Schadstoffmessungen im Schönbuch und Nordschwarzwald. Institut für Verfahrenstechnik und Dampfkesselwesen der Universität Stuttgart, Abt. Reinhaltung der Luft, Bericht Nr. 5 − 1987
28 Deutscher Wetterdienst, Aerologische Station Stuttgart: Ergebnisse der täglichen Radiosondenaufstiege. Persönliche Mitteilung, Stuttgart 1986
29 Bayerisches Landesamt für Umweltschutz, München, persönliche Mitteilung 1986
30 Landesanstalt für Umweltschutz, Karlsruhe, persönliche Mitteilung 1986
31 Amt für Umweltschutz der Stadt Stuttgart, Abt. Klimatologie, persönliche Mitteilung 1986
32 Schweizer, G.: Die Smog-Lage im Januar 1985 − Auswirkugen in Mittelbaden. Staub Reinhalt. Luft 45 (1985) Nr. 12, S. 587−590
33 Baumbach, G.; Baumann, K.; Dröscher, F.: Untersuchung der Verteilung von Luftverunreinigungen und ihres Eintrages in Waldbestände im Schwarzwald und Schönbuch. BMFT-Projekt Nr. 0339 112, Abschlußbericht, Universität Stuttgart 1990
34 Dröscher, F.: Vorkommen und Eintrag von atmosphärischen Partikeln in Waldbestände. Dissertation, Universität Stuttgart 1990
35 Baumbach, G.; Minner, G.; Konrad, G.: Luftverunreinigungen in einem Schwarzwaldtal bei Inversionswetterlagen, Teil 1. Institut für Verfahrenstechnik und Dampfkesselwesen der Universität Stuttgart, Abt. Reinhaltung der Luft, Bericht Nr. 3 − 1986
36 Petkovsek, Z.: Meteorologische Probleme der Luftverunreinigungen in Talbecken, Z. Meteor. 32 (1982) 1, S. 42−50
37 Dröscher, F.; Rauskolb, J.: Luftverunreinigungen in einem Schwarzwaldtal bei Inversionswetterlagen, Teil 2: Untersuchung der Staub-Immissionen. Institut für Verfahrenstechnik und Dampfkesselwesen der Universität Stuttgart, Abt. Reihaltung der Luft, Bericht Nr. 3 − 1986

3.5 Bibliography

38 Baumbach, G.; Göttlicher, R.; Winkelbauer, W.: Einfluß von Inversionen auf die Schadgasverteilung über einer Kleinstadt im Naturpark Schönbuch. Staub Reinhalt. Luft 45 (1985), Nr. 7/8, S. 365 – 368
39 Baumbach, G.; Baumann, K.; Dröscher, K.: Behaviour of air pollutants under inversion weather conditions. In: K. Grefen and J. Löbel (eds.): Environmental meteorology. Dordrecht: Kluwer Academic Publishers 1988
40 Paffrath, D.; Paffrath, M.; Peters, W.; Rösler, F.: Ergebnisse der Flugzeug-Messungen von Luftverunreinigungen über österreichischen Gebirgstälern vom 22.11.82 – 10.2.83. Interner Abschlußbericht SKAT Nr. IB-553-21-83, DFVLR, Institut für Physik der Atmosphäre, Oberpfaffenhofen 1983
41 Vonier, B.; Steisslinger, B.; Baumbach, G.: Luftbelastung in Tübingen. Institut für Verfahrenstechnik und Dampfkesselwesen der Universität Stuttgart, Abt. Reinhaltung der Luft, Bericht Nr. 12 – 1988
42 Becker, K.H.; Löbel, J. Hrsg.: Atmosphärische Spurenstoffe und ihr physikalisch-chemisches Verhalten. Berlin: Springer 1985
43 Fabian, P.: Atmosphäre und Umwelt. 3. Aufl. Berlin: Springer 1989
44 Christen, H.-R.: Grundlagen der allgemeinen und anorganischen Chemie. Frankfurt: Salle; Aarau: Sauerländer 1985
45 Umweltbundesamt: Luftreinhaltung '81 – Entwicklung – Stand – Tendenzen, Materialien zum 2. Immissionsschutzbericht der Bundesregierung an den Deutschen Bundestag. Berlin: Erich Schmidt 1981; Abschn. 1.2.2 Reaktionen von Luftverunreinigungen, S. 51 ff
46 Verein Deutscher Ingenieure, Kommission Reinhaltung der Luft: Säurehaltige Niederschläge – Entstehung und Wirkungen auf terrestrische Ökosysteme, Düsseldorf: VDI 1983
47 Guderian,'R. (ed.): Air pollution by photochemical oxidants. Ecological Studies 52, Berlin: Springer 1985
48 Löslichkeiten von Gasen in Wasser. In: Hütte, Taschenbuch der Werkstoffkunde (Stoffhütte), Berlin: Wilhelm Ernst & Sohn 1967, S. 124
49 Finlayson-Pitts, B.J.; Pitts, J.N. jr.: Atmospheric chemistry, fundamentals and experimental techniques. New York: John Wiley & Sons 1986
50 Beilke, S.; Gravenhorst, G.: Heterogeneous SO_2-oxidation in the droplet phase. Atmos. Environ. 12 (1978) 231 – 239
51 Schurath, U.; Ruffing, K.: Die Oxidation von NO durch Sauerstoff und Ozon in Abgasfahnen, Staub Reinhalt. Luft 41 (1981) Nr. 8, S. 277 – 281
52 Baumann, K.: Aufbau und Erprobung einer Meßstation zur Erfassung der Schadstoffausbreitung an einer Autobahn. Studienarbeit Nr. 2193 am Institut für Verfahrenstechnik und Dampfkesselwesen der Universität Stuttgart, Abt. Reinhaltung der Luft, 1985
53 Schurath, K.: Bildung von Photooxidantien durch homogene Transformation von Schadstoffen. Symposium Verteilung und Wirkung von Photooxidantien im Alpenraum, Gesellschaft für Strahlen- und Umweltforschung, München, GSF-Bericht 17/88, S. 136 – 151
54 Leighton, P.A.: The photochemistry of air pollution, New York: Academic Press 1961
55 Platt, U.: Oxidierte Stickstoffverbindungen in der Atmosphäre, IMA-Querschnittsseminar „Atmosphärische Prozesse" 22. – 24.1.86, Texte 13/86, Umweltbundesamt, Berlin 1986
56 Attmannspacher, W.; Hartmannsgruber, R.; Lang, P.: Verbesserung der Grundkenntnisse über die Klimatologie der vertikalen Ozonschicht durch verstärkte Ballonsondierung. Abschlußbericht BMFT-Vorhaben FKW 11, GSF München 1984
57 Kanter, H.J.; Reiter, R.; Munzert, K.-H.: Untersuchungen zur Frage der photochemischen Produktion von Ozon in Reinluftgebieten und ihrer vertikalen Verteilung, Forschungsbericht 104 02 800, Umweltbundesamt, Berlin 1982
58 Fricke, W.: Die Bildung und Verteilung von antropogenem Ozon in der unteren Troposphäre. Berichte des Instituts für Meteorologie und Geophysik der Universität Frankfurt, Nr. 44, Dezember 1980
59 Steisslinger, B.: Einfluß von Temperaturinversionen auf Konzentration und Verteilung von Luftverunreinigungen, Dissertation, Universität Stuttgart, 1993
60 Baumbach, G.: Fesselballonmessungen in Heilbronn – Vertikale Schadstoffverteilung und Austauschverhältnisse in den bodennahen Luftschichten. Abt. Reinhaltung der Luft im Institut für Verfahrenstechnik und Dampfkesselwesen der Universität Stuttgart, Bericht Nr. 28 – 1992

61 Wanner, H.; Künzle, T.; Neu, K.; Jhly, B.; Baumbach, G.; Steisslinger, B.: On the Dynamics of Photochemical Smog over the Swiss Middleland – Results of the First POLLUMET Field Experiment. Meteorol. Atmos. Phys. 51, 117–138 (1993)
62 Prinz, B.; Krause, G.H.M.; Stratmann, H.: Waldschäden in der Bundesrepublik Deutschland. LIS-Berichte Nr. 28, Landesanstalt für Immissionsschutz, Essen 1982
63 Seiler, W.; Fishman, J.: The distribution of carbon monoxide and ozone in the free troposphere. J. Geophys. Res., Vol 86 (1981) No. C8, p. 7255–7265
64 Fishman, J.; Seiler, W.: Correlative nature of ozone and carbon monoxide in the troposphere: implications for the tropospheric ozone budget. J. Geophys. Res. Vol 88 (1983) No. C6, p. 3662–3670
65 Conrad, F.; Seiler, W.: Influence of temperature, moisture and organic carbon on the flux of H_2 and CO between soil and atmosphere: field studies in subtropical regions. Geophys., Vol. 90 (1985) No. D3, p. 5699–5709
66 Inman, R.E. et al.: A natural sink for carbon monoxide. Science 172 (1971) 1229 f
67 Whitby, K.T.: The physical characteristic of sulphur aerosols. Atmos Environ. 12 (1987) 135–159
68 Klockow, D.: Analytical chemistry of the atmospheric aerosol. In: Georgii, H.W.; Jaeschke, W. (eds): Chemistry of the unpolluted and polluted troposphere. Dordrecht: Reidel 1982
69 Grosch, S.; Schmitt, G.: Experimental investigation of the deposition of atmospheric pollutants in forests. In: Grefen, K.; Löbel, J. (eds.): Environmental meteorology. Dordrecht: Kluwer Academic Publishers 1988
70 Umweltbundesamt, FRG: Daten zur Umwelt 1988/89. Berlin: Erich Schmidt 1989
71 Baumbach, G.; Vogt, U.; Hein, K.R.G.; Oluwole, A.F.; Ogunsola, O.J.; Olaniyi, H.B.; Akeredolu, F.A.: Air pollution in a large tropical city with a high traffic density – results of measurements in Lagos, Nigeria. The Science of the Total Environment 169 (1995) 25-31
72 Goldammer, J.G. (ed.): Fire in the Tropical Biota. Ecosystem Processes and Global Challenges. Ecological Studies 84. Berlin, Heidelberg, New York: Springer 1990
73 Crutzen, P.J.; Heidt, L.E.; Krasnec, J.P.; Pollock, W.H.; Seiler, W.: Biomass burning as a source of atmospheric gases CO, H_2, N_2O, NO, CH_3Cl and COS. Nature Vol. 282 (1979) 253-256
74 Kommission Reinhaltung der Luft im VDI und DIN (ed.): Typische Konzentrationen von Spurenstoffen in der Troposphäre. Düsseldorf 1992
75 Bridgman, H.A.: Global Air Pollution: Problems for the 1990s. London: Belhaven Press 1990
76 Graedel, T.E.: Chemical compounds in the atmosphere. New York: Academic Press 1978
77 Curran, T.; Fritz-Simons, T.; Freas, W.; Hemby, J.; Mintz, D.; Nizich, S.; Parzygnat, B.; Wayland, M.: National Air Quality and Emissions Trends Report, 1993. U.S. Environmental Protection Agency, Office of Air and Radiation, Office of Air Quality Planning and Standards. EPA report 454/R-94-026, North Carolina 1994
78 VDI nachrichten Umwelt-Index: Ozon-Werte in Deutschland.VDI nachrichten Nr. 82, Düsseldorf: VDI-Verlag, 14. Juli 1995
79 Der Rat von Sachverständigen für Umweltfragen: Umweltgutachten 1987. Stuttgart: W. Kohlhammer 1987
80 OECD: Control of Major Air Pollutants, Status Report. OECD, Organization for Economic, Co-Operation and Development, Environment Monographs No. 10, Paris 1987
81 Metropolitan Commission for Pollution Prevention and Control in the Valley of Mexico: Air Pollution in Mexico City – Present Situation. Technical Secretariat of the Metropolitan Commission, Mexico City 1994
82 Stroebel, R.: Air Quality in Urban, Industrial and Rural Areas of France. Proceedings of the 10th World Clean Air Congress, Espoo, Finland. The Finnish Air Pollution Prevention Society, Helsinki 1995. Volume 3, p. 520
83 Nyffeler, U.P.: Ambient Air Quality in Switzerland. Proceedings of the 10th World Clean Air Congress, Espoo, Finland. The Finnish Air Pollution Prevention Society, Helsinki 1995. Volume 2, p. 204
84 Schwela, D.H.: Public Health Implications of Urban Air Pollution in Developing Countries. Proceedings of the 10th World Clean Air Congress, Espoo, Finland. The Finnish Air Pollution Prevention Society, Helsinki 1995, Volume 3, P. 617

85 Larson, S.M.; Cass, G.R.: Characteristics of Summer Midday Low Visibility Events in the Los Angeles Area. Environ. Sci. Technol., Vol. 23, No. 3, 1989, 281-289
86 Viras, L.; Froussou, M. et al.: Air Quality in Athens, Annual Report. Ministry of Environment, City Planning and Public Works, Athens, Mai 1994
87 Moussiopoulos, N.: Air Pollution in Athens. Contribution to the book: Moussiopoulos, N. et al.: Urban Air Pollution: Computational Mechanics Publications, Southampton 1994
88 Gualdi, R.; Puglisi, F.: Introduzione, Studi per la valutazione della qualità dell'aria nella provincia di Milano. Aggiornamento al 31 marzo 1993, U.SS.L. 75/III di Milano, P. 9
89 Olaeta, I.: Environmental Health Service for Santiago-Chile, personal information, 1995
90 Baumbach, G.; Steierwald, G.; Wacker, M.; Mörgenthaler, V.: Kommunaler Luftreinhalteplan für die Stadt Heilbronn. Amt für Straßenverkehr und Umwelt der Stadt Heilbronn, Germany, 1992
91 Baumbach, G.: Verkehrsbedingte Schadstoffimmissionsbelastungen in Städten und an Autobahnen. Staub-Reinhaltung der Luft 53 (1993), 267-274
92 Romberg, E.; Bösinger, R.; Lohmeyer, A.; Ruhnke, R. und Röth, E.P.: $NO-NO_2$-Conversion Model for Application to Forecast Air Pollutions due to Automobile Exhaust. Staub-Reinhaltung der Luft, submitted paper 1995
93 Boden, T.A.; Kaiser, D.P., Sepanski, R.J.; Stoss, F.W. (eds.), Trends '93: A Compendium of Data on Global Change. ORNL/CDIAC-65. Carbon Dioxide Information Analysis Center, Oak Ridge National Laboratory, Oak Ridge, Tennessee, U.S.A., 1994
94 Keeling, C.D., and T.P. Whorf. 1994: Atmospheric CO_2 records from sites in the SIO air sampling network. pp 16-26. in T.A. Boden, D.P.Kaiser, R.J. Sepanski, and F.W. Stoss (eds.), Trends '93: A Compendium of Data on Global Change. ORNL/CDIAC-65. Carbon Dioxide Information Analysis Center, Oak Ridge National Laboratory, Oak Ridge, Tenn., U.S.A.
95 Neftel, A., H. Friedli, E. Moor, H. Lötscher, H. Oeschger, U. Siegenthaler, and B. Stauffer. 1994. Historical CO_2 record from the Siple Station ice core. pp. 11-14. In T.A. Boden, D.P. Kaiser, R.J. Sepanski, and F.W. Stoss (eds.), Trends '93: A Compendium of Data on Global Change. ORNL/CDIAC-65. Carbon Dioxide Information Analysis Center, Oak Ridge National Laboratory, Oak Ridge, Tenn., U.S.A.
96 Conway, T.J., P.P. Tans, and L.S. Waterman. 1994. Atmospheric CO_2 records from sites in the NOAA/CMDL air sampling network. pp. 41-119. In T.A. Boden, D.P. Kaiser, R.J. Sepanski, and F.W. stoss (eds.), Trends '93: A Compendium of Data on Global Change. ORNL/CDIAC-65. Carbon Dioxide Information Analysis Center, Oak Ridge National Laboratory, Oak Ridge, Tenn., U.S.A.
97 Fricke, W., and M. Wallasch. 1994. Atmospheric CO_2 records from sites in the UBA air sampling network. pp. 135-146. In T.A. Boden, D.P. Kaiser, R.J. Sepanski, and F.W. Stoss (eds.), Trends '93: A Compendium of Data on Global Change. ORNL/CDIAC-65. Carbon Dioxide Information Analysis Center, Oak Ridge National Laboratory, Oak Ridge, Tenn., U.S.A.
98 Colombo, T., and R. Santaguida. 1994. Atmospheric CO_2 record from in situ measurements at Mt. Cimone. pp. 169-172. In T.A. Boden, D.P. Kaiser, R.J. Sepanski, and F.W. Stoss (eds.), Trends '93: A Compendium of Data on Global Change. ORNL/CDIAC-65. Carbon Dioxide Information Analysis Center, Oak Ridge National Laboratory, Oak Ridge, Tenn., U.S.A.
99 Manning, M.R., A.J. Gomez, and K.P. Pohl. 1994. Atmospheric CO_2 record from in situ measurements at Baring Head. pp. 174-178. In T.A. Boden, D.P. Kaiser, R.J. Sepanski, and F.W. Stoss (eds.), Trends '93: A Compendium of Data on Global Change. ORNL/CDIAC-65. Carbon Dioxide Infromation Analysis Center, Oak Ridge National Laboratory, Oak Ridge, Tenn., U.S.A.
100 Houghton, J.T., Jenkins, G.J. and Ephraums, J.J. (Eds.): Climate Change. The Intergovernmental Panel for Climate Change (IPCC) Scientific Assessment. Cambridge University Press, 990
101 Stauffer, B., A. Neftel, G. Fischer, and H. Oeschger. 1994. Historical CH_4 record from the Siple ice core. pp. 251-254. In T.A. Boden, D.P. Kaiser, R.J. Sepanski, and F.W. Stoss (eds.), Trends '93: A Compendium of Data on Global Change. ORNL/CDIAC-65. Carbon Dioxide Information Analysis Center, Oak Ridge National Laboratory, Oak Ridge, Tenn., U.S.A.
102 Dlugokencky, E.J., P.M. Lang, K.A. Masarie, and L.P. Steele. 1994. Atmospheric CH_4 records from sites in the NOAA/CMDL air sampling network. pp. 274-350. In T.A. Boden,

D.P. Kaiser, R.J. Sepanski, and F.W. Stoss (eds.), Trends '93: A Compendium of Data on Global Change. ORNL/CDIAC-65. Carbon Dioxide Information Analysis Center, Oak Ridge National Laboratory, Oak Ridge, Tenn., U.S.A.
103 Prinn, R.G., R.F. Weiss, F.N. Alyea, D.M. Cunnold, P.J. Fraser, P.G. Simmonds, A.J. Crawford, R.A. Rasmussen, and R.D. Rosen. 1994. Atmospheric CFC-11 (CCl_3F), CFC-12 (CCl_2F_2) and N_2O from the ALE/GAGE network. pp. 396-420. In T.A. Boden, D.P. Kaiser, R.J. Sepanski, and F.W. Stoss (eds.), Trends '93: A Compendium of Data on Global Change, ORNL/CDIAC-65. Carbon Dioxide Information Analysis Center, Oak Ridge National Laboratory, Oak Ridge, Tenn., U.S.A.
104 Elkins, J.W., T.M. Thompson, J.H. Butler, R.C. Myers, A.D. Clarke, T.H. Swanson, D.J. Endres, A.M. Yoshinaga, R.C. Schnell, M. Winey, B.G. Mendonca, M.V. Losleben, N.B.A. Trivett, D.E.J. Worthy, V. Hudec, V. Chorney, P.J. Fraser, and L.W. Porter. 1994. Global and hemispheric means of CFC-11 and CFC-12 from the NOAA/CMDL flask sampling program. pp. 422-430. In T.A. Boden, D.P. Kaiser, R.J. Sepanski, and F.W. Stoss (eds.), Trends '93: A Compendium of Data on Global Change. ORNL/CDIAC-65. Carbon Dioxide Information Analysis Center, Oak Ridge National Laboratory, Oak Ridge, Tenn., U.S.A.
105 Fricke, W.; Dauert, U.: Luft kennt keine Grenzen. Umweltbundesamt, Berlin, 4. Auflage 1994, and personal information 1994
106 Landesamt für Umweltschutz Sachsen-Anhalt (ed.): Immissionsschutzbericht 1993. Berichte des Landesamtes für Umweltschutz Sachsen-Anhalt, Heft 12-1994
107 Senatsverwaltung für Stadtentwicklung und Umweltschutz, Referat Öffentlichkeitsarbeit (ed.): Luftverschmutzung in Berlin im Jahr 1991. Berlin 1993: Kulturbuch-Verlag
108 Pluschke, O.: Monitoring Local Air Quality: The Spatial and Temporal Pattern of Air Pollution Caused by Mobile Sources in the City of Nürnberg/Germany. Proceedings of the 10th World Clean Air Congress, Espoo, Finland. The Finnisch Air Pollution Prevention Society, Helsinki 1995, Volume 2, p. 220
109 Sächsiches Landesamt für Umwelt und Geologie (ed.): Luftqualität in Sachsen. Jahresbericht zur Immissionssituation 1993, Dresden 1994
110 Heida, H.; Coolen, M.; de Jong, A.; Bank, K.: Ambient Air Pollution Patterns in Amsterdam 1971-1991. Proceedings of the 10th World Clean Air Congress, Espoo, Finland. The Finnish Air Pollution Prevention Society, Helsinki 1995, Volume 2, p. 203
111 Durmaz, A.; Dogu, G.; Ercan, Y.; Sivrioglu, M.: Investigation of the Causes of Air Pollution in Ankara and Measures for its Reduction. Gazi University, Faculty of Engineering and Architecture, Ankara 1993
112 Gervat, G.P.; Air Quality in Hongkong 1992. Hong Kong Government Environmental Protection Department, Air Services Group, Report No. EPD/TR 7/94
113 Dutton, E.G. 1994. Aerosol optical depth measurements from four NOAA/CMDL monitoring sites. pp. 484-494. In T.A. Boden, D.P. Kaiser, R.J. Sepanski, and F.W. Stoss (eds.), Trends '93: A Compendium of Data on Global Change. ORNL/CDIAC-65. Carbon Dioxide Information Analysis Center, Oak Ridge National Laboratory, Oak Ridge, Tenn., U.S.A.
114 Landesanstalt für Umweltschutz Baden-Württemberg (ed.): Die Luft in Baden-Württemberg. Jahresbericht 1993. Berichte der Landesanstalt für Umweltschutz Baden-Württemberg, Karlsruhe 1994
115 Volz, A. and Kley, D.: Evaluation of the Montsouris series of ozone measurement made in the nineteenth century. Nature 332 (1988), 240-242
116 Bojkov, R.D.: Surface ozone during the second half of the nineteenth century. J. clim.appl. Met. 25 (1986), 343-352
117 Lisac, I. and Grubisic, V.: An analysis of surface ozone data measured at the end of the 19th century in Zagreb, Yugoslavia. Atmospheric Environment Vol 25A (1991), No. 2, pp. 481-486
118 Ozone Data for the World: Catalogue of ozone stations and ozone data for 1980-1983. AES Canada in Co-operation with WMO, Downsview Ontario, No. 18, 1983
119 Feister, U. and Warmbt, W.: Long-Term Measurements of Surface Ozone in the German Democratic Republic. Journal of Atmospheric Chemistry 5 (1987), 1-21
120 Low, P.S.; Davies, T.D.; Kelly, P.M.; Reiter, R.: Uncertainties in Surface Ozone Trend at Hohenpeissenberg. Atmospheric Environment Vol. 25A (1991), No. 2, pp. 511-515

121 Statistische Berichte: Umwelt, Immissions-Konzentrationsmessungen. Ed. by Statistisches Landesamt Baden-Württemberg, Stuttgart
122 Cox, W.M. and Chu, S.H.: Meteorologically Adjusted Ozone Trends in Urban Areas: A Probabilistic Approach. Tropospheric Ozone and the Environment II, Air and Waste Management Association, Pittsburgh, PA, 1992

4 Effects of Air Pollution

4.1 General Considerations

In our industrial age man releases a variety of trace substances into the atmosphere. Apart from the compounds primarily emitted, one also finds secondary components created by chemical transformations. All of them are integrated in manifold ways into substance cycles not only involving the atmosphere but also the soil (lithosphere), water (hydrosphere) and living space of man and animal (biosphere). Whether atmospheric trace substances become noxious for humans depends on both their direct effects on man's physical well-being and on his living conditions in his environment, with the food chain representing a particularly close connection to animal and plant life. Indirect damage, also, to climate and living space influence both physical health as well as mental well-being. Damage to material goods, on the other hand, merely causes material and psychological irritation.

4.1.1 The Range of Possible Types of Damage

The extent of the negative effects of air pollution can vary greatly. They range from an imperceptible basic load to irritation, sickness and death. The disorders caused do not usually occur immediately on exposure (acute damage). It is more frequently the case that only after prolonged exposure, often after enrichment, i.e., accumulation, that chronic ailments become visible. Particularly in complex organisms or biological communities, the effects become obvious after some time has elapsed:

One must distinguish between temporary, i.e., reversible, and permanent, i.e., irreversible damage. The search for clear-cut cause-and-effect patterns is often complicated by the simultaneous occurrence of several damaging factors with reinforcing effects, synergisms, by far exceeding the sum of the effects of single components. However, compensatory effects also occur. One then speaks of antagonisms.

In these generally highly complex effect relationships a clear understanding of the essential effect structures is required to be able to predict the behavior of larger, natural total systems. Therefore, a great deal of research is presently being concentrated on the effects of air pollution on humans, animals, vegetation, materials as well as climate and habitat. The effects on processes related to atmospheric chemistry and the radiation balance of the earth take up a special position in this list, as the processes are restricted to the atmosphere in these categories.

4.1.2 The Path of Air Pollutants to the Location where they Become Effective

In all areas the deposition of trace substances from the atmosphere is the precondition of air pollutants. In many cases special attention must therefore be paid to the conditions for deposition. They are the decisive factor governing which substances come into contact with the subsequent place of effect, the locations and the extent. These processes are fundamentally similar transport processes, regardless of whether biological surfaces, e.g., the lungs of a human being or a leaf, or abiotic surfaces, such as the facades of buildings are involved. The following transport mechanisms are the main processes of separation in varying degrees of frequency:
- turbulent diffusion due to atmospheric turbulence,
- molecular diffusion due to Brownian molecular motion,
- sedimentation due to gravitation,
- impaction (inertial precipitation of aerosols).

Gas deposition is mainly restricted to diffusion processes. Sedimentation and impaction are mass-dependent mechanisms and therefore pertain to fluid droplets and particles. Often, wind movement is decisive for the substance transport through the barrier layer between atmosphere and surface of the "target" object.

The extent of the deposition is additionally dependent on the structure and condition of the barrier area. They determine the sorption properties of the surface or its adhesive properties for aerosols.

Aqueous wetting films promote the solution of hydrophile gases or particle components which are inert in a dry condition. This is especially true of acidic gases such as SO_2, NO_2, HCl. In the case of sulfur dioxide, it is wet-chemical oxidation, e.g., which actually leads to the formation of the effective sulfuric acid.

Finally, air pollutants can penetrate the target object. Transport to the actual point of penetration is usually based on diffusion processes. In contrast to abiotic materials, plants, animals and even human beings have active transport mechanisms at their disposal. They themselves influence within their bodies the distribution of the air pollutants which have entered. In many cases specific detoxification mechanisms by which living beings discharge noxious substances or bind them so that they become harmless, i.e. immobilize them, take effect.

4.2 Climatic Changes Caused by Atmospheric Trace Substances

The term climate is generally understood to be the mean weather profile at a certain location measured over many years. The climate depends on atmosphere, hydrosphere and biosphere in a number of ways, with the radiation balance of the earth playing a special role.

Atmospheric trace substances influence this radiation balance considerably. E.g., the stratospheric ozone layer functions as a UV filter and thus makes life on earth possible. Then again, the naturally occurring infrared-active greenhouse gases temper the earth's surface to a mean of 15 °C; without them the mean temperature would be at approx. –20 °C [1, 2]. Therefore, changed trace substance concentrations can thus dramatically interfere with the radiation balance and thus modify the weather situation on earth in the long run.

The concentration development of some climate-relevant trace substances has been compiled in Table 4.1. Ozone, water vapor and aerosol particles do not show any readily apparent trends, whereas the anthropogenic greenhouse gases CO_2, CH_4, N_2O and fluorochlorinated hydrocarbons (CFC) show a clear increase in their concentrations.

Table 4.1. Climate-relevant atmospheric trace components and the changes in their occurrence (acc. to [1])

Trace component	Part of the atmosphere	Mixing ratio	Annual change of concentration	Reasons for change
Trace components without significant trend				
O_3	Troposphere	Variable	+ 1% (only lower troposphere of the mean latitudes of the northern hemisphere)	Burning of fossil fuels
	Stratosphere	Variable	– 0.3% (mean latitudes)	Photolytic decomposition of $CFCl_3$ and CF_2Cl_2
			– 40% (Antarctica, Sept. + Oct.)	Degradation products of $CFCl_3$ and CF_2Cl_2
H_2O vapor	Troposphere	Variable	Increase (tropical Pacific region)	
Aerosol particles	Troposphere	Variable	Increase (limited areas)	Air pollution
Greenhouse gases with rising trend				
CO_2	Troposphere	345 ppm	+ 0.5%=1.6 ppm/a	Fossil fuels, destruction of parts of the biosphere
CH_4	Troposphere	1.65 ppm	+ 1.2%=27 ppb/a	Cultivation of rice, cattle, dumps, fossil fuels, destruction of biosphere by burning
N_2O	Troposphere	0.32 ppm	+ 0.2%=0.9 ppb/a	Nitrogen fertilization, fossil fuels, destruction of parts of the biosphere
CF_2Cl_2	Troposphere	400 ppt	+ 4.5%=18 ppt/a	Refrigerants, spray cans
$CFCL_3$	troposphere	270 ppt	+ 5.2%=13 ppt/a	Refrigerants, spray cans, production of insulating foam

4.2 Climatic Changes Caused by Atmospheric Trace Substances

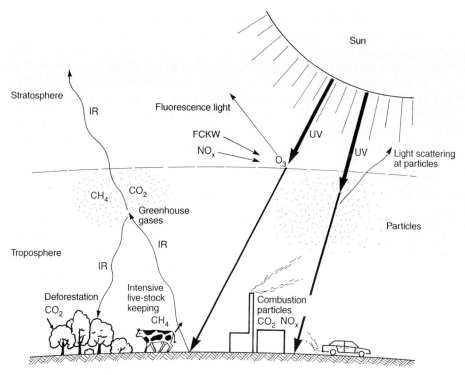

Fig. 4.1. Diagram showing the interactions of atmospheric trace gases with the radiation balance of the earth

The most important mechanisms leading to climatic changes due to atmospheric trace substances are shown schematically in Fig. 4.1. They can lead to both a rise in the mean surface temperature of the earth as well as to a drop.

4.2.1 Temperature Increase

Several effects can lead to a global warming of the earth's surface. Increased radiation of sunlight on the earth's surface, especially of high-energy UV light, warms the surface of the earth as also decreased light reflection from the area of the lower atmospheric layers. Therefore, the present points to be discussed are the destruction of the stratospheric ozone layer as well as the increase of the so-called greenhouse gases. A rise in temperature would have numerous, fearsome consequences, among them the shifting of climatic zones and with them the shifting of the inhabitable regions of the earth, the expansion of the deserts, as well as the flooding of larger areas of land due to the melting of polar ice masses.

4.2.1.1 Destruction of the Stratospheric Ozone Layer

In the upper stratosphere ozone occurs in high concentrations. This natural ozone layer acts as a filter for high-energy sunlight, as ozone has strong absorption bands in the UV range. The light energy absorbed by ozone then becomes independent of direction and is then reflected with a greater wave length. Only part of the light energy reaches the earth.

Owing to a number of anthropogenic air pollutants this ozone layer can be destroyed. The consequences would be a warming of the earth's surface, a decreased stability of the thermal layering of the stratosphere and increased UV radiation in the troposphere. Humans, many animals and plants are highly sensitive to hard UV-B and UV-C radiation which are even richer in energy.

The decomposition of the stratospheric ozone layer is attributed to two substance groups [7]:
- fluorochlorinated hydrocarbons (CFHC),
- oxides of nitrogen (NO_x).

CFHC, which stand out for their chemical stability, have very long lifespans (50-100 years) and therefore reach the upper air layers of the stratosphere after a migration period of approx. 10 years. Once in the ozone layer they are photolyzed by high-energy UV radiation ($\lambda < 220$nm). Split off chlorine atoms then can effect ozone decomposition [3, 5]. Conditions in the stratosphere were simulated in reaction chambers under laboratory conditions. Chlorine monoxide (ClO) was formed as the final product of complicated chain mechanisms.

Since 1979 a steady depletion of the stratospheric ozone layer has been observed with the help of satellite images during the winter months in Antarctica . During the first measuring flights in the Antarctic stratosphere, considerable ClO concentrations could be measured in the center of the "ozone hole", so that today the contribution of CFHC to ozone decomposition is considered a proven fact [5].

Ozone decomposition in the stratosphere is also brought about by the oxides of nitrogen released there. Air traffic which is still relatively infrequent in this elevated atmospheric layer will probably become busier in the future due to supersonic flights, above all military aircraft and space flights. Nuclear explosions in the atmosphere would, however, release even greater amounts of nitrogen oxide than aircraft and rocket engines. In this case tremendous changes in the composition of the atmosphere would have to be expected [6]. The rapid increase in air traffic with normal ultrasonic passenger aircraft (up to 10 % annually in the FRG) releases oxides of nitrogen and hydrocarbons in the region of the tropopause (at an approximate altitude of 8-12 km). According to current information these emissions tend to effect an increase of the ozone concentration similar to that close to the ground, according to model calculations an increase of approx. 20 % at an altitude of 10 km [7]. With the development of fuel-saving aircraft engines hydrocarbon emissions are dropping and emissions of oxides of nitrogen increasing further. This could entail an ozone decomposition at the operation range of passenger aircraft (at 8-12 km height).

Similar to CFHC the low-reactive dinitrogen monoxide N_2O (laughing gas), originating from microbial nitrification and denitrification in the soil, rises to stratospheric altitudes and can also contribute to ozone decomposition there.

Prognoses pertaining to developments in the stratospheric ozone layer are constantly being revised according to new findings from direct measurements and laboratory tests, for reaction kinetics and emission development in particular [7].

4.2.1.2 Greenhouse Effect of Infrared-Active Gases

Gases with strong absorption bands in the infrared (IR) light range are called greenhouse gases. These gases absorb the IR radiation reflected by the earth's surface in the troposphere and transmit it in no particular direction, so that it is partially reflected back to the earth. Like a glass-enclosed greenhouse, they do not impede the entry of short-wave sunlight.

The most significant IR-active gas is water vapor. It occurs in large amounts in the troposphere . Increased surface temperatures raise the rate of evaporation and

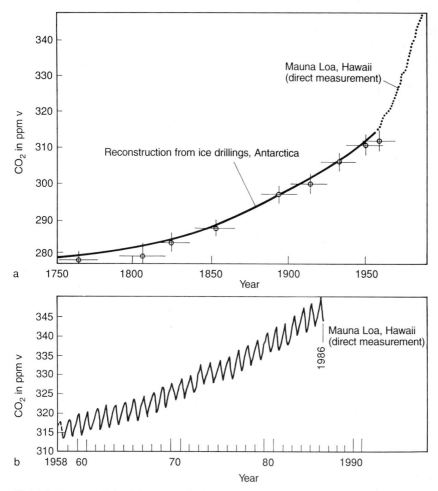

Fig. 4.2. Increase of the CO_2 content in the atmosphere since the beginning of industrialization (acc. to [11]). **a** The increase of the atmospheric CO_2 concentration is reconstructed from ice drillings in Antarctica acc. to Neftel, 1985; dots on upper right represent annual values; in reconstructions the crosses indicate the corresponding uncertainty range. **b** Monthly values of direct measurements on the Mauna-Loa, Hawaii, acc. to Keeling 1982, 1987

thus increase the water vapor concentration. The result is a reinforcement of the effect (so-called positive feedback).

The increase of anthropogenic greenhouse gases (s. Table 4.1) is linked to the increase in energy consumption in numerous ways. The particularly extreme increase of the trace gases CH_4 and CO_2 is additionally connected to the massive changes in agricultural utilization, above all in the Third World. This is evident from the example of the development of the atmospheric CO_2 content of the northern hemisphere during the last 100 years, s. Fig. 4.2.

Both the exponentially growing consumption of fossil fuels world-wide, especially in the industrial nations, (s. Fig. 1.5) as well as the accelerated deforestation of annually 1 % of the tropical forest area release enormous amounts of CO_2 into the atmosphere. Simultaneously, significant biomass volumes are destroyed world-wide which in turn decreases global photosynthesis performance. The carbon cycle of the earth loses its balance [8].

But the greenhouse effect of anthropogenically released CO_2 will be exceeded by that of the other greenhouse gases in the near future (s. Table 4.1). Although they only occur in small amounts they absorb IR radiation much more effectively than CO_2 [1, 2, 4].

Prognoses pertaining to the temperature development of the earth's surface due to greenhouse gases are, however, still very unreliable. Apart from model simplifications, changed frame conditions can lead to results which vary greatly. Even a diminution of the degree of cloudiness by a mere 4 % effectuates the same warming as a doubling of the CO_2 content in the troposphere [1].

4.2.2 Temperature Drop Due to Particle and Cloud Occurrence

The temperature on the earth's surface drops when incoming radiation energy is reduced by air pollution. Water droplets or dust particles in the atmosphere reflect, scatter and refract incoming sunlight and partially absorb it. Therefore only part of the radiated solar energy reaches the surface of the earth. The most significant natural sources of aerosol are volcanic eruptions. Their climatic impact can be far-reaching. The ice ages in the earth's climatic history are attributed to them.

Direct anthropogenic influence on particle occurrence and cloud formation has so far been essentially restricted to industrial areas (e.g., power plants, cooling towers). In the future, however, massive changes in agricultural utilization in the Third World (deforestation, forming of deserts) will probably have more widespread influence on particle and cloud occurrence in the troposphere.

Lately different scientists have drawn attention to the fact, that if nuclear weapons were used, enormous amounts of aerosols would be released into the atmosphere in a flash. Model calculations indicate that after a nuclear conflict aerosol concentrations would rise to such an extent even in the southern hemisphere, that there, too, surface temperatures would drop dramatically for a long time ("nuclear winter") [6, 10]. Along with that many forms of life would become impossible.

4.2.3 Prognostic Difficulties

It is difficult to make quantitative statements on the development of the earth's climate and to predict the complex interrelationships between the atmosphere and other parts of the earth, the hydrosphere with the enormous storing capacity of the oceans for substances and energy as well as the biosphere and lithosphere [1, 11-13]. At present a tendency of the mean surface temperature of the earth to increase by approx. 1 °C/100 years is to be observed [1]. This rise could quite possibly be caused by man's interference, but it could also lie within the range of natural climatic fluctuations, s. Fig. 4.3. Even if climatic research cannot answer this question with absolute certainty, it has, however, revealed that our present climate is definitely not a self-stabilizing system. On the contrary, it can react very sensitively to changes in climate-determining factors.

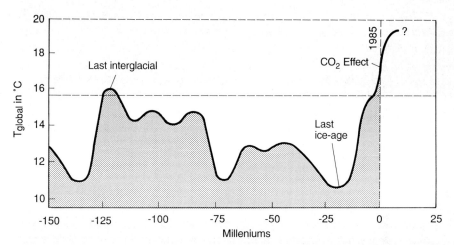

Fig. 4.3. Temperature reconstruction (in this case relative to the global mean value) based on ice drillings in Antarctica. A hypothetically possible anthropogenic influence on trace gases in the future has been included [1]

4.3 Effects on Materials

The effects of air pollution on material objects manifest themselves mainly in surface corrosion of varying degrees. The mechanisms of effect are usually relatively easy to grasp, the effects being more obvious than in biological systems. Nevertheless it is not always easy to differentiate between the effects of atmospheric pollutants and the influences of climate, particularly those of UV radiation and humidity, pollutant deposition conditions and bacterial activity. Synergistic effects of several simultaneously interacting pollutants are also known [14].

The building materials stone, steel, concrete and brick are the ones exposed to atmospheric conditions to a large extent. The damage to historical buildings made of

Fig. 4.4. Sandstone statue from the year 1702, Herten Castle near Essen, Germany. Condition in the year 1908 (**a**) and 1969 (**b**) [16]

natural stone is particularly serious. In these cases art treasures of historical value are being destroyed. Two photographs of a statue of Herten Castle near Essen show how rapidly this figure has decayed in a matter of only 60 years, s. Fig. 4.4.

Greater corrosion velocities by far than those caused by the basic pollution of the atmosphere occur in the vicinity of certain emission sources outside or inside buildings where diluting effects are minimal. In such an ambience, greatly varying emission-material combinations with specific corrosion effects occur. The natural corrosion of metal surfaces and mineral building materials is accelerated dramatically by the deposition of acid gases and dust particles in combination with humidity.

4.3.1 Mineral Building Materials

Sulfuric acid formed from sulfur dioxide plays a major role in the corrosion of mineral building materials. It corrodes the limestone ($CaCO_3$) which many kinds of natural stone as well as concrete contain and transforms it into gypsum $CaSO_4 \cdot 2H_2O$ [15]:

$$CaCO_3 + H_2SO_4 + 2H_2O \rightarrow CaSO_4 \cdot 2H_2O + H_2CO_3 . \tag{4.1}$$

The decisive factor for the corrosion effect is the extreme volume expansion of approx. 100 % during the formation of the gypsum structure. This, in combination with the presence of cracks, frequently leads to a laminar chipping off of the stone or masonry.

The carbonic acid forming at the same time can also effectuate the transformation of limestone:

$$CaCO_3 + H_2CO_3 \rightarrow Ca(HCO_3)_2. \tag{4.2}$$

During this process calcium hydrogen carbonate is formed. It is highly water-soluble, washes out well and, like gypsum, has poor mechanical strength.

Several authors have investigated the dependency of the corrosion velocity in natural stones on SO_2 pollution. For this, the same material samples were positioned at locations with different ambient air quality situations under defined conditions, and the weight losses finally determined. Such a relationship is shown in Fig. 4.5. In this situation, SO_2 pollution cannot be regarded as the sole cause but as the reference substance for many other urban air pollutants.

Corrosion is primarily a surface phenomenon. Building materials with alkaline reactions for the most part neutralise atmospheric acid depositions und thus protect deeper layers. In compact building materials such as reinforced concrete, the penetration of acids down to the steel reinforcements have nevertheless been observed [17-19]. In these cases metal corrosion prevails.

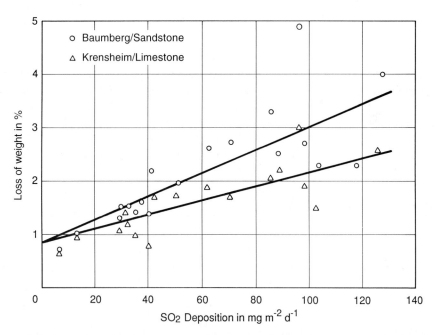

Fig. 4.5. Dependency of corrosion rate of different natural stones on SO_2 deposition, measured with IRMA passive samplers, Northrhine-Westphalia, Germany [20]

4.3.2 Metals

Metal corrosion, too, is caused largely by the effect of acid gases. In iron metals the main mechanism is electrochemical corrosion. Due to small spatial differences in the chemical and physical structure of the metal surface slight potential differences are created. Electrochemical cells, so-called local elements of corrosion with anodes and cathodes, are formed. These potential differences cause a corrosion current. The less precious metal of the anode is corroded.

Dry, clean surfaces show negligible corrosion rates. A series of climatic and pollution parameters promote corrosion. It has been tried again and again to weight these parameters as universally as possible. For this, empiric corrosion functions have been set up which are to enable the forecasting of corrosion losses. The following factors have mainly been taken into account in these functions [21, 22]:

- Duration of wetting: Aqueous wetting films owing to the formation of dew, deposition of hygroscopic particles or capillary action of pores offer reaction volumes for the local formation of acid from dissolved trace gases and are a medium for diffusive transport processes. They are formed when threshold values below 100 % humidity are exceeded.
- Pollutant concentration in the air: pollution caused by corroding trace substances.
- Wind velocity: a measure for the turbulent transport of gases to the material surface.
- Temperature: limits velocity of chemical reactions and diffusion processes.

Metal corrosion due to SO_2 pollution has been particularly well investigated. An example of this is shown in Fig. 4.6. The dependence of the corrosion rate of zinc,

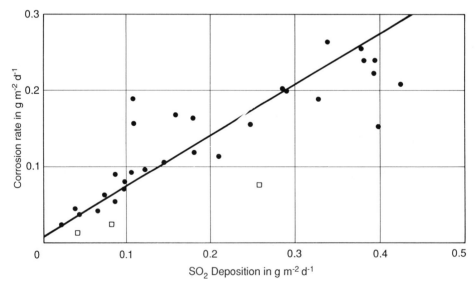

Fig. 4.6. Corrosion rate of iron in dependence of the SO_2 deposition, measured with an IRMA SO_2 passive samplers, FRG and UK [23]

established by the loss of mass in test plates on the SO_2 deposition rate is shown here [23]. Zinc is frequently used as corrosion protection for surface coating of steel parts. Similar dependencies were also established for steel and SO_2 pollution [24].

For one thing, the effect of SO_2 goes back to the destruction of passivating oxide layers forming on dry surfaces, for another to its function in electron transfer in electrochemical cells. Depositions of particles with a high metal content increase the inhomogeneity of the surface and with it the differences in potential which form here. It has been observed in industrial areas that heavy rainfall reduces the rate of corrosion by cleaning the surfaces of heavy metal and acid particles.

Similar to SO_2 other strong acid formers such as chlorine and fluorine vapors promote the corrosion of metals [17].

4.3.3 Other Materials

Damage to paper, leather and textiles is also caused by the effect of acid gases, particularly SO_2. If these materials are to be stored for many decades, e.g., in museums or libraries, preservation is required to prevent them from becoming brittle [21].

Synthetic materials such as varnishes and pigments which are increasingly used today, are, however, usually damaged by photooxidizing compounds, mostly characterized by the reference substance ozone. Moreover, polymeric materials are cor-

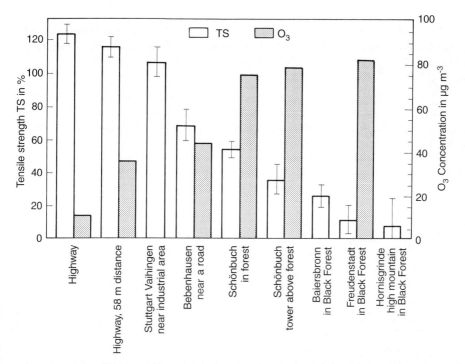

Fig. 4.7. Tensile strength of exposed NBR samples positioned in the shade with mean ozone concentration and their deviation at measuring sites in Southwest Germany [25]

roded by other trace gases, e.g., by nitrogen oxides [14]. The embrittlement of rubber products and elastomeres as well as the bleaching of dyes [14], e.g., automobile tyres and cable insulation, is damage known to be caused by photooxidizing compounds. Macromolecules with carbon double bonds are particularly prone to oxidation by ozone. Varying shielding degrees of these compounds explain the wide range of ozone sensitivity even within one substance group. Apart from chemical structure the power of traction also influences the predisposition to brittleness of synthetic materials. As a protection against ozone many synthetic substances are mixed with special additives.

The pronounced ozone sensitivity of certain synthetic compounds, can, on the other hand, be exploited as an indicator for air quality control. As synthetic indicators they are suitable for estimating the presence of photooxidizing compounds. Measurements in Southwest Germany revealed a clear correlation between the mean ozone concentration and the tensile strength of NBR elastomer samples (nitrile-butadien rubber without ozone protection), s. Fig. 4.7. With increasing brittleness caused by the exposure to photooxidizing compounds their tensile strength becomes poorer.

4.4 Effects on Vegetation

Wild and cultivated plants are both exposed to atmospheric trace substances in manifold ways. Gases or components of particles and precipitation have either a direct effect on aboveground plant parts, mainly on needles or leaves, or, after longterm exposure, an indirect effect on the roots via the soil. Changes in plant functions can thus occur which are either noxious for the plant or even beneficial. Under certain circumstances plants are even dependent on receiving certain trace substances from the atmosphere for their vital functions. Depositions of these substances then have a fertilizing effect. Particularly mineral elements and nitrogen compounds in precipitation and particles function as nutrients and important trace elements.

Increased contributions of atmospheric trace substances, however, can also lead to disorders in the natural plant functions which can interfere with the economic applications of the plants. Such damage can entail a loss of economic and idealistic values or also of human living quality. Moreover, on a larger scale, this type of damage can also upset nature's balance and reduce genetic diversity.

One of the most far-reaching effects of the impact of air pollution is the large-scale breakdown of entire ecological systems. The example of deforestation over large areas shows how humans are indirectly affected by air pollution. If the woods cannot continue to exert their protective function, erosion damage such as loss of soil and landslides, insufficient water retainment and filtration as also climatic changes are to be expected.

Table 4.2 compiles the most important consequences of varying degrees of vegetation damage which have been observed as effects of air pollution. Among them plant damage leading to economic losses in agriculture and forestry have been most closely investigated.

Table. 4.2. Possible effects of air pollutants on plants (acc. to [26])

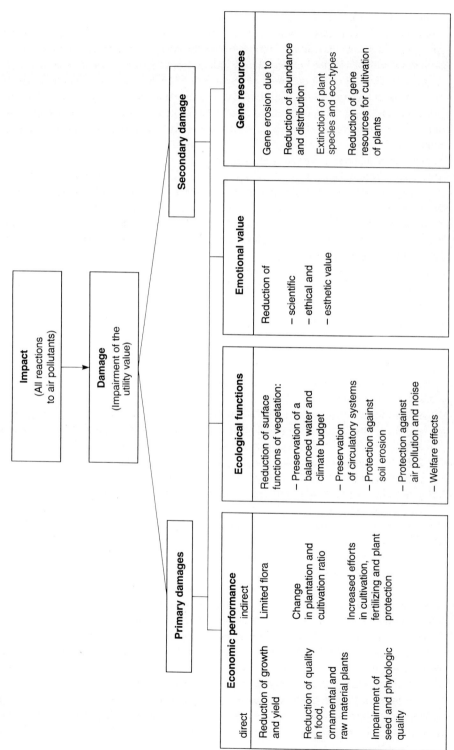

4.4.1 Plant Damage Caused by Air Pollution

Most information concerning vegetation damage has been derived from situations with an obviously higher occurrence of individual atmospheric pollutants in the vicinity of industrial emissions. Gas components SO_2, Cl_2, HCl, HF, NH_4 and C_2H_4 as well as particles containing heavy metals and alkalines are known for their phytotoxic, i.e., plant-damaging effect. As a rule, it is only near emission sources that their concentrations are high enough to cause acute damage. However, photochemical compounds such as ozone and peroxyacetyl nitrate (PAN) also occur on a large scale, although their formation takes place secondarily due to reactions involving atmospheric chemistry. In some regions of the USA their concentrations are high enough to cause considerable reductions of yields in particularly sensitive agricultural plants [26].

Concentrated acids formed from acid gases by atmospheric oxidation, usually bound in precipitation or on particles, are also to be found over large areas and can have a plant-damaging effect.

At longer distances from emission sources plant damage cannot usually be identified as the sole effect of certain air pollutants. Their damage profiles usually deviate from the ones obtained from controlled laboratory tests where one isolated damage factor was applied. A monocausal interrelationship between cause and effect does not apply in these cases.

4.4.1.1 Determination of Dose-Effect Relations

Whether and to what extent an air pollutant has a harmful effect on plants depends on a number of factors from the field of environmental pollution as well as on the sensitivity of the individual systems concerned. Some of the main factors are [26]:
– Pollutant situation: concentration, contact time: duration, frequency and sequence of exposure, simultaneous occurrence of different pollutants;
– external growth factors: light, temperature, humidity; water and nutrient supply;
– individual disposition: species, variety, specific individual sensitivity; state of development, germination, leaf age, physiological activity; prior development (development of defense mechanisms).

The multitude of different effect constellations can easily be imagined on looking at this list. As optimum growth conditions are aimed for in agricultural cultivation, the relationships here are more obvious than in extensive utilization. Based on pollution experiments in open, closed or air-conditioned greenhouses with generally good growth conditions and defined pollutant exposures, definite dose-effect relationships can be determined. The results, especially the symptoms produced, are compared with outdoor statistics in polluted areas, in which a larger number of influencing factors must be taken into account.

An entire life cycle can be followed merely by investigating cultivated annuals. It is difficult, however, to determine long-term effects on shrubs, bushes and trees. When investigating forest damage toxicological concepts are to be applied in which

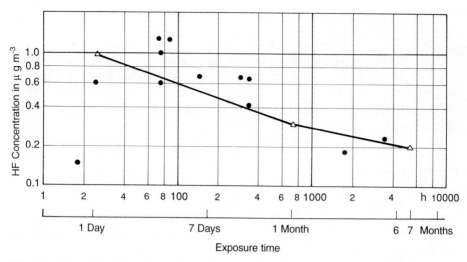

Fig. 4.8. Critical HF dosages at which highly sensitive plants show first visible damage. Test results (•), maximum concentration values (Δ) acc. to VDI guideline 2310, part 3 E (Δ = in µg/m³: 1.0; 0.3; 0.2) [27]

effects caused by short-term exposure doses are extrapolated to longer periods of time.

Threshold values of phytotoxicity are known for a series of gases harmful to agricultural plants. They take concentration and contact time into account. Using HF as an example, such a dose-effect relationship is shown in Fig. 4.8.

At the first proximation critical exposure and dose values (= concentration x time) at which particularly sensitive plant species first show damage profiles are often used [26]. Based on these effect relationships, threshold limits result in high values for short exposure periods, but low values for long exposure periods.

Threshold values such as these are listed in Table 4.3 for selected air pollutants. They take into account first visible plant damage, e.g., discolorations (chloroses) or drying (necroses) of leaves or needles or else a distinct growth depression of plant species especially sensitive to the particular pollutant. Indirect effects via soil are not taken into account any more than latent, "invisible" damage, as these can be inadequately simulated in short-term experiments. Combination effects of several air pollutants also remain unconsidered.

Synergisms occurring during an impact involving two components have been known in a series of cases [26, 28-30]. The concurrence of three and more harmful atmospheric factors has been insufficiently investigated so far. Efforts are presently being undertaken as part of the research into the recent forest damage in Germany ("neuartige Waldschäden").

4.4.1.2 Damage Mechanisms and Profiles of Single Air Pollutants

Absorption is the first step of an air pollutant to its place of effect within a plant. There are two paths of entry into the leaf or needle:

- via stomata mainly on the leaf or needle undersides; they open and close according to the CO_2 required for photosynthesis and evaporation for temperature regulation;
- via the epidermis which is covered on the outside by a protective wax film (cuticle).

Most substances enter the plant via the stomata, particularly gases and the dissolved components in aqueous films on the leaves.

Photochemical compounds

From the point of entry these dissolved substances are relocated in the intercellular space of the plant. Some of them destroy cell structures, others interfere with the plant's metabolism. Table 4.3 lists in a brief outline the main damaging mechanisms of the most important air pollutants. The strong radical formers O_3 and PAN already have an effect in the vicinity of the stomata. They react particularly fast with unsaturated fatty acids and sulfonyl groups, e.g., of amino acids. To a limited extent the plant is protected against these strong oxidizing agents by its own enzymes. E.g., the

Table 4.3. Plant damage caused by air pollution (compiled acc. to [26, 27, 31-33])

Damaging pollutants	Most sensitive plants	Visible damage to foliage	Damage mechanisms	Damage threshold
Ozone – O_3	Tobacco, wheat rye, bean, potato, vine, tomato, pine (pinus silvestris)	Reddish-brown pigmentation on upper leaf chloroses and necroses at tips and margins of needles, premature leaf shedding	Absorption via stomata, oxidative destruction of lipoproteins of the surface, cell membranes ($\rightarrow$leaching), destruction of products of metabolism such as unsaturated fatty acids and sulfunyl groups, disruption of photosynthesis	40 ppb/7-8 h 30 ppb long-term exposure
Peroxyacetyl-nitrate – PAN	Tobacco, bean, leafy lettuce, species of pine, poplar, oak	Bronze-colored stains on underside of leaves, discoloration of needles, premature aging	Similar to O_3, in particular inhibition of fat synthesis, inhibition of enzymes	10 ppb/8 h
Sulfur dioxide- SO_2	Wheat, rye, barley, cotton, species of fruit trees, species of pine and birch	Necroses between leaf nerves and on rims (deciduous trees), reddish-brown striped discoloration of needle tips	absorption via stomata, formation of H_2SO_3 or H_2SO_4 in the intercellular space, disruption of assimilation functions, changes of enzymes and proteins, rigid stomata	180 ppb/ 8 h 8...17 ppb/ vegetation period

4.4 Effects on Vegetation

Damaging pollutants	Most sensitive plants	Visible damage to foliage	Damage mechanisms	Damage threshold
Hydrogen fluoride – HF	Fruit trees, pine, species of spruce, vine, tulip	Necroses on tips and margins	Aerosol and caustic damage on epidermis; gas: absorption via stomata, transport to margins and tips; storage, damage to cell structures, disruption of assimilation functions	1.2 ppb/14 h 0.18 ppb/ vegetation period
Chlorine, hydrogen chloride – Cl_2, HCl		Similar to SO_2		500...1500 ppb/ 0.5...3 h
Nitrogen oxides NO, NO_2	pine, species of spruce	Grayish, brownish discolorations starting on the leaf margins going inward, reddish-brown to red-violet discoloration starting on the tips of the needles	Absorption through stomata, pH decreasing at the point of entry, disruption of longterm assimilation functions, possibly formation of nitrosamine	20000 ppb/1 h 1000 ppb/ longterm
Ammonia NH_3		Similar to SO_2	Collapsed tissue without loss of chlorophyll	55000 ppb/1 h
Hydrogen sulfide H_2S		Necroses between leaf veins		100000 ppb/ 5 h
Acid precipitation		Small, stain-like necroses on vascular bundles and stomata	External attack on leaf areas without hydrophobic wax layer (reinforced by O_3 exposition), necroses, leaching, absorption via stomata, buffering by plants, ion exchange (H^+ against mineral substances from plant), loss of assimilates	pH<3.1...3.2 days to months
Metals	Various	Different	Disruption of assimilation functions and cell membrane structure	Cd, Cr, Co, Ta, V: 5...10ppm[a] Hg, Ni: 25...40ppm[a] Cu, Pb: 125...140 ppm[a] Zn: 740 ppm[a]

[a] mass concentration in the ashes of leaves or needles

enzyme peroxidase which also occurs in the cell walls inactivates secondarily formed hydrogen peroxide before it can penetrate into the cell.

Despite similar effect mechanisms the symptoms of O_3 and PAN in leaves differ greatly, as the two substances damage different organs. Ozone typically damages cells on the upper side of the leaf, whereas PAN destroys cells on the underside of the leaf (spongy parenchyma). When combined with acid precipitation O_3 can effect a premature aging of the cuticle. As a rule, large-leaved field and ornamental plants are considerably more at risk from photochemical compounds than coniferous plants [30].

Acid-forming gases and heavy metals

Comparatively weaker oxidisers are products of acid gases. It is only in very high concentrations that they have an effect externally, cauterizing the leaf surface in the same way as high-alkaline particles (e.g., spent limestone from cement factories), or directly at the point where they enter into the plant. In some cases they accumulate in certain organs of the plant. E.g., the plant relocates fluorides absorbed by the roots into the margin and tip zones of needles and leaves. That is why, after slight HF exposures over long periods of time, the highest fluoride concentrations were to be found in older needles. Deciduous trees are less sensitive to fluorine. Damage symptoms can be observed only after certain concentrations in needles or leaves are exceeded [27].

Heavy metals are enriched in plants in a similar way, then acting at their point of accumulation [31]. In these cases it is therefore more sensible to indicate damage thresholds in the form of mean contents of these substances in the leaves and needles.

Sulfur compounds, too, accumulate in leaves and needles, so that in respect to SO_2 coniferous trees are more at risk than plants shedding their leaves. The sulfuric and sulfurous acid forming from SO_2 interferes in many ways with the metabolic functions. This is reflected in generally poorer photosynthesis as well as in an increase in breathing (respiration) which leads to a lack of assimilate in the plant. Growth starts slowing down. Even the slightest SO_2 concentrations are sufficient to interfere with the closing mechanisms of the stomata. This effect results in an excessive loss of water [29].

Leaf and needle wetting with acid precipitation

In the context of the recent forest damage effects of acid precipitation have been investigated more closely. Both strong and weak acids have an effect on leaf organs via aqueous films (mainly H_2SO_4, HNO_3). High proton concentrations damage the epidermis in the leaf and needle areas where the hydrophobic cuticle is particularly thin. In combination with the impact of O_3 chloroses and necroses are formed [28]. They lead to an increased leaching of nutrients and assimilates. Even low proton concentrations have an effect on the mineral balance of the plant. The protons contained in the wetting water on leaves displace mineral substances such as magnesium, potassium and manganese in the cation exchanger places of the cuticle and in the area of the stomata [34]. As a result the leaves are drained of these essential minerals. If these substances are not replenished via the roots in sufficient amounts, then the lack of magnesium, the central atom in the chlorophyll complex, e.g., will show a typical brownish-yellow discoloration of needles.

4.4.2 Forest Damage

Classic smoke damage

Until a few years ago extensive damage to trees could usually be traced back to single causes. Weather-related events, pests, faults in forestry and also increased pollutant loads often showed very specific effects which either appeared only in certain species of trees or were limited to certain locations and times. Among them was forest damage near emission sources such as foundries and cement factories, or large furnaces which have been known since the last century particularly from the highly industrialized highland regions of the Ore Mountains, the Harz and the Thuringian and Franconian Forest.

The biggest incidence of "classic smoke damage" of this type in Europe can at present be found on the ridges of the Ore Mountains and the Giant Mountains. Both are directly influenced by the Saxon and North Bohemian brown coal and industrial areas with their heavy discharges of SO_2 and particulate matter containing heavy metals. Since 1960 collapses of entire areas of spruce forest have been observed in the Ore Mountains which have led to the dying of approx. 600 km^2 of the total forest area of 3500 km^2 on the territory of the former CSSR alone. The leaf-shedding tree species beech, larch and rowan, however, which are less sensitive to SO_2 have hardly at all been affected [35]. Similar extensive forest damage is known from the area near the largest Canadian nickel foundry in Sudbury [29, 33].

Most recent forest damage

Since the early eighties forest damage has occurred in many areas in Central Europe and in the Eastern United States in many widely differing coniferous and deciduous forests – damage which cannot be attributed to to a single one of the hitherto known causes either in kind or in extent. In contrast to classic smoke damage they are designated as "recent forest damage".

At first only firs were affected, but then the damage quickly spread to spruce and other conifers and later to beech, oak and other deciduous trees. Having started out in the high altitudes of the South German Highlands, particularly the Black and Bavarian Forest, the damage also quickly spread to Middle and North German forest areas and the Alp region, so that sick trees are to be found in the widely differing soil, climate, stock and silvicultural conditions today. Generally, the forest damage increases with the elevation of the stock location. The high altitudes of the Highlands are particularly badly affected.

West-facing stock is generally more prone to damage, even if the main pollutant load comes from other directions. This is even true of the Bavarian Forest in which the eastern slopes are more exposed to the impact of the emissions of the neighboring North Bohemian brown coal area [28].

Sites exposed to light and air are very much at risk, i.e., primarily single trees at the edge of a forest, trees standing free or those towering above the canopy.

4.4.2.1 Damage Profiles

There is no uniform damage profile. Even when looking at one tree species symptoms can vary greatly. A systematic differentiation of damage profiles is at present being developed. With its help individual characteristics can be classified better. Nevertheless, the indicators of sickness of the trees most affected, fir and spruce, can be summarized in the following characteristics [28]:
- growth anomalies: in spruces: branches hang down limply, 2nd degree (lametta syndrome);
- in firs: reduced growth of the top shoot (stork's nest);
- formation of young shoots, branches and twigs in the lower trunk area (water sprouts);
- needle discoloration: needles of the upper side of branches exposed to light show brownish-yellow discoloration, usually with the exception of the new year's growth of needles; as a rule, there is a lack of magnesium and potassium with unproblematic concentrations of heavy metals and sulfur in the needles;
- premature shedding of needles: loss of needles begins with the oldest needles and proceeds from the trunk to the periphery, from the base of the crown to the top;
- disorders in the root area: reduced growth of fine roots and impaired mycorrhization (symbiosis of soil fungi and tree in fine root area, enabling the tree to absorb nutrients from the ground);
- increment losses: in part abruptly since the sixties without any external indicators, in part gradual decline leading to the death of the tree.

Apart from these there are specific regional characteristics, such as reddish needles near the limestone Alps.

In deciduous trees, the most uniform damage profile is that of the beech. The main characteristics are [28]:
- growth anomalies: whip-like short shoots between leaf veins, reduction in leaf size, curling up of leaf edges;
- leaf discoloration: brownish-yellow discoloration of the upper crown areas exposed to light;
- premature shedding of leaves: defoliation, mainly in the upper crown, starting at the shoot tips;
- increment losses: continuous decline.

At a later stage of sickness secondary parasites generally appear, adding to the syndrome. Among them are wood fungi which can lead to a watercore in the center of the trunk and bark beetles which suck on the vascular bundles conducting nutrients.

4.4.2.2 Assumed Mechanisms of Effect

A series of hypotheses which permit a more complete overall picture have been suggested to explain this recent forest damage, with a great variety of debilitating factors and possible trigger factors coinciding.

The initially propounded theory of soil acidification forms the basis for this. According to this theory, acid precipitation destroys the nutrient and water balance of a tree in several ways.

Soil acidification

On the one hand, acids can directly damage the stomata of leaves and needles and thus lead to nutrient leaching. On the other hand, continuous exposure to acids harms the buffering capacity of the soil and can finally reduce the pH value of the soil solution. As a consequence nutrients are leached from the soil to a much greater extent. In addition, when pH values fall below critical thresholds, metals such as aluminum are dissolved and are then biologically available. Particularly threatened by this are mycorrhiza fungi living in symbiosis in the fine root area of many trees species and supporting the tree in its nutrient supply. If, as has been frequently observed, there are only few active fine root tips, then such a tree is poorly supplied with water and nutrients, s. Fig. 4.9.

In addition to this basic loading (predisposing factors), triggering factors such as frost and dryness make things worse. They, too, have a disadvantageous effect on the nutrient and water balance of the trees. Increased activity of nitrifying bacteria in warm, dry summers leads to further soil acidification through the formation of nitric acid.

One limitation of this hypothesis is that it does not explain damage on well buffered soils (limestone). Also, toxic aluminum concentrations have so far not been proven even on acid soil sites.

Nitrogen supply

Another explanatory concept leads on from the soil adification theory. It takes into account the nutritient quality of the nitrogen in the atmosphere. In most forest areas this contribution exceeds the amount necessary for a balanced nutrient supply. With these originally favourable growth conditions an initially healthy plant development can quickly exhaust the low reserves of other nutrients and trace substances. Insufficient adaptation to this situation leads to – as is known from agriculture – greater susceptibility to being damaged by frost and pests.

The possible effect relationships of this hypothesis are presented schematically in Fig. 4.9.

Ozone and acid precipitation

The discoloration of the light-exposed upper sides of the needles of spruce and fir occurring mainly in the higher elevations of the highlands seems to explain the ozone

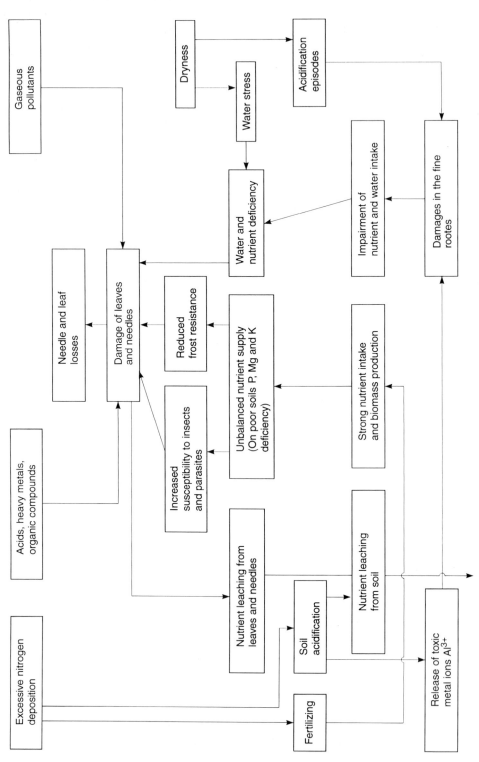

Fig. 4.9. Estimated effect relations due to excessive acid and nitrogen depositions (soil acidification) (acc. to [36])

4.4 Effects on Vegetation

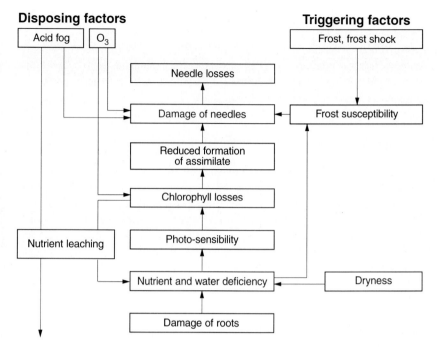

Fig. 4.10. Estimated effect relations due to ozone and acidic precipitation (acc. to [36])

hypothesis [28]. These forest areas with mostly low-nutrient soils are characterized by a high acid load due to precipitation of rain, but also by considerably more polluted fog water as well as by high atmospheric concentrations of photochemical compounds, e.g., ozone. The direct effect of ozone on the stomata of the needles is intensified by exposure to acids. Due to this, nutrients are leached from the needles to a much higher degree. The yellow discoloration observed in needles is the consequence of an – apparently light-induced – chlorophyll decomposition and is particularly linked with a lack of magnesium.

Reduced photosynthesis can impair the supply of assimilates to the root space and thus interfere with the formation of fine roots so that the nutrient and water balance is disturbed. This damage profile goes hand in hand with an increased susceptibility to frost. Here, too, climatic conditions are added to basic air pollution as triggering factors. It is assumed that extreme dryness in the years 1976, 1982 and 1983 as well as the high recurrence of frost shocks in the years 1978, 1981, 1982 and 1983 with precipitous drops in air temperature of up to 30 Kelvin (New Year's Eve 1978) led to a simultaneous occurrence of forest damage in the Central European highlands.

The effect relationships of exposure to ozone in combination with acid precipitation are shown schematically in Fig. 4.10.

Organic substances

So far investigation of recent forest damage has concentrated on the effects of inorganic atmospheric pollutants. The effect of many organic substances on forest trees has hardly been researched at all. They, too, must be taken into consideration as causes for the recent forest damage.

Current information on the presence of such possible plant poisons at forest sites and their deposition on plant surfaces as also the effect mechanisms of such substances in plant physiology is as yet entirely insufficient to be able to explain the effects of organic components on the health of trees.

Recent forest damage – a complex of diseases

In spite of several years of intensive research in the field of recent forest damage none of the above-mentioned hypotheses have been able to entirely explain the great variety of observed symptoms within the most manifold internal and external frame conditions. According to current information a series of damaging mechanisms must work together so that in each case the weighting of the influence of single factors must be questioned. Unfortunately, special measures for emission reduction cannot be derived from such a complex of causes.

However, there is sound evidence to show that under unfavorable soil and climatic conditions, there are a number of anthropogenic trace substances which lead to latent plant damage in forest areas even if they are present only in low concentrations as is the case in locations far from emission sources. Any additional load (stress) can then trigger visible forest damage to individual trees or even endanger the stock of entire forests.

4.5 Impact on Human Health

Apart from food and drink whose daily intake does not exceed a few kilograms and which can be chosen more or less freely, the air we breathe represents a medium which is taken in by all people of a certain area in the same way. On an average the air volume coming into contact with the organism of an adult comes to approx. 20 m^3 per day, which corresponds to a mass of 25 kg. Humans can live without solid food up to 40 days; without air to breathe, however, only a few minutes. One must therefore devote special attention to the quality of the air inhaled.

The pollutants contained in the air reach the inside of the body through respiration and remain there to be transformed or stored and enriched over a period of many years.

The harmfulness of air pollutants for humans depends on the following factors:
- toxicity of the individual pollutants,
- concentration of the pollutants x length of exposure = pollutant dose,
- combined effect of several pollutants,
- ambient conditions such as temperature, radiation, air movement, humidity a.s.o.,
- age and health condition of the individual.

Air pollutants can cause irritation or physical and mental impairment.

There are varying types and degrees of damage, s. sect. 4.1.1. In the worst of cases mortality can increase (note the casualties during smog episodes, Table 1.2). Annoyance can, e.g., be caused by
- visible clouds of smoke,
- odors,
- combinations of both.

Visible clouds of smoke convey to many people negative information on environmental pollution. If this situation recurs often, mental well-being can be impaired due to the annoyance about this phenomenon. Unpleasant odors have much the same effect, with the personal perception of the individual being even more intense than with the visual impression. In this respect the sense of smell exerts a natural warning about harmful substances. When headaches or respiratory irritation follow the perception, then the lack of psychological well-being becomes a physical one.

4.5.1 Possibilities and Difficulties of Recording Harmful Effects

The effect of atmospheric pollutants on human health is the main argument for limiting pollutant emissions. The symptoms are, however, manifold, and simultaneous exposure along with other irritants often makes it very difficult to determine the specific effects of air pollutants. The findings available are from the research fields of occupational medicine, epidemiology, clinical studies and toxicology, s. Table 4.4. To obtain exact information on the effects of individual air pollutants, experiments on human beings would have to be carried out with all other influencing factors excluded (clinical studies). Such tests on voluntary test persons have to remain restricted to acute influences with reversible effects. Even then strictest precautions would have to be taken so as not to expose those involved to any danger of risking their health.

Effects of long-term exposure cannot be investigated in this manner as one would have to expect permanent damage. However, an extensive long-term experiment has been taking place for many years: voluntary exposure of many people to combustion exhaust gases when smoking or involuntarily as passive smokers.

According to the definition of the World Health Organization (WHO) *epidemiology* deals with the investigation of the distribution of diseases, physiological variables and social consequences of diseases in human population groups as well as with the factors influencing this distribution [38]. As is known from epidemiological studies, 25,000 persons die of lung cancer annually in the FRG and 80-90 % of these cases are ascribed to smoking [39, 40]. Not only lung cancer but also numerous other diseases, e.g., in the cardiovascular system, are caused by inhaling cigarette smoke. Even passive smoking is not without its dangers and often leads to an increased cancer risk, as the so-called side-stream smoke contains much higher pollutant concentrations in part than main stream smoke [41, 42].

The effects of smoking are so severe that they often mask the effects caused by polluted ambient air. E.g., epidemiological studies have shown that there is an increased risk of cancer in urban areas as compared to rural districts; this, however,

Table 4.4. Scientific disciplines researching the effect of air pollution on human health and their strengths and weaknesses on the basis of [37]

Discipline	Population groups	Strengths (advantages)	Weaknesses (disadvantages)
Occupational medicine (as a branch of epidemiology)	Certain professions	Limited disease factors, possibly long exposure times	Affects primarily only healthy persons
Epidemiology	Population groups (e.g., urban and rural population) Affected groups	Actual exposure, long-term low level effects No extrapolations necessary, Inclusion of sensitive groups	Difficult quantification of exposure, Many covarities, Minimal results in dose-response relations Assumed instead of causal correlations
Clinical studies	Experiments (human experiments) diseased persons	Controlled exposure, few covarities, examination of susceptible persons, causal correlations	Artificial exposure, only acute effects, no long-term effects, risk, ethical scruples (public acceptance)
Toxicology	Animal experiments	Much dose-response data,	Applicability to humans? ethical scruples
	Cell tissue	fast accumulation of data	Threshold of human response?
	Biochemical systems	Causal correlations Mechanisms of response	Extrapolation

cannot be primarily attributed to atmospheric pollution, as people who live in cities smoked more ten to thirty years ago than those living in the country [43, 44].

A particular difficulty with epidemiological studies is that the persons are not only exposed to outside air but predominantly breathe indoor air whose quality can vary greatly depending on the individual case. All this makes the deduction of causal interrelationships between pollutant concentrations in the air and the effects on human beings more difficult.

Thanks to occupational medicine which looks for interrelationships between occupational diseases and the exposure to certain pollutants there are a number of findings on the effects of air pollutants [45, 46]. In outside air, however, atmospheric pollutant concentrations are mostly one scale smaller than in work-places so that the effects of being exposed to low concentrations over long periods of time can still be considered a gap in our knowledge of medicine.

While it is taken for granted that healthy persons of employable age who are exposed to pollutants at their work-places must be protected, it is the following groups of people who are particularly sensitive to pollutants in ambient air:

- small children whose respiratory and cardiovascular system are still at the developmental stage,
- old people with weak heart, circulatory and respiratory functions,
- sick people, e.g., those with asthma, bronchitis or cardiovascular problems.

When setting up guidelines for pollutants in outside air these groups of people have been particularly kept in mind [47].

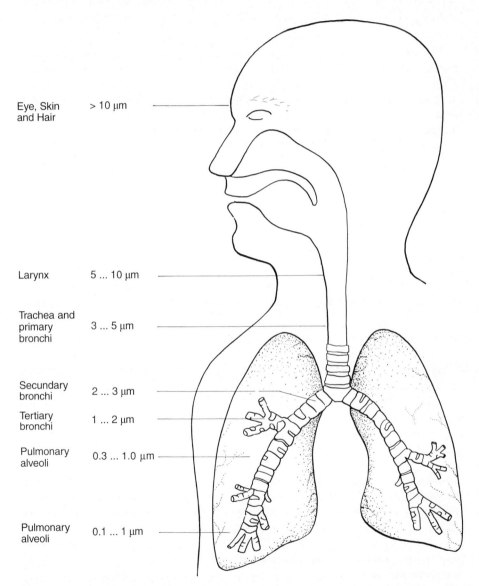

Fig. 4.11. Particle deposition in the different areas of the respiratory tract in dependence of the mean diameter of the particles (acc. to [37])

4.5.2 Paths Air Pollutants Take in the Human Body

Apart from pollutant intake via the skin or the mucous membranes of the eyes, air pollutants enter the human body mainly via the respiratory tract. Particle-shaped pollutants are eliminated or absorbed at different locations of the respiratory tract, fundamentally depending on the aerodynamic properties of the particles. An overview is given in Fig. 4.11. Whereas coarse particles (> 5-10 μm mean diameter) remain in the nasopharyngeal space and some of the smaller particles settle in the ciliated epithelium of the bronchial tubes (cilia), the major part is deposited in the lungs.

The meati of the nose and the upper bronchi are coated with a mucous layer. This layer not only absorbs particles but also water-soluble gases. The ciliated epithelium of the bronchial tubes causes a constant upward movement of the mucus. In this way the gases and particles absorbed are transported into the pharyngeal space, then swallowed and finally discharged via the digestive tract. The particles absorbed in the "pre-filter" nasal passages are also discharged again by mucus secretion.

With gases their water solubility determines their location of deposition in the respiratory tract. Gases with good solubility such as SO_2 are absorbed early in the upper respiratory passages. When inhaling SO_2, e.g., there is a characteristic sour taste. Gases with poorer water solubility such as CO, NO or O_3 penetrate into the lung area and from there reach the lymphatic or blood system by molecular diffusion via the pulmonary alveoli. Irritant gases are assumed to excite nerve cells on respiratory system walls and thus cause manifold effects, e.g., sneezing, coughing or rapid, shallow breathing [37].

4.5.3 Effects of the Most Important Air Pollutants

The effects of air pollutants on humans range from slight irritations of the eyes and mucous membranes up to death. However, the locations of effect are not only restricted to the respiratory tract. Many different organs in the whole body are affected. The manifold effects of the indivi dual air pollutants are compiled in Table 4.5.

4.6 Ambient Air Guidelines and Standards

Casualties of smog catastrophes, vegetation damage in the vicinity of emitting factories and the manifold consequences of air pollution caused by all the different emittent groups call for pollution reduction measures. These will continue to require great efforts as the necessary minimum level of emissions has not yet been achieved in many fields and emission reductions have partly been offset by the increasing output of emission sources. Besides, new effects, which were hitherto unknown, are continually being discovered.

Apart from emission reduction measures monitoring air quality on the output side is necessary for the protection of the population and for man's animate and inanimate

Table 4.5. Important air pollutants and their effects on human health

Air pollutants	Effects	References
SO_2	Colorless gas, sour taste in pure air from 0.6 mg/m^3, irritant gas for respiratory system, dissolves in mucous membranes of eyes, mouth, nose and bronchi; increase of flow resistance in respiratory system; the main effect is caused by repeated peak-like exposure	[49-51]
SO_2 + fine particulate	Increase of toxic SO_2 effect; penetration of acidic aerosol into the inner respiratory organs and formation of H_2SO_4; increased susceptibility to chronic bronchitis; higher risk of acute diseases of the respiratory system; increased frequency of broncho-pathic symptoms;	[50]
NO_2	Brown color, oppressive odor, odor threshold 0.2 mg/m^3, habituation to odor, irritant gas for respiratory system, dissolves in mucous membranes; increased susceptibility to germs in the respiratory system;	[49,51,52]
NO	Colorless, odorless, poor aqueous solubility; no irritation of mucous membranes (but rapid transformation into NO_2 in air), toxic effect approx. 20% of NO_2, formation of methemoglobin in the blood which is not capable of transporting O_2	[49, 51]
CO	Odorless gas, 200 to 300 times greater affinity to the blood pigment hemoglobin than O_2: formation of carboxyhemoglobin (COHb), obstruction of O_2 transport in the blood and thus of the O_2 supply to the body; points of attack: central nervous system (brain) and cardiovascular system; headaches, fatigue, drowsiness, weakening will-power, interference with sleeping-waking behavior;	[49,51,53]
O_3	Colorless (blue in higher concentrations), oppressive odor, odor threshold from 0.05 mg/m^3, strongest oxidant; strong irritant to respiratory system; penetrates into the inner lung due to poor aqueous solubility; impairment of lung functions by oxidation of enzymes, proteins, amino acids, lipids etc., causes breaks in chromosomes, greater susceptibility to infections; tussive irritation, eye irritation; O_3 is a component of photochemical smog, combination effect with other photooxidizing compounds;	[49,54,55]
H_2S	Colorless gas, strong rotten-egg odor, odor threshold 1.6 to 5 µg/m^3, habituation to smell; irritant gas for eyes and mucous membranes, cellular or ferment toxin, damage of nerve tissue, fatigue, easy irritability, headaches, disorders of balance and sleep etc.;	[51,56,57]
Cl_2, HCl	Cl_2: green color, odor threshold approx. 0.15 to 0.3 mg/m^3; HCl: colorless; Strong irritant gases for respiratory system, absorbed by mucous membranes as easily water-soluble; relevant in proximity of particular emission sources and failures;	[49]
HF	Strong oxidant; disordered calcium metabolism and thus damage to bones and teeth; strong plant toxin; relevant in proximity of particular emission sources;	[27, 49]

Table 4.5. (continued)

Air pollutants	Effects	References
HCHO (formaldehyde)	Colorless gas, strong pungent odor, odor threshold 0.05 to 1 mg/m^3, irritant gas for eyes and respiratory system, suspected of having a carcinogenic and mutagenic effect;	[51,58]
Aldehydes, ketones, phenols, mercaptanes	Strong odorous substances, irritations of eyes and respiratory system (oxidized hydrocarbons, e.g., from incomplete combustion);	[59]
C_6H_6 (benzene)	Colorless liquid with characteristic odor; as air pollutant benzene vapor; absorption via respiratory system, in the case of chronic exposure accumulation in fatty tissue and bone marrow, metabolizing in the liver, damage of the blood-forming systems, carcinogenic (leukemia) when occupationally exposed;	[60]
PAH polycyclic aromatic hydrocarbons	More than 100 different PAHs as the result of incomplete combustion of fossil fuels, frequently adsorbed to particulate matter; e.g., in cigarette smoke or diesel soot; some of the PAH are carcinogenic, the best known representative: benzo[a]pyrene;	[61-63]
Soot	Carbon particles (agglomerates) to which PAH can be adsorbed: carriers of PAH into the respiratory system; diesel soot is suspected of being carcinogenic;	[51]
Halogenated hydrocarbons	Highly diverse substances, most of them are relatively non-reactive, some are highly toxic; possibly affect skin, metabolism and excretory organs (liver, kidneys), the germ track and the central nervous system, possible accumulation in fatty tissue;	[51]
Pb	Occurrence in particles; affects metabolic activity and brain, particularly in small children, accumulates in bones;	[51]
Cd	Occurrence in particles; increased resorption during vitamin and mineral deficiencies; impairment of kidney function during longer and increased absorption;	[51]
Asbestos	Fibrous particles; increased risk of mesothelioma and lung cancer during prolonged exposure	[51]

environment. Various institutions have worked out or passed air pollution guidelines and standards for the assessment of air quality . Comparing pollutant concentrations with the given maximum values helps to determine where protective measures and emission reduction are particularly needed (e.g., setting up clean air plans). When certain threshold values are exceeded, smog warning and smog alarm plans entailing immediate and drastic measures of emission reduction become effective.

An overview of the most important air quality guidelines of the World Health Organization (WHO) and of the Association of German Engineers (Verein Deutscher Ingenieure) as well as the national ambient air standards of the USA and Germany (identical with the EU guidelines) are listed in Table 4.6 (for human health protection). Guideline values for individual substances based on effects on terrestrial

4.6 Ambient Air Guidelines and Standards   205

Table 4.6. List of air quality guidelines, air quality standards and smog alarm values of the most important air pollution substances for the protection of human health

Substance	WHO air quality guidelines for individual substances [1] [45]				U.S. National ambient air quality standards [48]		German ambient air quality standards [67]		German VDI guidelines [3] [47, 50, 52, 54, 69]		German smog alarm values [66]	
	Time-weighted average	Averaging time	Carcinogenic risk [2]	Site of tumour	Concentration	Averaging time	Concentration	Averaging time	MIK values	Averaging time	Concentration	Averaging time
Arsenic As			4×10^{-3}	lung								
Benzene C_6H_6			4×10^{-6}	blood, leukemia			$10\ \mu g/m^3$	1 year				
Asbestos			10^{-6}–10^{-5} [4] 10^{-3}–10^{-4} [4]	lung cancer [5] mesothelioma	HAP [6]							
Cadmium Cd	1–5 ng/m^3 10–20 ng/m^3	1 year (rural areas)							50 ng/m^3	24 h		
Carbon disulfide CS_2	100 $\mu g/m^3$	24 h										
Carbon monoxide CO	100 mg/m^3 60 mg/m^3 30 mg/m^3 10 mg/m^3	15 min [7] 30 min [7] 1 h [7] 8 h [7]			40 mg/m^3 10 mg/m^3	1 h [8] 8 h [8]			50 mg/m^3 10 mg/m^3 10 mg/m^3	30 min 24 h 1 year	30 [9] 45 [10] 60 [11]	3 h 3 h 3 h
Chromium (VI) Cr			4×10^{-2}	lung								

Table 4.6. (continued)

Substance	WHO air quality guidelines for individual substances [1) [45]				U.S. National ambient air quality standards [48]		German ambient air quality standards [67]		German VDI guidelines [3) [47, 50, 52, 54, 69]		German smog alarm values [66]	
	Time-weighted average	Aver-aging time	Carcino-genic risk [2)	Site of tumour	Concen-tration	Aver-aging time	Concentration	Averaging time	MIK values	Aver-aging time	Concen-tration	Aver-aging time
Formaldehyde HCHO	100 µg/m³	30 min							70 µg/m³ 30 µg/m³	30 min 1 year		
Hydrogen fluoride HF									200 µg/m³ 100 µg/m³ 50 µg/m³	30 min 24 h 1 year		
Hydrogen sulfide H₂S	150 µg/m³	24 h										
Lead Pb	0,5–1,0 µg/m³	1 year			1.5 µg/m³	1 year			3 µg/m³ 1.5 µg/m³	24 h 1 year		
Mercury Hg	1 µg/m³ [12)	1 year										
Nitrogen dioxide NO₂	400 µg/m³ 150 µg/m³	1 h 24 h			100 µg/m³	1 year	200 µg/m³	98% of all 1 h values over 1 year	200 µg/m³ 100 µg/m³	30 min 24 h	0.6 mg/m³ [9) 1.0 mg/m³ [10) 1.4 mg/m³ [11)	3 h 3 h 3 h

4.6 Ambient Air Guidelines and Standards

Pollutant	Value	Time	Value	Time	Value	Time	Value	Time	Value	Time
Ozone O_3	150-200 µg/m³	1 h	100-120 µg/m³	8 h	235 µg/m³	1 h [8] daily max.	110 µg/m³	8 h	120 µg/m³	30 min
									180 µg/m³ [3,9]	1 h
									240 µg/m³ [10]	1 h
Particulate Matter (PM 10)	150 µg/m³	24 h [8]			300 µg/m³ [16]	95% of all 24 h values over 1 year	500 µg/m³ [16]	1 h		
	50 µg/m³	1 year			150 µg/m³ [16]	1 year aver. of 24 h values	250 µg/m³ [16]	24 h		
							75 µg/m³ [16]	1 year		
Polynuclear aromatic hydrocarbons PAH [15]	9×10^{-2}	lung								
Sulfur dioxide SO_2	500 µg/m³	10 min	1300 µg/m³	3 h [8]	80 µg/m³ [17]	1 year aver. of 24 h values	1000 µg/m³	30 min	0.6 mg/m³ [3,9]	3 h
	350 µg/m³	1 h	365 µg/m³	24 h [8]	120 µg/m³ [18]		300 µg/m³	24 h	1.2 mg/m³ [10]	3 h
			80 µg/m³	1 year	130 µg/m³ [18]	winter aver. of 24 h values			1.8 mg/m³ [11]	3 h
					180 µg/m³ [20]	98% of all 24 h values of 1 year				
					250 µg/m³ [21]					
					350 µg/m³ [22]					
Sum of SO_2 + 2×Total Susp. Part. Matter									1.1 mg/m³ [3,9]	24 h
									1.4 mg/m³ [10]	24 h
Toluene C_7H_8	8 mg/m³	24 h				98% of all				
Vinyl chloride	1×10^{-6}	liver and other sites								

Remarks:

1) Information of this table should not be used without reference to the rationale given in the WHO Air Quality Guidelines book [45]
2) Cancer risk estimates for lifetime exposure to a concentration of 1 µg/m^3
3) Ambient Air Guideline Values of the Association of German Engineers (Verein Deutscher Ingenieure VDI): Maximum Immission Concentrations (MIK)
4) Cancer risk estimates for lifetime exposure to a fibre concentration of 500 fibres/m^3 (fibres measured by optical methods)
5) Lung cancer in a population with 30% smokers
6) Belonging to the 189 hazardous air pollutants (HAPs) identified in the U.S. Clean Air Act Amendments [68] with several health effects
7) Exposure at these concentrations should be for no longer than the indicated times and should not be repeated within 8 hours
8) Not to be exceeded more than once per year
9) Early warning
10) First alarm degree
11) Second degree
12) The guideline value is given only for indoor air pollution; no guidance is given on outdoor concentrations (via deposition and entry into the food chain) that might be of indirect relevance
13) 4 periods per day: 0-8 hours, 8-16 hours, 12-20 hours, 16-24 hours
14) With traffic restrictions
15) Expressed as benzo(a)pyrene (based on benzo(a)pyrene concentration of 1 µg/m^3 in air as a component of benzene-soluble coke-oven emissions)
16) Total Suspended Particulate Matter TSM
17) with > 150 µg/m^3 total suspended particulate matter (TSM), (average of 24 h values of 1 year)
18) with ≤ 150 µg/m^3 TSM (average of 24 h values of 1 year)
19) with > 200 µg/m^3 TSM (average of 24 h values in winter)
20) with ≤ 200 µg/m^3 TSM (average of 24 h values in winter)
21) with ≥ 350 µg/m^3 TSM (98% of all 24 h values over 1 year)
22) with ≤ 350 µg/m^3 TSM (98% of all 24 h values over 1 year)

vegetation and different sensitive plants are described in Table 4.7. The following sections will provide a brief explanation of the individual guidelines and standards.

4.6.1 Nature of WHO Guidelines [45]

The primary aim of the air quality guidelines[1] is to provide a basis for protecting public health from the adverse effects of air pollution and for eliminating, or reducing to a minimum, those air contaminants that are known or likely to be hazardous to human health and wellbeing.

The guidelines are intended to provide background information and guidance to governments in making risk management decisions, particularly in setting standards, but their use is not restricted to this. They also provide information for all those who deal with air pollution. The guidelines may be used in planning processes and various kinds of management decisions at the community or regional level. When guideline values are indicated, this does not necessarily mean that they must take the form of general countrywide standards, monitored by a comprehensive network of control stations. In the case of some agents, guideline values may be of use mainly for carrying out local control measures around point sources.

It should be emphasized that when air quality guideline values are given, these values are not standards in themselves. Before standards are adopted, the guideline values must be considered in the context of prevailing exposure levels and environmental, social, economic and cultural conditions. In certain circumstances there may be valid reason to pursue policies which will result in pollution concentrations above or below the guideline values.

Ambient air pollutants can cause several significant effects which require attention: irritation, odor annoyance, acute and longterm toxic effects (including carcinogenic effects). Air quality guidelines either indicate levels combined with exposure times at which no adverse effect is expected concerning noncarcinogenic endpoints, or they provide an estimate of lifetime cancer risk arising from those substances which are proven human carcinogens or carcinogens with at least limited evidence of human carcinogenicity.

It is believed that inhalation of an air pollutant in concentrations and for exposure times below a guideline value will not have adverse effects on health and, in the case of odorous compounds, will not create a nuisance of indirect health significance (see definition of health, Constitution of the World Health Organization). Compliance with recommendations regarding guideline values does not guarantee the absolute exclusion of effects at levels below such values. For example, highly sensitive groups especially impaired by concurrent disease or other physiological limitations may be affected at or near concentrations referred to in the guideline values. Health effects at or below guideline values can also result from combined exposure to various chemicals or exposure to the same chemical by multiple routes.

---

[1] Guidelines in the present context are not restricted to suggested numerical values, but also include any kind of recommendation or guidance in the relevant field.

Table 4.7. List of air quality guidelines of the most important air pollution substances for the protection of vegetation

Substance	WHO guidelines [45]			German VDI Guidelines [26, 27, 64, 65]		
	Guideline values	Averaging time	Remarks	MIK values [1]	Averaging time	Remarks
Hydrochloric acid HCl				800 µg/m³	1 day	very sensitive plants
				100 µg/m³	1 month	
				1200 µg/m³	1 day	sensitive plants
				150 µg/m³	1 month	
Hydrogen fluoride HF				1 µg/m³	1 day	very sensitive plants
				0.25 µg/m³	1 month	
				0.15 µg/m³	7 months	
				2 µg/m³	1 day	sensitive plants
				0.6 µg/m³	1 month	
				0.4 µg/m³	7 months	
				6 µg/m³	1 day	less sensitive plants
				1.8 µg/m³	1 month	
				1.2 µg/m³	7 months	
Nitrogen dioxide NO_2	95 µg/m³	4 hours	In the presence of SO_2 and O_3 levels which are not higher than 30 µg/m³ (arithmetic annual average) and 60 µg/m³ (average during growing season) respectively	600 µg/m³	0.5 hours	sensitive plants
	30 µg/m³	1 year		350 µg/m³	7 months	
Total nitrogen deposition	3 g/m²	1 year averaged over growing season	Sensitive ecosystems are endangered above this level			
Ozone O_3	200 µg/m³	1 hour		320 µg/m³	0.5 hours	very sensitive plants
	65 µg/m³	24 hours		70 µg/m³	8 hours	
	60 µg/m³	averaged over growing season		480 µg/m³	0.5 hours	sensitive plants
				800 µg/m³	0,5 hours	less sensitive plants

Peroxyacetyl-nitrate PAN	300 µg/m³	1 hours		
	80 µg/m³	8 hours		
Sulfur dioxide SO₂	30 µg/m³	1 year	insufficient protection in the case of extreme climatic and tropographic conditions	
	100 µg/m³	24 hours		
	250 µg/m³	0.5 hours	very sensitive plants	
	50 µg/m³	7 months		
	400 µg/m³	0.5 hours	sensitive plants	
	80 µg/m³	7 months		
	600 µg/m³	0.5 hours	less sensitive plants	
	120 µg/m³	7 months		

[1] Maximum Immission Concentration

It is important to note that guidelines have been established for single chemicals. Chemicals, in mixture, can have additive, synergistic or antagonistic effects; however, knowledge of these interactions is still rudimentary.

Risk estimates for carcinogens do not indicate a safe level; they are presented so that the carcinogenic potencies of different carcinogens can be compared and an assessment of overall risk made.

The guidelines do not differentiate between indoor and outdoor exposure (with the exception of the exposure to mercury) because, although the sites influence the type and concentration of chemicals, they do not directly affect the basic exposure-effect relationship.

4.6.2 National Ambient Air Quality Standards

Many countries have set up ambient air quality standards which require that clean air efforts be initiated when these standard values are exceeded. Monitoring adherence to these values generally requires continuous measurements over long periods of time. Most values can only be checked after measurements of one year. Several years of measurements in many sites are required to identify trends. The standards of the USA and Germany are listed in Table 4.6 as examples.

4.6.3 MIK² Values of the Verein Deutscher Ingenieure (VDI = Association of German Engineers)

MIK values of the VDI were set up by a team of scientists from different fields and experts from industry and administration and were subjected to a hearing with counter-arguments from the public. It is the aim of the VDI's MIK values to prevent detrimental effects of air pollutants on man and his environment, i.e., animals, plants, materials, soil, water and atmosphere, including their functioning interrelationship such as in, e.g., ecosystems [47]. MIK values for the protection of human health and well-being are included in Table 4.6, those for the protection of the earth's vegetation in Table 4.7.

The relevant reference times are also indicated when determining maximum immission values. These reference times have been adapted both to the nature of the pollutants' effects and to the reactions of the relevant subjects. The MIK values for the protection of human health and well-being, e.g., have fixed time bases of 0.5 or 24 hours. Exceeding the maximum concentration thus does not depend on the total of all data measured, but on the momentarily occurring individual half-hourly or diurnal mean values.

---

[2] MIK stands for "Maximale Immissions-Konzentration" = Maximum Immission Concentration)

4.6.3 Smog Alarm Values

Based on earlier smog catastrophes involving numerous cases of illness and death the smog warning plans of the federal states of the FRG [66] serve the purpose of restricting air pollution during periods of high contaminant concentrations and of warning the population. However, due to the emission reduction measures of the past years pollutant concentrations have not been as high during low-exchange weather conditions as they used to be in the past.

When the values of the smog pre-warning stage are exceeded and there is a likelihood of the low-exchange weather situation continuing for an extended period, population and industry are asked to avoid pollutant emissions wherever possible. In the cases of alarm 1 and 2, e.g., emission restrictions take effect in the form of automobile operation prohibition and operation restrictions for industrial furnaces and other emitting plants.

4.7 Bibliography

1. Graßl, H.: Klimaveränderung durch Spurengase. Energiewirtschaftliche Tagesfragen 37 (1987) 127−133
2. Flohn, H.: Treibhauseffekt und Klima. GRS-Fachgespräch „Betriebserfahrung und Reaktorsicherheit", Gesell. f. Reaktorsicherheit, Köln 1987
3. Fabian, P.: Antarktisches Ozonloch: Indizien weisen auf Umweltverschmutzung. Phys. Bl. 44 (1988) 2−11
4. Ehhalt, D.H.: Kreisläufe klimarelevanter Spurenstoffe. In: VDI-Kommission RdL: Globales Klima. Schriftenreihe der VDI-Kommision RdL Bd. 7, Düsseldorf: VDI (1987) 78−96
5. Stolarski, R.S.: Das Ozonloch über der Antarktis. Spektrum der Wissenschaft 1988
6. Crutzen, P.; Birks, J.: The atmosphere after the nuclear war, Ambio 11 (1982), 116−125
7. Fabian, P.: Atmosphäre und Umwelt. 3. Aufl., Berlin: Springer 1989
8. Junge, C.: Kreisläufe von Spurenstoffen in der Atmosphäre. In: Jaenicke, R. (ed.): DFG Sonderforschungsbereich Atmosphärische Spurenstoffe. Weinheim: Chemie 1987, S. 19−30
9. Schönwiese, C.D.: Das natürliche Klima und seine Schwankungen. In: VDI-Kommission RdL: Globales Klima, Schriftenreihe der VDI-Kommission RdL, Bd. 7, Düsseldorf: VDI 1987, S. 4−26
10. Budyko, M.I.; Golitsyn, G.S.; Izrael, Y.A.: Global climatic catastrophes. Berlin: Springer 1988
11. Schönwiese, C.D.; Dickmann, B.: Der Treibhauseffekt − Der Mensch ändert das Klima. Stuttgart: Deutsche Verlagsanstalt 1987
12. Jäger, J.: Climatic changes: floating evidence in the CO_2-debatte. Environment 28 (1986), S. 6
13. Bolle, H.-J.: Die Bedeutung atmosphärischer Spurenstoffe für das Klima und seine Entwicklung. In: [9], S. 27−77
14. Schreiber, H.: Wirkungen von Photooxydantien auf Materialien. In: Guderian, R. (ed): Luftqualitätskriterien für photochemische Oxidantien. UBA Berichte 5/83. Berlin: Erich Schmidt 1983, S. 481
15. Gauri, L.K.; Holdren, G.C.: Pollutant effects on stone monuments. Environmental Science and Technology 15 (1981) 386−390
16. N.N.: Westfälisches Amt für Denkmalpflege, Salzstraße 38, D-4400 Münster, Bildarchiv
17. Wangerang, E.: Freiheitsstrafen nach dem Deckeneinsturz im Hallenbad. VDI-Nachrichten 18.12.87, S. 23
18. Isecke, B.: Einfluß von Luftverunreinigungen und Umwelteinflüssen auf das Korrosionsverhalten von Stahl- und Spannbeton. In: VDI-RdL: Materialkorrosion durch Luftverunreinigungen, VDI Berichte 530. Düsseldorf: VDI 1985, S. 137

19 Lindner, W.: Betonschutz und Betonsanierung mit dem Disbocret System 500. In: VDI-Rdl: Materialkorrosion und Luftverunreinigungen. VDI-Berichte 530. Düsseldorf: VDI 1985, S. 219
20 Luckat, S.: Quantitative Untersuchung des Einflusses von Luftverunreinigungen bei der Zerstörung von Naturstein. Staub 41 (1981) 440–442
21 Yocom, J.E.; McCaldin, R.O.: Effect of air pollutants on materials and the economy. In: Stern, A.C. (ed): Air pollution. Vol. 1: Air pollution and its effects. New York: Academic Press 1968
22 Topol, L.E.; Vijauakumar, R.: Material demage from acid deposition. In: Proceedings of the 6th World Congress on Air Quality, 16–20. May 1983, Paris, Vol. 3, p. 51–58
23 Schikerr, G.: Die Bedeutung des Schwefeldioxids für die atmosphärische Korrosion der Metalle. Werkstoffe und Korrosion 15 (1964) 457–463
24 Upham, J.B.: Materials deterioration and air pollution. JAPCA 15 (1965) 265
25 Schliep, M.: Kunststoffe als Indikatoren für Luftverunreinigungen – Ergebnisse von Expositionsversuchen. IVD-Bericht Nr. 4–1986, Universität Stuttgart 1986
26 Verein Deutscher Ingenieuere: Guideline VDI 2310, part 6: Maximum Immission Values to Protect Vegetation. Maximum Immission Concentrations for Ozone. Berlin: Beuth April 1989
27 Verein Deutscher Ingenieure: Guideline VDI 2310, part 3: Maximum Immission Values to Protect Vegetation. Maximum Immission Concentrations for Hydrogen Fluoride. Berlin: Beuth December 1989
28 Prinz, A.; Krause, G.H.M.: Waldschäden in der Bundesrepublik Deutschland, Staub 47 (1987) 94
29 Guderian, R.: Air pollution – phototoxicity of acidic gases. Berlin: Springer 1977
30 Guderian, R.; Tingey, D.T.; Rabe, R.: Wirkung von Photooxidantien auf Pflanzen. In: Guderian, R.: Luftqualitätskriterien für photochemische Oxidantien. UBA-Berichte 5/83. Berlin: Erich Schmidt 1983, S. 205
31 Hindawi, I.J.: Air pollution injury to vegetation. US Dept. Health Education and Welfare. Publ. AP 71 Raleigh, N.C. 1970
32 Brandt, C.S.; Heck, W.W.: Effects of air pollutants on vegetation. In: Stern, A.C. (ed): Air pollution. Vol. 1: Air pollution and its effects. New York: Academic Press 1968
33 Smith, W.H.: Air pollution and forests. New York: Springer 1981
34 VDI: Säurehaltige Niederschläge – Entstehung und Wirkung auf terrestrische Ökosysteme. VDI-Kommission Reinhaltung der Luft. Düsseldorf: VDI 1983
35 Exkursion anläßlich der XIII. Internationalen Arbeitstagung forstlicher Rauchschadensachverständiger (INFRO) 27.8.–19.1984, Most, CSSR
36 Cowling, E.; Krahl-Urban, B.; Schimansky, C.: Wissenschaftliche Hypothesen zur Erklärung der Ursachen der neuartigen Waldschäden. In: Papke, H.E., Krahl-Urban, B.; Peters, K.; Schimansky, C. (ed): Waldschäden. KFA Jülich, Jülich 1987
37 Stern, A.C. et al.: Fundamentals of air pollution. New York: Academic Press 1984
38 Wichmann, H.-E.: Umweltepidemiologie – wozu?, Staub Reinhalt. Luft 48 (1988) 175–176
39 Bundesministerium des Innern: Was sie schon immer wissen wollten. Luftreinhaltung. Stuttgart: W. Kohlhammer 1983
40 Ergebnisse der Epidemiologie des Lungenkrebses, Umweltbundesamt. Berichte 3/86. Berlin: Erich Schmidt 1986
41 Blot, W.J.; Frummeni, J.F. jr.: Passive smoking and lung cancer. JNCl, Vol. 77, No. 5, November 1986, p. 933–1000
42 Remmer, H.: Gefährdung durch Passivrauchen. Studie, Universität Tübingen 1988
43 Becker, N.; Frentzel-Beyme, R.; Wagner, G.: Krebsatlas der Bundesrepublik Deutschland, 2. Auflage. Berlin: Springer 1984
44 Klein, R.G.: Krebs durch die Atemluft – Fragestellungen und Probleme. In: Krebsforschung heute, S. 106–110. Darmstadt: Steinkopf 1986
45 Air quality guidelines for Europe. Copenhagen: World Health Organization, Regional Office for Europe. WHO regional publications, European series No. 23, ISBN 92-890-1114-9
46 Henschler, D. (Hrsg.): Gesundheitsschädliche Arbeitsstoffe, Toxikologisch-arbeitsmedizinische Begründung von MAK-Werten. Loseblattsammlung. Weinheim: VCH Verlagsgesellschaft 1987
47 Verein Deutscher Ingenieure: Guideline VDI 2310, part 1: Aim and Significance of the Guidelines Maximum Immission Values. Berlin: Beuth October 1988

48 Measuring Air Quality, The Pollutant Standards Index. EPA-451/K-94-001, U.S. Environmental Protection Agency, Office of Air Quality Planning and Standards, Research Triangle Park, NC, February 1994
49 Leithe, W.: Die Analyse der Luft und ihrer Verunreinigungen. Stuttgart: Wissenschaftliche Verlagsgesellschaft 1974
50 Verein Deutscher Ingenieure: Guideline VDI 2310, part 11: Maximum Immission Values Referring to Human Health. Maximum Immission Concentrations for Sulfur Dioxide. Berlin: Beuth August 1984
51 Der Bundesminister für Umwelt, Naturschutz und Reaktorsicherheit (Hrsg.): Auswirkungen der Luftverunreinigungen auf die menschliche Gesundheit. Informationsschrift, Bonn 1987
52 Verein Deutscher Ingenieure: Guideline VDI 2310, part 12: Maximum Immission Values Referring to Human Health, Maximum Immission Concentration for Nitrogen Dioxide. Berlin: Beuth June 1985
53 Berichte des Kohlenmonoxid-Kolloquiums der VDI-Kommission Reinhaltung der Luft 28.–29.10.1971. Staub Reinhalt. Luft 32 (1972) Nr. 4
54 Verein Deutscher Ingenieure: Guideline VDI 2310, part 15: Maximum Immission Values Referring to Human Health. Maximum Immission Concentration for Ozone (and Photochemical Oxidants). Berlin: Beuth April 1987
55 Umweltbundesamt: Luftqualitätskriterien für photochemische Oxidantien. Berichte 5/83. Berlin: Erich Schmidt 1983
56 Kappus, H.: Die Toxikologie des Schwefelwasserstoffs. Staub Reinhalt. Luft 39 (1979) Nr. 5, S. 153–155
57 Hettche, H.O.: Wirkungen von Luftverunreinigungen auf Menschen und Tiere. VDI-Bildungswerk BW 616, Düsseldorf 1966
58 Bundesgesundheitsamt, Bundesanstalt für Arbeitsschutz und Umweltbundesamt: Formaldehyd. Schriftenreihe des Bundesministers für Jugend, Familie und Gesundheit, Band 148. Stuttgart: W. Kohlhammer 1984
59 Cheremisinoff, P. N.; Young, R.A.: Industrial odor technology assesment. Ann Arbor: Ann Arbor Science Publishers 1975
60 Umweltbundesamt: Luftqualitätskriterien für Benzol. Berichte 6/82. Berlin: Erich Schmidt 1982
61 Seidenstücker, R.; Wölcke, U.: Krebserrregende Stoffe, Chemische Kanzerogene im Laboratorium – Struktur, Wirkungsweise und Maßnahmen beim Umgang. Nr. 3 Schriftenreihe Arbeitsschutz, Bundesanstalt für Arbeitsschutz und Unfallforschung Dortmund, 1979
62 Umweltbundesamt: Luftqualitätskriterien für ausgewählte polycyclische aromatische Kohlenwasserstoffe. Berichte 1/79. Berlin: Erich Schmidt 1979
63 Verein Deutscher Ingenieure: Luftverunreinigung durch polycyclische aromatische Kohlenwasserstoffe. VDI-Berichte 358. Düsseldorf: VDI 1979
64 Verein Deutscher Ingenieure: Guideline VDI 2310, part 2 (draft): Maximum Immission Values to Protect Vegetation. Maximum Immission Concentrations for Sulfur Dioxide. Berlin: Beuth August 1978
65 Verein Deutscher Ingenieure, Guideline VDI 2310, part 5 (draft): Maximum Immission Values to Protect Vegetation. Maximum Immission Concentration for Nitrogen Dioxide. Berlin: Beuth Verlag 1978
66 Verordnung der Landesregierung, des Ministeriums für Umwelt und des Innenministeriums zur Verhinderung schädlicher Umwelteinwirkungen bei austauscharmen Wetterlagen (Smog-Verordnung-Smog VO) vom 27.6.1988 Gesetzblatt für Baden-Württemberg 1988, S. 214 f
67 Zweiundzwanzigste Verordnung zur Durchführung des Bundes-Immissions-Schutzgesetzes (Verordnung über Immissionswerte - 22. BImSchV) vom 26. Oktober 1993. Deutsches Bundes-gesetzblatt I S. 1819, geändert durch Verordnung vom 27.5.1994, Bundesgesetzblatt I S. 1095
68 Clean Air Act Amendments of 1990, U.S. Code, vol. 42 sec. 7412 (b) (2), 1990
69 Verein Deutscher Ingenieure, Guideline VDI 2310, part 19: Maximum Immission Values Referring to Human Health. Maximum Immission Concentrations for Suspended Particulate Matter. Berlin: Beuth April 1992

5 Measuring Techniques for Recording Air Pollutants

5.1 General Criteria

5.1.1 Applications of Measuring Techniques

The choice of a measuring process for air pollutants depends on the substance to be measured, the properties it has, and the information to be gained from the measured values. Measurements must be carried out both at the location of formation of the air pollutants to determine emissions and at the location of effect to determine air quality. Pollutants occur in the different states: gaseous, liquid and solid. Accordingly, manifold measuring techniques are required. An overview is presented in Fig. 5.1.

Concentrations of air pollutants are primarily measured. *Mass flows* are also relevant for emissions. Measuring techniques for them will be dealt with later. For the subject of air quality, too, *mass flow* or *deposition measurements* are of interest. Measuring techniques for these must be tailored for this special purpose.

5.1.2 Discontinuous or Continuous Measurements

In measuring technology one must distinguish between discontinuous and continuous measuring methods. Apart from the question of the available technical equipment it mainly depends on the object of the measurements whether discontinuous single measurements are to be carried out or whether a continuously operating measuring device has to be employed. Whereas continuously operating measuring devices reflect the temporal profile of the measured quantity, discontinuous methods provide the measured quantity as a mean value over the sampling period. When using discontinuous methods knowledge on the temporal profiles of the air pollutants to be measured should be available so that sensible sampling periods can be set. The applicability of discontinuous single measurements and continuous measuring is compared in Fig. 5.2 by using various emission profiles as examples.

Ambient air pollutants also have fluctuating concentration profiles, as pollutant emissions change constantly along with atmospheric dispersion conditions. Here the measuring method to be applied is essentially determined by the possible effect of the air pollutants. For long-term effects it is sufficient to determine mean values which can be fixed quite well by a series of individual discontinuous measurements. To determine peak values, however, one must employ continuously operating measuring instruments, reflecting the temporal course of the concentrations as completely

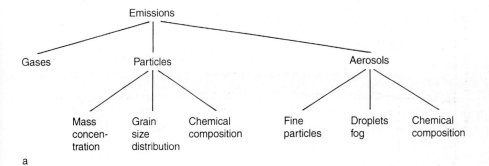

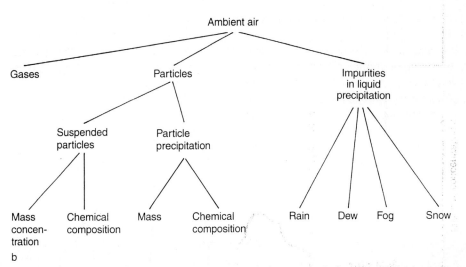

Fig. 5.1. Fields of application for measurement techniques, **a** emission control, **b** air quality control

as possible. E.g., conditions where concentrations are high cannot be predicted most of the time, as they depend on many factors.

Apart from the profiles of the measured quantities the use of continuous or discontinuous measuring procedures is determined by the technical possibilities of the methods themselves. The characteristics of the two measuring procedures are compared in the following:

Continuously operating measuring instruments

- automatic measuring instruments based on physical, chemo-physical or chemical measuring principles,
- constant recording and indication or electric output of the measured values,
- the instruments must be calibrated, e.g., with calibration gases or by comparative measurements with manual processes,
- automatic instruments for all measured components have not been developed and are therefore not available for all of them.

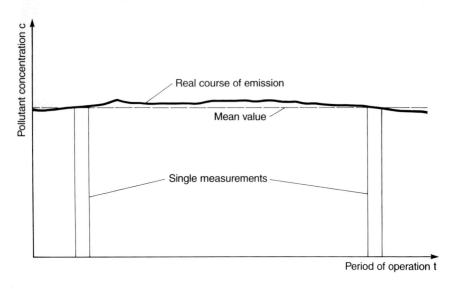

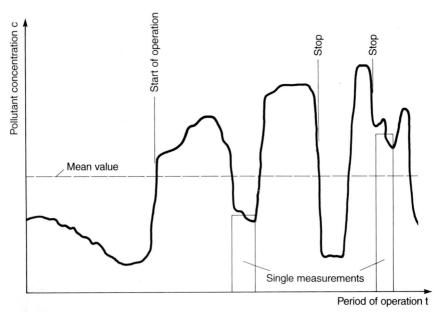

Fig. 5.2. Application of continuous and discontinuous measuring methods for different emission profiles. **a** approximately constant emission concentrations, e.g., of a continuously operated firing, emissions are recorded well by random or single measurements; **b** fluctuating emission concentrations due to the production plant's batch operation; actual concentrations can only be determined by continuous measurements

Discontinuous measuring methods

- mostly manual processes or automated discontinuous processes,
- generally, measurements are carried out in two stages:
  1. collecting of samples on location in different sampling devices: in collecting vessels, e.g., in gas sampling tubes or syringes, in absorbing solutions, on filters or by adsorption on solid substances, e.g., activated carbon,
  2. analysis of the samples in the laboratory,
- the samples taken can be analysed thoroughly in the laboratory so that it is also possible to determine substances which cannot be accessed by continuously operating measuring instruments,
- by accumulation over long sampling periods, very low concentrations to which continuously operating instruments almost do not respond can be detected,
- main disadvantage: the results measured are only available after a certain period of time after the samples were taken,
- measured quantities are frequently based on weight and volume determinations, e.g., wet-chemical methods. A calibration with external standards, e.g., with test gases, is not necessary in these cases. For this reason, these types of processes are used as *reference measuring methods*. E.g., the precise concentration of the test gases being used for the calibration of continuously working measuring instruments can be determined with wet-chemical reference measuring methods.

5.1.3 Physical and Chemical Measuring Principles

In *physical measuring methods* a specific physical property of the pollutant is made use of as quantity to be measured; in these methods the air sample does not change materially during its measurement. Specific physical properties of the substances to be investigated are applied to which other components of the sample do not contribute.

In *chemical measuring methods* the quantity to be measured is transformed into a condition with characteristic and measurable properties by a chemical reaction; during this chemical reaction the measured quantity changes.

Measuring processes based on a physical principle can generally be automated better for continuous processes, chemical methods usually being suitable for discontinuous measurements. Chemo-physical measuring principles are also applied.

An essential principle which is primarily applied in the continuous measurement of gaseous pollutants, is the *excitation of molecules* by adding energy. Excitation can be caused by exposure to radiation in different wave lengths, by generating high temperatures, e.g., via combustion, or by chemical reactions. Either the energy used for excitation or the energy released in another form is exploited for measurement. So physical quantities are measured, while excitation can also take place chemically.

There are still further methods of excitation, e.g., excitation by electric, magnetic or nuclear forces. Methods of this type can be used for laboratory analyses of air pollutant samples collected. They are used less frequently, however, for direct measuring.

Table 5.1 shows examples of different types of excitation and the measuring effects thereby applied, s. also Fig. 5.3.

Table 5.1. Different types of molecular excitation, measuring effects and measuring principles used in emission and air quality control

Molecular Excitation	Measuring Effect	Measuring Principles Used
Rotary spectrum – microwaves – infrared radiation (IR)	Specific absorption of the radiation or the microwaves IR-photometer,	
Vibrational spectra due to infrared radiation	Specific absorption of the radiation	e.g., non-dispersive or laser photometer
Electron spectra	Specific absorption of visible (VIS) or ultraviolet (UV) radiation;	UV and VIS photometer
	Resulting radiation emission: UV fluorescence Scattered light	Fluorescence measurement Raman spectroscopy
Chemical reaction	Intensity of the resulting light emission	Chemiluminescence
Flames	Specific light emission Formation of ions	Flame photometry Flame ionisation

5.1.4 Different Requirements for Emission and Air Quality Measurements

The different requirements for emission and air quality measurements are compared in table 5.2. The main difference lies in the varying ranges of concentration. Gener-

Table 5.2. Different requirements in emission and air quality measurements

	Emission	Air Quality
Concentration range		Factor 1000···100000 lower, low detection thresholds, small drifts
Interfering parameters	High concentrations e.g., in water vapor, particulates, CO_2, aggressive gases Work-intensive sampling and gas preparation, high selectivity of measuring methods, frequent maintenance	lower
Pollutant flows	Determination of carrier gas volume flows with auxiliary parameters	Volume flows: work-intensive meteorological measurements

ally, air quality measuring devices must have a considerably lower detection limit than emission devices. In contrast to this, emission measurements often require time- and work-intensive sample preparation.

5.1.5 Emission Components to be Recorded

Using exhaust gases from combustion processes as an example, table 5.3 lists the most important emission components to be recorded with the help of measuring techniques.

Table 5.3. Tasks and present state of technology for the detection of pollutants emitted by combustion processes

Pollutant	Task of Measurements				State of Technology			
	Continuous	Discontinuous	Prescrib.[a]	Interest, Research	Commercially avail. instruments	Manual methods	Laboratory methods	Sophisticated research methods
Leading parameters for combustion								
O_2	×		×		×			
CO_2	×	×	×		×			
H_2O	(×)	×	(×)		(×)	×		
(CO)	×		×		×			
(Soot)	×	×	×		×			
H_2SO_4-dewpoint	×	×			×		×	
Products of incomplete combustion								
CO	×	×	×		×	×		
Hydrocarbons Alkanes Alkenes Alkines		× (sum)	×	× (sum)	×	× (sum)	×	×
Aromatics		×	(×)	×		×	×	×
Oxidized hydrocarbons, e.g., aldehydes		×	×		×	×	×	
Odorous substances		×	(×)	×		×		
Soot	×	×	×		×			
Sulfur compounds								
SO_2	×	×	×		×	×		
SO_3		×	×		×		(×)	
H_2S	(×)	×	×	(×)	×			
Nitrogen oxides								
NO_x (NO+NO_2)	×	×	×		×	×		
single NO, NO_2	×	×	×		×	×		
NH_3	×	×	×	×	×	×	(×)	

Table 5.3. (continued)

Pollutant	Task of Measurements				State of Technology			
	Continuous	Discontinuous	Prescrib.[a]	Interest Research	Commercially avail. instruments	Manual methods	Laboratory methods	Sophisticated research methods
Other inorganic gases								
Cl$_2$ or HCl	(×)	×	×	×	(×)	×	×	
F$_2$ or HF	(×)	×	×	×	(×)	×	×	
Particulate matter content of exhaust gases								
Total part. matter	×	×	×		×	×		
Particle size distribution, aerosols		×	(×)	×	(×)	×	×	
Particulate components								
Sulfate, nitrate, chloride, fluoride			×	(×)	×		×	×	(×)
Heavy metals								
E.g., V, Ni, Cd, Hg, Pb		×	(×)	×		×	×	×
Organic components		×	(×)	×		×	×	×

[a] Different plants are subject to different regulations relative to measuring technology

5.1.6 Ambient Air Components to be Recorded

Table 5.4 lists those ambient air components which, on the one hand, have been released into the air as emissions and which, on the other hand, can be formed as secondary pollutants via chemical transformations in the atmosphere and for which measuring techniques are to be provided.

5.2 Measuring Methods for Gaseous Pollutants

The the most important measuring principles for gaseous air pollutants are described in the following paragraphs.

An overview of the most important measuring methods and the gases to be detected in each case is shown in table 5.5.

Table 5.4. Tasks and state of technology for air quality control

Pollutant	Task of measurements				State of technology			
	Continuous	Discontinuous	In-measur. grids	Special cases, Research	Commercially avail. instruments	Manual methods Research	Prod. in lab.	Sophisticated research methods
Classic pollutants								
SO_2	×	×	×		×	×		
NO	×		×		×			
NO_2	×	×	×		×	×		
O_3	×	×	×		×	×		
CO	×	(×)	×		×	×		
C_nH_m (total, methane-free)	×		×	×	×	×		×
CO_2	×		×	×	×	×		
Suspended particulate matter	×	×		×	×	×		×
Sedimented particulate matter		×	×		×	×		
Air-chemically formed gases, special gases								
H_2S	(×)	×		×	(×)	×		
NH_3	(×)	×		×	(×)	×	×	
HNO_3 (gaseous)		×		×	(×)	×		×
NO_3, N_2O	(×)	×		×	(×)			×
OH	×	×		×	(×)			×
HO_2, RO_2	×	×		×	(×)			×
HCHO	×	×		×	(×)			×
higher aldehydes		×		×	(×)			×
PAN	×	×		×	(×)			×
CH_4 C_nH_m	×	(×)	(×)	×	(×)			×
Individual components		×	(×)	×	×	×	×	×
Precipitation								
Amount of rain	×	×	×		×	×		
Duration of rain	×	×	×		×			
Snow		×	×			×		
Fog		(×)		×		×		×
Dew		(×)		×		×		×
Components:								
pH value	×	×	(×)		×	×		
conductivity	×	×	(×)		×	×		
Sulfate, nitrate, chloride, fluoride, ammonia	×	×	×		×	×	×	

Table 5.4. (continued)

Pollutant	Task of measurements				State of technology			
	Con-tinuous	Discon-tinuous	In-measur. grids	Special cases, Re-search	Commer-cially avail. struments	Manual methods Re-search	Prod. in lab.	Sophisticated research methods
Metals:								
alkali, alkaline earth		×	(×)	×	×	×	×	
Heavy Metals		×	(×)	×	×	×	×	
Particulate components								
Salts		×		×	×	×	×	
Metals		×		×	×	×	×	
Org. substances		×		×	×	×	×	

Table 5.5. Overview of the most important measuring techniques for gaseous air pollutants

Measuring technique	Emission	Air quality	Con-tinuous	Discon-tinuous	Examples of gases to be detected
Radiation absorption in the infrared range (IR Photometer)	×	only CO	×		CO, CO_2, SO_2, NO, NH_3, H_2O, CH_4, C_2H_2, C_2H_4, C_2H_6, a.s.o.
Radiation absorption in the ultraviolet range (UV photometer)	×	×	×		O_3, NO, NO_2, SO_2, Cl_2, H_2S, $HCHO$, (NH_3)
Radiation absorption in the visible range	×		×		NO_2, Cl_2
UV fluorescence	×	×	×		SO_2
Chemilumin-escence	×	×	×		NO, $NO+NO_2$, O_3
Flame photometry	×	×			Total sulfur or SO_2
Flame ionization	×	×	×		Total hydrocarbons
Conductometry (elect. con-ductivity)	×	×	×		SO_2, H_2S, CO_2, Cl_2, HCl
Amperometry		(×)	×		NO_2, H_2S, O_2, CO, SO_2
Coulometry	×	×	×		SO_2, Cl_2
Potentiometry	×	×	×		pH of solutions, HF, HCl, O_2
Paramagnetism	×		×		O_2
Heat conductivity	×		×		CO_2, H_2
Gas chromato-graphy (GC)	×	×		×	Organic components, individual determination
High-performance liquid chroma-tography (HPLC)	×	×		×	Organic components, individual detemination, ions in solutions

Table 5.5. (continued)

Measuring technique	Emission	Air quality	Con-tinuous	Discon-tinuous	Examples of gases to be detected
Wet-chemical methods					
Colorimetry	(×)	×	(×)	×	SO_2, NO_2, F^-, HCl, Cl_2, H_2S, NH_3, O_3, $HCHO$
Titration	×		×		SO_2, SO_3, $NO+NO_2$, CO, Cl_2, H_2S, NH_3

5.2.1 Photometry

Photometry uses the absorption of infrared (IR), visible (VIS) or ultraviolet (UV) radiation by the gases as measuring effect. Wave length ranges are:

IR: 1 000... 10 000 nm,
VIS: 400... 800 nm,
UV: approx. 200... 400 nm.

In the visible and UV range shell electrons are excited by radiation, in the IR range predominantly molecule vibrations but also rotations are. During this, the gases absorb energy in certain wave length ranges (absorption bands). The loss of radiation intensity caused by this is measurable.

Fig. 5.3 shows the types of excitation for the different wave length ranges as well as the "absorption bands" of some gases in the IR range (s. also [1-3]).

A photometer (s. Fig. 5.4) consists of a radiation source whose focused beam of light falls through a cell on a radiation detector; the latter transforms the beam into an electric signal of proportional intensity. The loss of radiation intensity due to the absorption of the measuring gas is – when frame conditions are constant – a measure for its concentration.

The interrelationships of radiation absorption are described by the Lambert-Beer law:

$$I = I_0 \cdot e^{-\varepsilon \cdot \varrho \cdot l}. \tag{5.1}$$

When pressure and temperature are constant the following applies:

$$I = I_0 \cdot e^{-\varepsilon \cdot c \cdot l}. \tag{5.2}$$

Transformed, the following results:

$$\frac{I}{I_0} = T = T, \tag{5.3}$$

$$E = \ln \frac{1}{T} \tag{5.4}$$

$$\frac{1}{T} = \frac{I_0}{I} = e^{\varepsilon c l} \tag{5.5}$$

$$E = \ln \frac{I_0}{I} = \varepsilon c l; \tag{5.6}$$

where I_0 stands for intensity of entering radiation (intensity of light for the reference), I for intensity of exiting radiation (intensity of light for the sample), T for transmis-

226   5 Measuring Techniques for Recording Air Pollutants

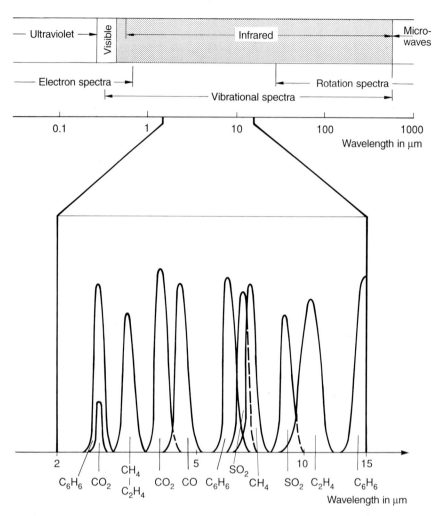

Fig. 5.3. Types of excitation in the individual wave length ranges and selected IR absorption bands

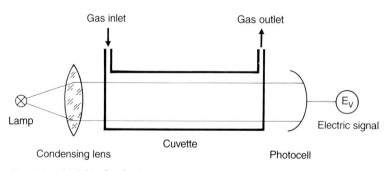

Fig. 5.4. Principle of a photometer

sion, ε for the extinction coefficient (dependent on wave-length), ρ for gas density, c for the concentration of gas or pollutant, l for optical pathlength of the cell measurement, E for extinction (non-dimensionalized) of the absorbing substance inverse logarithm of the transmission T.

5.2.1.1 IR Photometer

In contrast to diatomic elementary gases with symmetrical electron arrangement in the molecule all heteroatomic gases are somewhat dipolar in character. This dipole character is the reason why the molecule, when exposed to IR radiation, is set vibrating and thus absorbs radiation. Therefore, depending on their molecular structure heteroatomic gases have absorption bands of varying strengths which are separated from each other relatively well. The IR absorption bands of some gases are shown as examples in Fig. 5.3. The relatively wide bands are the result of vibrating molecules. If there is a sufficiently high resolution a fine structure becomes visible in the bands which results from the excitation of rotation.

Most IR photometers function as so-called non-dispersive instruments (NDIR), i.e., radiation is emitted in the entire IR range; there is no spectral splitting of th-e IR radiation emitted by the radiation source. Selectivity is achieved by installing a radiation detector filled with the component to be measured. This type of detector is possible only in the IR range, as the lifespan of the molecules excited by IR radiation is so long that the excitation energy can be released via molecule collisions as thermal energy.

Extinction E as a measure of the radiation absorption of a gas (or also of a liquid) is thus dependent on the properties of the gas (extinction coefficient ε), on the concentration c and on the optical pathlength l which the beam of light must pass through. If ambient conditions are constant, ε for a gas is constant. If l is also kept constant, then extinction E is directly dependent on the concentration of the gas to be measured.

In practice it is not sufficient to form the logarithm of the ratio of intensity I_0 in and intensity I out and to thus determine extinction E. Even without the presence of the component to be measured the instruments absorb radiation, e.g., via optical windows and the gases to be investigated. Thus, even without the presence of the component to be measured, radiation in is not equal to radiation I_0 out. This blank absorption must generally be determined experimentally.

For small extinctions (up to 1) the calibration curve is usually a linear when pressure is constant, s. Fig. 5.5. With increasing concentrations and uniform cuvette length l extinction gradually undergoes a saturation (curve b). Usually calibration curves are determined experimentally.

So far, single-beam photometers have been treated. In this type of photometry fluctuations of radiation intensity directly influence the stability of zero point. To compensate for the instability of the zero point one either operates with a reference beam through a blank cell (dual beam principle), or zero gas and the gas to be measured are fed through the same cell alternately, or the measuring beam and a reference beam of a neighboring wave length are sent alternately through one and the same

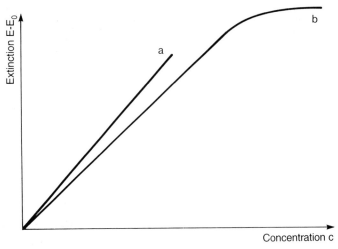

Fig. 5.5. Correlation between extinction and concentration (calibration curve); ε, σ and l are constant. *a* linear correlation at low extinctions; *b* saturation at high extinctions and concentrations

cell. The unknown measured component may not absorb any radiation from the neighboring wave length. What is measured is always the relationship of intensity between measuring beam and reference beam. Photometers operating with only one cell have the advantage that not only fluctuations of source radiation intensity but also soiled cell windows and in part even interferences by other gases are compensated for. To raise sensitivity and zero line stability, the radiation is usually modulated with chopper wheels. In this case the signals appear as alternating voltage after the detectors and can easily be amplified [4].

Nowadays instruments operating with infrared and ultraviolet radiation have become well established in trace gas analyses. Instruments operating with visible light have not been able to prevail as their uses remain restricted to few colored gases such as NO_2 (brown), chlorine (green) and other halogens, and their absorption coefficients are not very large. Photometers operating with visible light are relatively insensitive and unsuitable for air quality measurements unless the atmospheric air masses are directly passed through using very long path lengths (s. sect. 5.2.1.3).

The following sections will deal more closely with infrared and ultraviolet photometry and its use in trace gas analyses.

The arrangement of an IR gas analyzer is shown in Fig. 5.6.

Usually electrically heated filaments serve as radiation source. Radiation which is modulated by a rotating chopper wheel is passed through the measuring and reference cell. To reduce the influence of interfering gases whose absorption bands slightly overlap those of the measured component (s. Fig. 5.3), filter cells filled with the interfering component are generally placed before the measuring and reference cell in the radiation path. In this way, only the radiation part which corresponds to the interfering components is absorbed in advance. None of the different absorption bands of the measured components are cut off. In combustion gases the main interfering component is CO_2; therefore the filter cells are filled with CO_2.

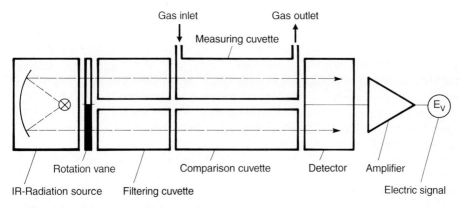

Fig. 5.6. Diagram of a non-dispersive IR gas analyzer

The radiation detector is placed behind the cells. It consists of two chambers, both of which are filled with the gas to be measured. The measuring effect results from the fact that when the unknown gas to be measured is present in the measuring cell, a pre-absorption of the IR radiation takes place there. Due to this less radiation reaches one chamber of the receiver than the other. Behind the reference cell this leads to varying degrees of warming and thus to a varying increase in pressure in the two chambers of the detector. The increase in pressure takes place periodically at the frequency of the chopper wheel which modulates the IR radiation. The amplitude of this pressure fluctuation depends on the extent of the pre-absorption of IR radiation in the measurement cell and is therefore a measure of the concentration of the gas to be measured. The individual types of instruments essentially differ in the construction of the detector which transforms the pressure fluctuations into electric signals. Thus, the classical detector type consists of two chambers separated by a very thin metal membrane. Together with a fixed electrode the metal membrane forms an electric condenser which changes its capacity constantly with oscillation of the membrane. The electric signal is received from this [5, 6]. Other instruments work with microflow sensors placed between the two detector chambers [7, 8]. These microflow sensors are electrically heated filaments and are part of a Wheatstone bridge. Here, these sensors transform the different pressure fluctuations in the detector chambers into electric signals.

NDIR instruments are primarily used for emission measurements; analyzers are mainly suitable for the determination of the gases CO, CO_2, NO, SO_2, H_2O, CH_4, C_2H_6 and many other hydrocarbons. For CO and CO_2, NDIR photometry is the most commonly used measuring technique, which is also unrivalled in its application for the measurement of these gases in the ambient air range.

Infrared gas analyzers are being constantly updated in their development. There is a trend towards smaller cells in the instruments, and with this towards shorter response times, less sensitivity to vibrations und higher stability of the signal. E.g., the latest development in the field of CO measurement is the so-called in-line measurement, i.e., an infrared measuring path is inserted directly into the exhaust gas channel or into the stack and radiation absorption is then measured on location without having to suction off a part of the exhaust gases and passing it into the measuring instrument [9].

For dispersive IR analyzers mention must be made of an instrument for the measurement of different organic components in which the wavelength of the emitted radiation can be adjusted by a system of filters: Miran [10]. The lower sensitivity which results – in contrast to NDIR photometry – from the fact that only one absorption band at a time is used for the measurement is compensated for by the use of extra long cells (up to 20 m, "White Cell"). In a minimum of space the long path is achieved by multiple reflection [10].

Nowadays, instruments with tunable IR lasers are used for multicomponent measurements [11, 12]. As these instruments are highly complex and correspondingly expensive, apart from laboratory and research applications, they are not yet available for general emission and air quality measurements. Instruments for emission measurements derived from laboratory equipment are now being offered which are based on Fourier transformation spectroscopy [13].

5.2.1.2 UV Photometer

A diagram of the UV absorption spectra of some air polluting gases is shown in Fig. 5.7. It is to be seen that there are gases with relatively narrow UV absorption bands,

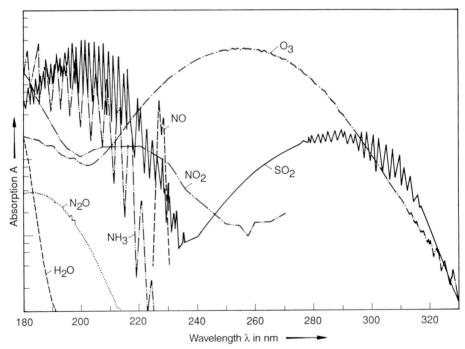

Fig. 5.7. UV absorption spectra of some air polluting gases [14]; absorption A versus wavelength λ

e.g., NO and NH_3. Other gases, however, absorb UV radiation widebandedly; these wide bands, however, may contain a finer structure. To be able to obtain selective measurements certain strategems are to be applied. A first selection is made by choosing the right UV lamp. Depending on their metal or gas fillings UV lamps have certain emission spectra. Therefore one uses lamps with emission bands which best correspond to the absorption bands of the measuring components or which lie within the absorption bands as clearly defined lines. However, gases absorbing widebandedly are still an interfering factor. Besides soiled cells and lamps with fluctuating intensities, they also lead to high and fluctuating reference values I_0 of the radiation and thus to high blank values of extinction. There are different possibilities of compensating for the blank extinctions and of receiving corrected reference radiation intensity I_0; two of them will be outlined using the example of UV-photometric measurement of NO and O_3.

UV absorption measurement with blank value compensation by wavelength comparison

A UV absorption measuring instrument for NO was developed whose lay-out is shown schematically in Fig. 5.8 [14, 15].

In a hollow cathode lamp filled with nitrogen and oxygen at reduced pressure excited NO molecules are formed in an electrical discharge. The energy of the excited molecules is dissipated by emission of characteristic luminescence radiation. The source of radiation is selective; it produces an emission range which corresponds precisely to the absorption range of NO in the measuring cell. This is called resonance absorption. One peculiarity of the radiation excited by electrical discharge is that two groups of NO-specific lines are emitted, i.e.:

1. "cold" emission lines – this is the group absorbed by the NO to be determined in the measuring cell (measuring radiation),
2. "hot" emission lines – that group of radiation showing lines in the neighboring range and meeting the detector not influenced by NO (reference radiation).

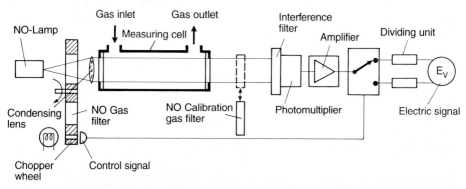

Fig. 5.8. Diagram of a UV gas analyzer for the detection of NO with blind value compensation by wavelength comparison [15]

The radiation is modulated by a chopper wheel and passed through the measuring cell via a condensing lens. It reaches the radiation detector, a photomultiplier, via an interference filter where interfering radiation is removed. If NO is present in the measuring cell, then the radiation is reduced by resonance absorption (extinction E) according to the Lambert-Beer law.

In this measuring technique the blank value of extinction E_0 is compensated for by alternately setting the chopper wheel to a position where all radiation (hot and cold emission lines) is passed through and to a position with a gas filter. The gas filter contains NO in high concentrations which completely absorbs the cold emission lines. The hot emission lines, however, which are in the immediate neighborhood range, pass through it as reference radiation. Just like the measuring radiation they are influenced by the in-line optics, by the cell windows but mainly by the wide-banded interfering components to produce the intensity reference value I_0 at the photomultiplier. The photomultiplier's signals are converted by a division unit controlled by the chopper wheel ($\ln\frac{I_0}{I}=E$), so that the signal is shown being proportional to the NO concentration. Intensity fluctuations of the lamp are also compensated for in this manner. This blank value compensation of a neighboring wave length is called a wave length comparison [4]. For calibration purposes a defined NO calibration gas filter can be inserted into the radiation path either manually or by automatic control.

The main difficulty with this instrument was to develop sufficiently stable UV lamps for continuous use, something which has obviously been successfully achieved subsequently [14].

UV absorption measurement with blank value compensation by substance comparison

UV photometers have been developed for O_3 measurements which are currently in widespread use in air quality measuring networks. The underlying principle is shown in Fig. 5.9.

The UV lamp, a mercury low-pressure lamp (high absorption by O_3 with the mercury resonance line at 253.7 nm, s. Fig. 5.7), sends its radiation unmodulatedly as constant light through the measuring cell to the radiation detector, a photomultiplier. Controlled by a solenoid valve, the air to be measured sucked in by the pump alternately enters the measuring cell either directly or via the converter. The converter contains several sieves covered with manganite (MnO_2). Its task is to quantitatively remove O_3 from the air to be measured, but to let all other gases, which possibly absorb UV radiation wide-bandedly, pass.

The solenoid valve switches, e.g., every 10 seconds. While air cleaned of O_3 reaches the measuring cell via the converter, reference intensity I_0 is measured. The intensity I measured during the next cycle is related to the output intensity I_0 and then delogarithmatized. The measuring signal is thus proportional to the O_3 concentration. Lamp aging, cell polluting and interfering gases are in this way compensated for. However, the lamp may not decrease in its intensity too much, otherwise one is be-

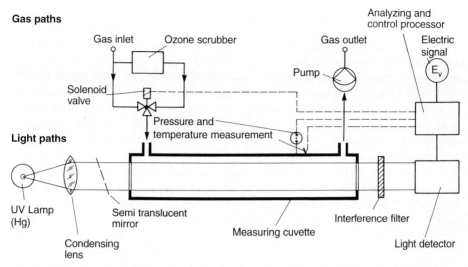

Fig. 5.9. Diagram of a UV gas analyzer for the detection of O_3 with blind value compensation by substance comparison [16, 17]

yond the valid range of the calibation curve. For this purpose the lamp is checked by a control detector which signalizes a fault when intensity gets too low. Radiation absorption in the measuring cell depends on O_3 concentration and measuring gas density. To eliminate the dependency on gas density, a pressure and temperature compensation has been added.

As blank value compensation is, in this case, carried out at the same wave length but with a different gas (same gas but with O_3 removed) one speaks of a substance comparison [4].

The precise functioning of this UV photometer succeeds or fails with the converter's efficiency. If it leaks O_3, the indicated value is too low; if it absorbs interfering gases, then the indicated value is too large. Due to the converter's sensitivity such an O_3 measuring instrument is only suitable for use in relatively pure air (air quality measurements). In furnace exhaust gases (e.g., for monitoring O_3 formation in electric precipitators) the converter would be instantly damaged.

UV photometry is being increasingly used for trace gas analysis and is constantly being further developed. E.g., there is an SO_2 emission measuring instrument working on the principle of UV absorption [8]. For analyzing formaldehyde emissions, e.g, UV photometry is presently being developed, with the greatest problems here caused by wide-bandedly absorbing gases [19-21].

5.2.1.3 Long-Path Photometry

For long-path photometry highly dispersive sources of radiation, e.g., tunable lasers, can be utilized. For this, radiation is only emitted in the wavelength range where the absorption bands of the required gases are situated. Measuring paths can be several kilometers in length here. The instruments serve for the measurement of mean values over extended sources such as refineries, chemical factories, other industrial complexes or entire cities. E.g., in Vienna NO_2 concentrations were measured over a distance of 3 km with a long-path photometer [22].

Another possibility is to emit light with a lamp in the widest possible wavelength range. At the detector site which can be located at distances of hundreds of meters up to several kilometers radiation absorption of different wave lengths can be measured with a tunable monochromator. In this way concentrations of trace gases and even radicals (OH radicals, HCHO, HNO_2, N_2O_5 etc.) which are normally inaccessible for any measuring technique have been measured in Germany between the Feldberg and the Schauinsland (path distance 8 km) without any direct contact [23]. A new, commercially available instrument for noxious gas components such as SO_2, NO, NO_2, O_3, formaldehyde (HCHO) and with some limitations for aromatic hydrocarbons such as toluene (C_7H_8) and p-xylene (C_8H_{10}) operates according to the same principle [24]. In this process called DOAS (differential optical absorption spectroscopy) visible light and UV radiation are emitted, detected after a certain distance (300 m to several km) and then transmitted to an evaluation device via an optical glass fiber cable. A very fine spectral decomposition of the received light takes place now at differing wavelength ranges. The evaluation of the received spectra is carried out by comparison with stored ones with the help of modern PC (personal computer) technology. By subtracting the spectra of the different gases the process becomes selective [24, 25]. This integrally measuring method offers advantages where results of point measurements are strongly influenced by local flows, e.g., in streets with heavy traffic. In addition there is the advantage that several components can be registered by one instrument.

Another method for local detection of air-polluting substances over longer distances uses the scattering of laser light on polluting particulate (small suspended particles of aerosols or pollutant molecules) as measuring effect. Distance and particle density of a pollutant cloud or a smoke plume can be calculated from travel time and intensity of the backscattered laser light impulses (Laser-Lidar). Gaseous pollutants can be analyzed qualitatively and quantitatively by evaluating the Raman-spectrum or fluorescence spectrum of the light backscattered by them [26].

5.2.2 Colorimetry

In colorimetry the color intensity of an absorbing solution, which has entered a chemical color reaction with the gaseous component of interest, is measured. The measuring gas to be investigated must be brought into close contact with the smallest possible amount of a reaction solution specific to the pollutant of interest, so that the chemical reaction can take place as completely as possible. For this, the gas sample is passed through washing bottles with fritted disks containing the reaction solution, s. Fig. 5.17c. Within certain limits the resulting coloring obeys the Lambert-Beer law, i.e., color intensity is proportional to the concentration of the measuring component of interest in the reaction solution. Color intensity is measured with a photometer. Colorimetric measuring processes have been developed for numerous air polluting gases and are described in VDI guidelines (s. sect. 5.2.12). Many color reactions are highly sensitive, thus being suitable for detecting low concentrations, and they react mostly selectively, i.e., the reaction solutions are insensitive to other interfering gases.

Generally, colorimetric measuring methods are discontinuous samplings followed by evaluation in the laboratory. But the process of the gas-fluid reaction and its subsequent photometric measurement has also been achieved with automatic measuring instruments.

5.2.3 UV Fluorescence

UV fluorescence as a measuring technique is related to photometry. The measuring gas is also exposed to radiation. However, it is not the radiation absorption which is measured but a luminous phenomenon (fluorescence) which is caused by the excitation of molecules through UV radiation of a certain wavelength. The measuring principle is, e.g., applied in SO_2 ambient air measurement, s. Fig. 5.10.

The air sample is exposed to UV radiation in the wavelength range of 190-320 nm. If present, SO_2 gives off a fluorescence radiation of 320-380 nm. Due to an interference filter only a radiation of this wavelength is recorded by the detector (photomultiplier); thus, the measuring principle is strictly selective. The higher the SO_2 concentration, the greater the fluorescence.

One interference in this measuring technique is that other components can absorb the energy of the excited SO_2 molecules, thus reducing the fluorescence yield. This interfering effect is known particularly from water vapor and hydrocarbons. By interposing a permeation gas exchanger an elimination of the interfering components

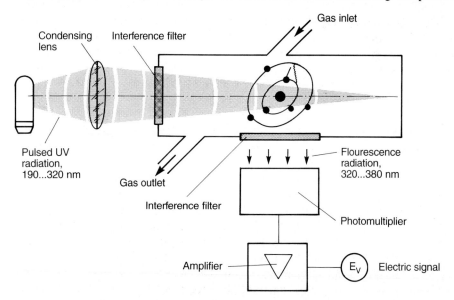

Fig. 5.10. Principle of UV fluorescence measurement (acc. to [27])

from the measuring gas is attempted. Measuring instruments operating according to this principle are used for both SO_2 air quality measurements as well as for SO_2 emission measurements [27, 28]. In emission measurements the interferences are higher due to the higher concentrations of the interfering components. Air quality measurement instruments operate very stably as far as zero point and sensitivity drift are concerned and as long as the intensity of the UV lamp remains constant.

5.2.4 Chemiluminescence

Chemiluminescence is related to UV fluorescence. The difference between the two is that in chemiluminescence molecules are not excited by UV radiation, but are excited by a chemical reaction. Thus, the measuring principle is a chemo-physical one. The intensity of the radiation created is a measure for the concentration of the reacting gas in a mixture of gases, if the external conditions (pressure, temperature and volume flow of the measuring gas) are kept constant. Just as is the case in UV fluorescence, the radiation created is recorded by a photomultiplier acting as radiation detector and is transformed into an electric signal. This method is used mainly for measuring NO, NO+NO_2 (i.e., NO_x) and O_3. It is equally suitable for emission and air quality measurements, whereby O_3 is usually investigated as ambient air pollution.

5.2.4.1 NO_x Measurement

When measuring NO O_3 is required as an auxiliary gas. The following reactions take place in the measuring instrument:

$$NO + O_3 \rightarrow NO_2^* + O_2 \tag{5.7}$$

$$NO + O_3 \rightarrow NO_2 + O_2 \tag{5.8}$$

$$NO_2^* \rightarrow NO_2 + hv \tag{5.9}$$

$$NO_2^* + M \rightarrow NO_2 + M, \tag{5.10}$$

where hv is emitted light energy at 600-3200 nm with a radiation maximum of 1200 nm, M triple collision partner which absorbs energy but otherwise does not take part in the reaction.

The reactions proceed at different reaction velocities. A constant part of the NO (approx. 10 % [4]) reacts with ozone to an excited NO_2^* molecule. When the excited NO_2^* molecule changes into its basic state, according to (5.9) light energy hv (chemiluminescence) is emitted whose intensity can be used as a measuring effect. Part of the excited NO_2^* molecules, however, give off their energy to collision partner M and in doing so change into their basic state. This radiation-free energy discharge is called quenching. The lower the pressure und with it the gas density in the measuring chamber, the smaller the probability becomes that an NO_2^* molecule can discharge its energy via a collision partner without giving off radiation.

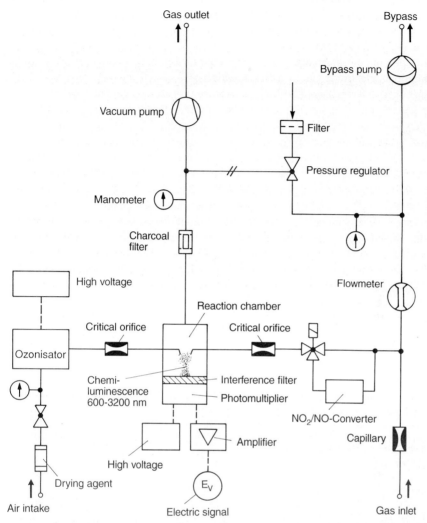

Fig. 5.11. Diagram of a chemiluminescence analyzer for the detection of NO and NO_x (NO + NO_2)

Fig. 5.11 shows the diagram of a chemiluminescence instrument for the determination of NO or of NO+NO_2 (i.e., NO_x).

The ozone required for the reaction is produced via an electric gas discharge in an air or oxygen flow. During this, approx. 2 % ozone is formed which must be abundantly available for the reaction with the NO in the reaction chamber. When NO concentrations are high, pure oxygen is required for a greater ozone yield in the ozonizator. To sustain low pressure in the reaction chamber the major part of the measuring gas is conducted through a bypass system with only a small part being conducted into the reaction chamber via fine capillaries or a critical nozzle. If merely the gas required by the reaction chamber were sucked in by the measuring instrument, then the small volume flows in the intake lines would cause too great a delay in response. In addition, the bypass system upholds constant pressure conditions just before the capillary or the critical nozzle.

After passing an interference filter the radiation emitted by the reaction is transformed into an electric signal by a photomultiplier.

NO_2 can also be measured with this technique. For this, it must be reduced to NO before the chemiluminescence reaction takes place. This is done in a converter by reduction with hot metals, e.g., molybdenum. By setting up and controlling the the solenoid valve SV efficiently, both the total nitrogen oxide concentration (NO + NO_2) as well as the NO concentration can be measured. From this, NO_2 concentration is determined by subtraction. There are also devices with two reaction chambers of which one is supplied directly with measuring gas and the other one via the converter. In this way NO and NO + NO_2 can be measured simultaneously.

There are several manufacturers of chemiluminescence devices for NO_x measurement. Depending on their design the measuring instruments are suitable for air quality measurements down to measuring ranges of 0 – 1000 ppt NO_x and for emission measurements up to 10000 ppm NO_x. The indication is very linear and owing to the specific reaction, there are generally only very slight interferences towards other gases.

5.2.4.2 O_3 Measurement

According to the same chemiluminescence reaction as in the NO measurement, ozone could be measured by its reaction with NO. A better and more inexpensive reaction partner for ozone, however, is ethene (C_2H_4):

$$O_3 + C_2H_4 \rightarrow O_2 + C_2H_4O^* \tag{5.11}$$

$$C_2H_4O^* \rightarrow C_2H_4O + h\nu \ (300 \ldots 600 \text{ nm}) \tag{5.12}$$

C_2H_4O, $H_2C\underset{O}{\overset{}{\diagdown\!\!\!\diagup}}CH_2$ Ehtylenoxid, Oxiran.

During this reaction chemiluminescence radiation is once again formed to be measured analogous to NO determination. The sole disadvantage of this ozone measuring technique is that ethene is required which is only available from a gas cylinder. As it is a flammable gas, this measuring technique is regarded with disfavor in air quality measuring stations and has given way increasingly to UV photometry. In the matter of interference and susceptibility to faults the chemiluminescence method is superior to UV photometry.

5.2.5 Flame Photometry

In flame-photometry atoms are excited in a flame and made to luminesce. The spectral line of the atom of interest is filtered out from the radiation of the flame via an interference filter and measured with a photomultiplier. In gas analyses this process is used mainly for sulfur measurements, but it is also suitable for measuring phosphorous compounds. In sulfur measurement, however, the flame-photometric effect is not based on an atom emission but on a recombining of sulfur atoms whereby excited S_2^* molecules are formed which pass into their basic state under a light emission of approx. 320 nm-460 nm. With an optical filter a wave length of 394 nm is chosen for sulfur detection [4].

The total sulfur content of the air, mainly H_2S and SO_2, is primarily measured. If individual compounds are to be identified, then single gases must be removed by absorption and adsorption filters prior to measuring. This process is distinguished by a high sensitivity (low detection limit!) and by a very brief response time. Therefore measuring devices working on this principle are used, e.g., for air quality measurements with aircraft [29]. Owing to the fact that hydrogen is required as an auxiliary gas for generating the flame inside the device the flame photometer is used less frequently in stationary air quality measuring stations. It is not common practice to use it for emission measurements as the concentrations to be measured are too high and there are too many interfering components (quenching).

5.2.6 Flame Ionization

Gases can be ionized more or less easily by the addition of energy. For gas analyses the ionization of organic molecules in flames (flame ionization) has gained the greatest significance. Ionization by radiation of radioactive substances in detectors, e.g., in gas chromatography, is also applied [30].

The so-called flame-ionization detector (FID) was originally developed for gas chromatography. Nowadays, it is also used as the most important measuring device for the continuous recording of organic substances in exhaust gases or in ambient air.

The measuring principle of the FID has been described many times, e.g., in [30-32]. It will be summarized here with the help of Fig. 5.12 .

The hydrogen flame burns out of a metal nozzle which simultaneously represents the negative electrode of an ionization chamber. The positive counter-electrode is fixed above the flame, e.g., as a ring. Between the two electrodes direct voltage is applied. The ion current is measured as a voltage drop above the resistor W. The measuring gas is added to the burning gas shortly before entering the burner nozzle. The air required for combustion flows in through a ring slot around the burner nozzle.

For stable measuring conditions it is essential that all gases – combustion gas, combustion air and measuring gas – are conducted into the flame in constant volume flows. For this, all gas flows are conducted via capillaries. Constant pressures before the capillaries ensure a constant flow. Sensitive pressure regulators for combustion gas and combustion air are used to achieve this fine-tuning. The measuring gas is pumped past the capillary in the bypass in a great volume flow. Pressure is kept constant by the back pressure regulator, so that a constant partial flow reaches the flame via the capillary. Most FIDs operate with overpressure, i.e., the measuring gas pump is located before the capillary.

To avoid condensation of the hydrocarbons to be measured almost all instruments can be heated to 150-200 °C. Heating includes the particle filter and the measuring

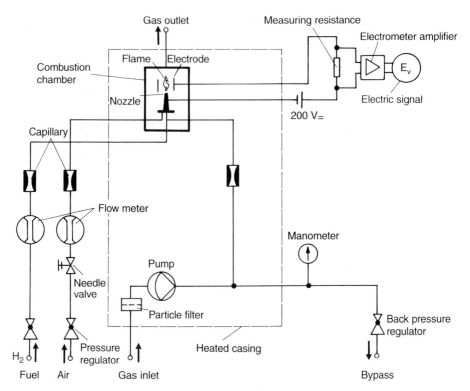

Fig. 5.12. Diagram of a flame ionization detector (FID)

gas pump; in most cases, particularly with warm exhaust gases, a heated sampling line is also used from measuring gas sampling to the measuring instrument.

Hydrocarbon compounds are oxidized in the flame with ions being formed as an intermediate product. In a certain range of the accelerating voltage the strength of the ionization current is in first approximation directly proportional to the amount of C atoms of the burned substance. Thus, an FID basically responds to all hydrocarbons and measures their total sum. Corresponding to the number of carbon atoms, larger molecules with many C atoms produce a higher signal than smaller molecules with a small number of C atoms. Ionization energy does not only stem from the flame's energy, but mainly from the oxidation energy of the carbon. Accordingly partially oxidized hydrocarbons provide a weak detector signal, completely oxidized hydrocarbons no signal at all; HCHO, CO and CO_2, e.g., are not detected.

If exhaust gases predominantly consist of mixtures of pure, i.e., non-oxidized or halogenated hydrocarbons, the FID provides a signal nearly proportional to the carbon mass content of the exhaust gas. With the help of special technical controlling, e.g., adding helium to burning gas, operating conditions can be stabilized and proportionality of the indication to the C content of the measuring gas be improved [33]. Flame ionization detectors are characterized by very short response times (<1 s).

FID is the standard measuring method for the determination of unburnt hydrocarbons in automobile exhaust gases and is used in test stands to determine HC emissions of the different driving cycles. Within certain limits the FID can also be used for total C monitoring which is mandatory in Germany for certain plants according to Technical Directive On Air Pollution Control [34]. Application possibilities and limitations are described in the VDI guideline 3481, part 6 [35]. The FID can also be used for monitoring volatile organic substances, e.g., solvents. It can be directly calibrated for the substances emitted in which case so-called response factors are used. The response factor shows the ratio of the FID's signal between the substance of interest and a reference substance (when both substances are of the same concentration). The calibration procedures for the determination of the response factors are described in detail in [36]. If information on the composition of the emitted organic substances is to be gained, then samples must be taken discontinuously and analyzed, e.g., gas-chromatographically. Knowing the composition alone, however, is of no use, if nothing is known about the temporal course. In batch processes, e.g., emissions can be subject to considerable fluctuations, s., e.g., [37]. In such cases a total HC measurement of the course with an FID is to be recommended with samples being taken at specific periods for a differentiated analysis.

5.2.7 Conductometry

In conductometric methods the gas to be investigated is brought into contact with a reaction solution specific for the measuring component, and the change in electrical conductivity of this solution is measured. It is a chemical measuring method. All gases causing a change in conductivity in a reaction solution can in principle be measured with such a method, e.g., SO_2, CO_2, H_2S, HCl, NH_3 and others. CO, e.g., can be oxidized to CO_2 by precombustion and can then also be measured via conductivity change [32, 38, 39].

Measuring a change in conductivity is the longest established continuous measuring method for the determination of SO_2 for both the fields of emission as well as air quality investigation. The reaction solution used is an aqueous hydrogen peroxide solution (H_2O_2) with which SO_2 reacts to form sulfuric acid (H_2SO_4):

$$H_2O_2 + SO_2 + 2H_2O \rightarrow H_2SO_4 + 2H_2O \rightarrow 2H_3O^+ + SO_4^{2-} . \tag{5.13}$$

In the aqueous solution the sulfuric acid formed occurs in a virtually completely dissociated form. The change in conductivity of the absorption solution caused by the ions formed is a measure of the SO_2 concentration. The change in conductivity is measured via an electrode measuring section which is part of a Wheatstone bridge.

The principle underlying a measuring instrument of this type is shown as a diagram in Fig. 5.13.

The measuring gas is sucked in by the membrane pump in the bypass. A smaller part of the gas flow is sucked out of the bypass into the measuring cell. The reaction solution is pumped from a reservoir into the reaction section where it mixes with the measuring gas. The SO_2 contained in the measuring gas reacts completely with the reaction solution. After the reaction the gas is separated from the fluid as the latter must be free of gas for the conductivity measurement. After the conductivity meas-

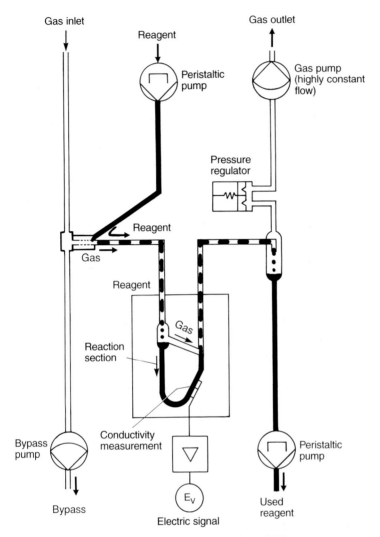

Fig. 5.13. Principle of a conductometric gas analyzer [40]

urement in the electrode measuring section, the fluid is sucked out of the measuring instrument together with the measuring gas. Constant gas and fluid flows are essential for an exact measurement.

With emission measurements it is sufficient to measure the conductivity of the solution after the reaction with SO_2.

For SO_2 air quality measurements, however, the method is somewhat more complicated. True, the measuring principle is analogous to the one for emission measurements, but the H_2O_2 solution is acidulated with H_2SO_4. In the measuring instrument the difference between the conductivities before and after the reaction with SO_2 is determined. Applying the conductivity method many different instruments have been used for SO_2 measurement [42-45].

Temperature directly influences the conductivity of the solution. Therefore, a thorough thermostatting of the measuring cells is essential. Problems frequently occur due to clogging of the fine capillaries, the formation of algae or soiling. Also, purity of the electrodes in the conductivity measuring section is of great importance. Flawless functioning of the conductivity measuring instruments therefore requires relatively high maintenance. When properly cared for, SO_2 conductivity measuring instruments can work very well [46]. They were, e.g., formerly used in all air quality measuring networks of Germany [47, 48], but they have been partly replaced by the physically operating UV fluorescence instruments nowadays.

5.2.8 Amperometry

Amperometry is the application of a galvanic current produced in an electrolyte after a reaction with the measuring component. The measuring cell represents a galvanic element which only supplies current via an external closing circuit if its electrolyte comes into contact with the gas of interest and in doing so produces ions. The ions migrate to the electrodes and charge them negatively or positively, so that a current can flow in the external circuit. What is important here is that the measuring component alone sets off the current-producing process and determines its intensity. Then the current intensity is proportional to the measuring component in the gas. Possible interfering components are generally removed by absorption filters before the measuring cell. The electrodes and the electrolyte are gradually exhausted. The electrochemical measuring cells must therefore be replaced from time to time.

Formerly there were cells with flowing electrolytes but these were never generally accepted. Nowadays, electrochemical cells are used in hand-held or portable measuring instruments, e.g., for spot sampling determination of O_2, CO and other components in workplaces and small furnaces, s., e.g., [51-54]. These instruments are much less expensive than, e.g, IR gas analyzers but they are not suitable for continuous measurements. They do not operate stably enough and would also wear out too fast in continuous operation.

5.2.9 Coulometry

Coulometry measures the quantity of electricity required for dissociating an electrolyte. The electrolyte is formed by introducing the gas component in question into a suitable solvent. For this, the air to be analyzed is continually pumped through a measuring cell containing the reagent solution. The component to be determined must be completely dissolved and must form the ions necessary for the electrolytic reaction. The electrolytic current caused by a voltage between two electrodes is directly proportional to the number of ions formed and thus to the measuring components' mass flow.

The best-known measuring instrument working on this measuring principle was a Philips SO_2 monitor for air quality measurement [55]. Its production was discontinued, however. Current instruments for measuring extremely low SO_2 concentrations work on a similar measuring principle developed by Rumpel using the polarographic-coulometric process. This polarographic-coulometric process is based on the following principle [4]:

If a very easily polarizable and an unpolarizable electrode of suitable potential are submerged into a suitable electrolytic solution, a current flows if a substance acting as depolarizer is present in the electrolyte. The unpolarized counterelectrode keeps its potential even if an external voltage is applied.

The set-up of the SO_2 measuring instrument developed by Rumpel is shown in Fig. 5.14. A very slowly flowing electrolytic solution (1 % H_2SO_4) is saturated with iodine in cell 1, then subsequently purified from possible iodide (J⁻) by electron emission in cell 2 at a polarizable Pt-electrode. This electrolyte solution then reaches cell 3, where a continuously controlled flow of measuring gas bubbles through it. If SO_2 is present in the measuring gas, then it is oxidized with iodine, with the iodine being reduced to iodide:

$$SO_2 + J_2 + 2H_2O \rightarrow SO_4^{2-} + 4H^+ + 2J^-. \tag{5.15}$$

The iodide ions act as depolarizer and again oxidize to iodine at the polarizable Pt-electrode (indicator electrode) of cell 3. The electrolytic current generated there is determined by the SO_2 concentration. The current measured between polarization electrode and non-polarizable reference electrode in cell 4, through which the solution subsequently flows, represents the measuring signal. An external cell voltage of 200mV keeps the indicator and reference electrode at a potential which is constant in reference to each other. After the solution has passed through the measuring cells, it is collected in a receptacle.

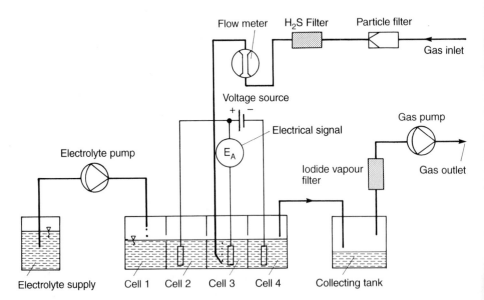

Fig. 5.14. Principle of a coulometric SO_2 gas analyzer [56, 57]

This measuring instrument was used by the German Federal Environmental Agency for detecting SO_2. It permits measurement of the frequently very low SO_2 concentrations at the clean air stations. It is said that its detection threshold is less than 1 µg/m³ SO_2 [56].

5.2.10 Potentiometry

If electric direct current is applied to an electrolytic solution its chemical composition changes. Inversely, electrical energy is to be gained from chemical processes. This fact is utilized in potentiometric measurements.

5.2.10.1 PH Measurement

The potentiometric measuring principle is the one applied the most frequently in the pH measurements of aqueous solutions. For this, glass electrodes which are filled with a buffered reference electrolyte are generally used [58]. This reference electrolyte is separated from the measuring solution by a glass diaphragm which is permeable to H ions. However, although a reduction-oxidation reaction (redox reaction) with an electron or ion exchange does not occur, at the phase boundaries potentials are created due to a charge exchange which are conducted through metal electrodes (reference and measuring electrode) or whose difference can be determined by highly resistant voltmeters. The voltage created is described by the Nernst equation [59]. Nernst relies on the principle of the glass electrode:

If a glass sphere with the properties of a glass membrane filled with a buffered solution of known pH value pH_i is submerged in a solution with the pH value pH, voltages of different strengths occur at both phase boundaries. The potential difference is called membrane voltage and can be measured. The Nernst equation determines the relationship between the membrane voltage and the activities of the determining ion type

$$U_E = \frac{RT}{nF} \ln \frac{a}{a_i}, \tag{5.16}$$

in which U_E is membrane voltage, T absolute temperature, R gas constant=8.31439 J · (K · mol)$^{-1}$, n the number of elemental charges (in hydrogen $n=1$), F Faraday constant (charge of one mol of a substance), 1 F=96486 As mol^{-1}, a the activity of the ion to be measured and a_i the activity of the internal electrolyte.

For potentiometric pH measurement the activity of the monovalent hydrogen ion H$^+$ is decisive when glass electrodes are used. As the same H$^+$ ions exist inside and outside, the activities a and a_i are proportional to the concentrations c and c_i of the H$^+$ ions.

The Nernst equation shows that when the pH value is measured, the temperature of the solutions or glass electrode also has an influence.

In literature pH measurement has been described in detail and has been largely standardized in German industrial standards [60-68]. In air pollution analysis pH measurement is used mainly to determine the acid content of precipitation. Apart from discontinuous analysis of precipitation samples, automatic precipitation monitors which are now commercially available have also been developed for this purpose [69-71]. Apart from automatic pH measurement, precipitation sampling in the automatic monitors is also highly labor-intensive. At present, precipitation monitors still require a lot of maintenance.

5.2.10.2 HF and HCl Measurement with Ion-Sensitive Electrodes

Electrodes developed not only for hydrogen ions but also for other ions are commercially available as so-called ion-sensitive electrodes. Potentiometric methods with ion-sensitive electrodes have been used to measure hydrogen chloride (HCl) and hydrogen fluoride (HF) in exhaust gases [72]. HCl and HF measurements are, e.g., required in waste incineration plants.

5.2.10.3 O_2 Measurement with Solid-State Ion Conductor Zirconium Dioxide

The same principle of potentiometric measurement forms the basis for solid-state ion conductors, e.g., the so-called λ-sensor for the measurement or control of oxygen in exhaust gases. If gases with different oxygen contents are separated by a zirconium dioxide (ZrO_2) membrane at high temperatures, then oxygen ions diffuse through vacancies in the lattice of the ZrO_2 membrane and generate an electrical charge. If both sides of the zirconium dioxide membrane are covered with a platinum grid an electrical charge can be collected if the two separate gas mixtures have different oxygen concentrations. Just as is the case in pH measurement according to the Nernst equation (5.16), membrane voltage depends logarithmically on the partial pressure ratio or the concentration ratio of the oxygen on both sides of the membrane. Furthermore, temperature has an influence, too,

$$U = \frac{R \cdot T}{n \cdot F} \cdot \ln \frac{p_{O_2}(\text{Luft})}{p_{O_2}(\text{Abgas})} \:, \tag{5.17}$$

where p_{O2} is partial oxygen pressure.

It must be noted, that O_2 ion conductivity of the ZrO_2 starts only at more than approx. 400 °C. Fig. 5.15a shows a diagram of a ZrO_2 exhaust gas probe.

On the inside of the thimble-shaped ZrO_2 membrane there is air with 20.93% volume percent of O_2. Exhaust gases are passed on the outside. In λ measurement (air excess measurement) of automobile exhaust gases the sensor is exposed directly to exhaust gases. The smaller the O_2 content of the exhaust gas, the greater the O_2 ratio, as there is always the constantly high O_2 content of the air on the inside of the sensor. With increasing O_2 concentration ratio, the sensor voltage goes up exponentially, i.e. the sensitivity of the measuring method increases as the quantity to be measured becomes smaller, a phenomenon not frequently observed in measuring techniques. Fig. 5.15b shows the dependencies of the sensor voltage on the O_2 concentration in the exhaust gas for different temperatures. The voltage cannot increase indefinitely with decreasing O_2 content; if voltage increases, at a certain voltage (depending on the internal resistance) a discharge via the ZrO_2 membrane itself takes place.

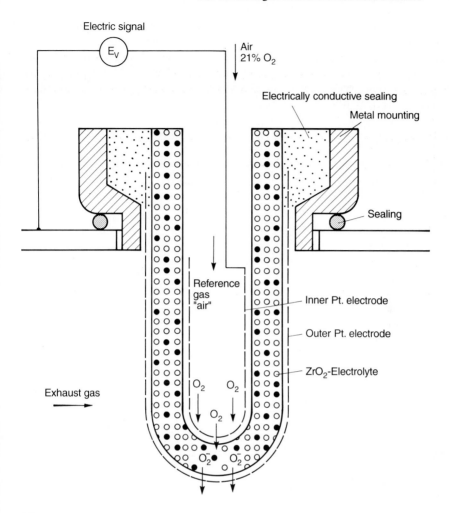

Fig. 5.15a. Diagram of a ZrO_2 probe [73]. With decreasing O_2 content the measured voltage increases

More and more frequently ZrO_2 sensors are used for O_2 measurement in furnace flue gases. There the sensors are heated to a constant temperature, so that usually only one of the curves shown in Fig. 5.15b actually applies. In automobile exhaust gases, however, usually no exact O_2 measurement takes place. Rather the air-fuel ratio is controlled such that the O_2 content approaches zero at the sensor, during which the sensor voltage rises sharply. This steep voltage rise is used for controlling. Voltage switch limits are set such that differing temperatures at the sensor do not exert any great influence.

If there are any appreciable concentrations of CO and hydrocarbons in the exhaust gases then a catalytic afterburning takes place at the sensor (catalyzed at the

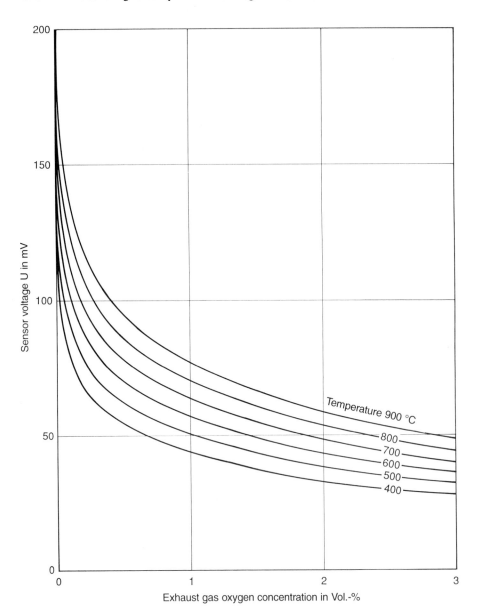

Fig. 5.15b. Dependency of sensor voltage on O_2 concentration in the exhaust gas for different temperatures [74]

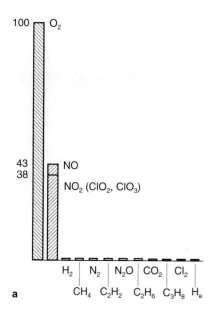

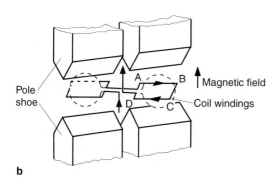

Fig. 5.16. Paramagnetic Oxygen Measurement. a. Magnetic susceptibility of some gases referred to O_2 (= 100%) [5]. b. Principle of a paramagnetic oxygen measuring instrument

platinum grid of the sensor) with residual oxygen being used up, i.e., O_2 content indicated by the sensor would be too low. In automobile exhaust gases this is no problem as residual oxygen is, e.g., needed in the succeeding catalytic converter for the afterburning of CO and hydrocarbons. Thus, not the O_2 content is controlled by the sensor, but the air ratio λ approaches 1, i.e., only as much O_2 or air is supplied as is needed for the complete oxidation of all unburned components. Therefore the name λ sensor.

5.2.11 Paramagnetic Oxygen Measurement

One principle of oxygen measurement uses the specific paramagnetic features of this gas. Fig. 5.16a shows the magnetic susceptibility of some gases referred to O_2 (= 100 %). Most of the gases are showing no magnetic susceptibility. The concentrations of gases of about 40 % susceptibility relative to oxygen (NO, NO_2, ClO_2, ClO_3) are so low (compared to O_2) that no interference has to be feared.

The principle of a paramagnetic oxygen measuring instrument is shown in Fig. 5.16b. In the measuring cell a "dumb-bell" is situated comprising two quartz spheres filled with nitrogen. This dumb-bell is suspended on a thin platinum band under tension. In the center of the dumb-bell a small mirror is placed. the dumb-bell is surrounded by a magnetic coil. Outside the measuring chamber there are permanent magnets. The magnets generate a powerful inhomogenous magnetic field. When oxygen molecules enter the measuring chambertogether with the sample gas, they are displaced in the area of the strongest magnetic field and thus exert differential forces on the two spheres of the dumb-bell, which is rotated away from its stationary position. As soon as the mirror is diverted away from its stationary position, the signal produced by a detector generates a current in the connectd amplifier. This current is transmitted through the coil which then generates its own magnetic field to bring the dumb-bell into the stationary position again. The current generated is thus a direct measure of the oxygen concentration present, and can be measured and displayed by a galvanometer.

The whole measuring instrument consists of the measuring cell itself, a permanent magnet, a power supply, the evaluation electronics including the display and a housing. The sample gas is sucked through the instrument by an external pump.

5.2.12 Measurement of Thermal Conductivity

The different thermal conductivity of gases is primarily used for measuring CO_2 and H_2. H_2 has a considerably higher thermal conductivity than other gases; CO_2 occurs in relatively high concentrations in furnace flue gases and is different from other gases in this respect. The principle of thermal flow measuring instruments is based on the fact that heated filaments exposed to the sample are cooled to different extents depending on the gas composition and thus change their electric resistance in the process. Thermal conductivity is not suitable, however, for the measurement of trace gases.

As the gases CO_2 and H_2 are not pollutants but auxiliary variables, e.g., for the determination of air excess in combustion processes, we will refrain from a more detailed description here. Relevant literature is offered in [5, 7, 32].

5.2.13 Manual Analysis Methods

In manual analysis methods generally a sample is first taken and then further processed in the laboratory. During sample taking an enrichment of the measuring component usually takes place, so that even very low pollutant concentrations can be detected. Fig. 5.17 shows different possibilities of manual analysis sampling. The choice of method depends on the species of interest and on the type of laboratory method with which the sample taken is further evaluated.

The simplest type of sample taking is done with gas-sample collectors (Fig. 5.17a and b) – in plastic bags for large volumes, in gas-samplingtubes made of glass for small volumes. In the laboratory the sample taken can, e.g., be fed in an automatic measuring instrument or it can be investigated by manual analysis.

Wet-chemical measuring methods in which the sample is collected in absorbing solutions and bound chemically are widespread. The simplest of this type of measurement is *volumetry*. In this method a defined exhaust gas volume is taken and flushed several times through a specific absorption fluid. The remaining residual volume is subsequently measured; the difference to the initial volume corresponds to the volume concentration of the absorbed gas in the % range. The method is well-known as Orsat Analysis. It is used for manual CO_2 determination in exhaust gases.

Another method of determining the absorbed pollutant content in a reaction solution is by *color change titration*. A sample is collected in the wash bottle according to Fig. 5.17c. Then, in the lab a reagent is gradually added to the absorbing solution until the bound pollutant is transformed quantitatively or the excess reagent in the absorption solution is used up quantitatively. The end of the reaction is indicated by appropriate color indicators.

Colorimetric methods show a highly sensitive reaction. The reaction solution in the wash bottle either reacts directly with the pollutant or after further chemicals have been added to form a color complex. Here, color intensity is a measure of the pollutant content in the reaction solution. Color intensities are calibrated with known pollutant amounts. The pollutant content of the reaction solution in proportion to the gas volume sucked through which is indicated on the gas meter shows the concentration of the gas in the investigated sample.

Table 5.6 lists the wet-chemical measuring methods for different components described in the VDI guideline [76].

5.2.14 Chromatographic Techniques

Chromatography – translated from Greek it means "color writing" – is an analytical procedure based on physical principles. One of the simplest chromatographic analytical procedures is paper chromatography. If a solution of different dyes is dropped on blotting or filter paper, then the individual substances in the solution spread to varying extents due to their different affinities to the absorption substance, the blotting paper. On the paper they can be identified by color bands or concentric color rings.

In modern chromatography, however, so-called separating columns through which the substance mixture is passed, are used. Depending on their different affini-

252   5 Measuring Techniques for Recording Air Pollutants

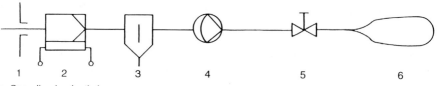
a Sampling in plastic bag

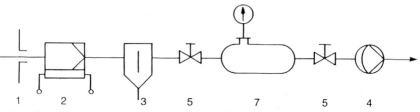
b Sampling with gas-sampling tube

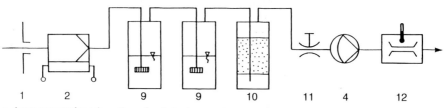
c Arrangement for adsorption of pollutant gases in reagents

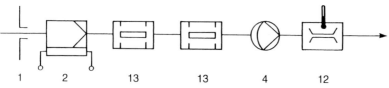
d Arrangement for adsorption of pollutant gases on solid adsorbents, e.g. activated charcoal

1 Sampling probe
2 Particle filter, heated if necessary
3 Condensate precipitator
4 Gas pump
5 Cock
6 Plastic bag
7 Glass gas sampling tube
8 Manometer
9 Wash bottle with sintered insert
10 Flask with drying agent
11 Throttle valve
12 Gas meter
13 Adsorption tube

For ambient air measurement parts 1, 2, 3 are left out

Fig. 5.17a-d. Sampling methods for manual measurements with evaluation in the lab

5.2 Measuring Methods for Gaseous Pollutants

Table 5.6. Wet-chemical measuring methods for air and exhaust gas analysis described in VDI guidelines [76]

Component to be measured	Measuring Method Name	Principle	Emission (E) Immission (I)	VDI Guideline/ISO standard d=draft	Date
SO_2	Iodine thiosulfate	Titration	E	2462, part 1	02/1974
	Hydrogen peroxide/barium perchlorate/thorin method	Titration	E	ISO 7934 d	09/1987
	Hydrogen peroxide method	Gravimetry	E	2462, part 3	02/1974
	Silica gel method	Adsorption/ colorimetry	I	2451, part 1	08/1968
	TCM	Colorimetry	I	2451, part 3 d	11/1994
	Thorin spectrophotometric method	Colorimetry	I	ISO 4221	05/1990
SO_3	2-propanol method	Titration	E	2462, part 7	03/1985
$NO + NO_2$	Phenoldisulfonic acid method	Colorimetry	E	2456, part 1	12/1973
	Hydrogen peroxide method	Titration	E	2456, part 2	12/1973
	Sodium salycilate method	Colorimetry	E	2456, part 8	01/1986
NO_2	Saltzman	Colorimetry	I	2453, part 1	10/1990
CO	Iodine pentoxide method	Titration	E	2459, part 2	02/1994
Total F^-	NaOH absorption	Colorimetry/ potentiometry	E	2470, part 1	10/1975
F^-	Sorption method with silver balls	Potentiometry or photometry	I	2452, part 3	07/1987
HCl	Absorption with water	A) Titration, B) potentiometry, C) colorimetry	E	3480, part 1	07/1984
Cl_2	Methyl orange method	Colorimetry	E	3488, part 1	12/1979
	Bromide-iodide method	Titration	E	3488, part 2	11/1980
	Methyl orange method	Colorimetry	I	2458, part 1	12/1973
H_2S	Iodimetry	Titration	E	3487, part 1	11/1978
	Molybdenum blue sorption	Colorimetry	I	2454, part 1	03/1982
	Methylene blue impinger	Colorimetry	E	2454, part 2	03/1982
NH_3 (total N)	Sulfuric acid absorption	Colorimetry/ titration	E	3496, part 1	04/1982
NH_3	Indophenol method	Colorimetry	I	2461, part 1	03/1974
	Nessler method	Colorimetry	I	2461, part 2	05/1976

Table 5.6. (continued)

Component to be measured	Measuring Method		Emission (E) Immission (I)	VDI Guideline/ISO standard	
	Name	Principle		d=draft	Date
Aliphatic aldehydes (C_1 to C_3)	MBTH method	Colorimetry	E	3862, part 1	12/1990
	Sulfite-pararosaniline	Colorimetry	I	3484, part 1	01/1979
Phenols	p-nitraniline method	Colorimetry	I	3485, part 1	12/1988
TOC (Total Organic Carbon)	Silica gel adsorption	CO_2 coulometry	E	3481, part 2	04/1980
	Silica gel adsorption	CO_2 coulometry	I	3495, part 1	09/1980
Polychlorinated dibenzodioxins and dibenzo-furans	Dilution method, filtering or absorption	Extraction GC/MS	E	3499, part 1 d	03/1990
Polycyclic aromatic hydrocarbons (PAH)	Filter/condenser method	Extraction GC/MS	E	3499, part 2 d	03/1993
	Filter condenser method	Extraction GC/MS	E	3872, part 1	05/1989
	Dilution method	Extraction GC/MS	E	3873, part 1	11/1992

ties to the individual adsorbents, substances with great affinity migrate more slowly through the separating column than those with a small affinity, so that they exit the column at different times. At the column's outlet the individual substances are recorded by means of physical detectors.

5.2.14.1 Gas Chromatography

Gas chromatography is a method for separating gas mixtures. Fluids can also be separated, only they must be vaporized beforehand and may not decompose. The principle has been frequently described, s., e.g., [30, 31]. Fig. 5.18 shows the schematic diagram of a gas chromatograph.

A carrier gas, the so-called mobile phase – hydrogen, nitrogen or helium – flows through the separating column. The gas or fluid mixture to be separated is injected into the carrier gas flow with an injection syringe. The injection block is heated to such an extent that even substances injected in a fluid state vaporize instantly up to a certain boiling point. The sample mixture is then flushed into the separating column by the carrier gas. To enlarge the inner surface the "packed" columns are filled with a porous inert material, e.g., silicagel. In so-called gas-fluid chromatography this material is coated with a fluid phase, the stationary phase, to improve the separating performance. When flowing through the stationary phase the mobile phase, the carrier gas, carries the components of the sample along with it according to the distribu-

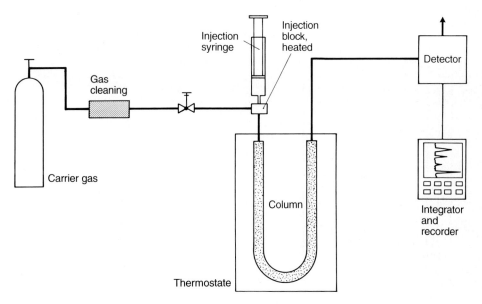

Fig. 5.18. Principle of gas chromatography

tion equilibrium between mobile and stationary phase. Therefore, the individual gas or vapor components flow through the separating column at different velocities, so that they reach the column outlet at characteristic intervals of time. Subsequently they flow through the detector where their concentrations are determined individually. The detector's signals are registered by a recorder as peaks. The record showing the succession of the individual peaks is called "gas chromatogram", s. Fig. 5.19.

The distribution equilibrium of the individual substances between the mobile and stationary phase is temperature-dependent. The separating column is therefore kept at a constant temperature in the so-called column oven or is heated with defined temperature programs, so that even nonvolatile substances can be expelled from the column and be analyzed.

The separating performance of the packed columns is limited. For this reason capillaries have been developed as separating columns in the past years which are coated with the stationary phase as an extremely thin film of fluid on the inside wall. With these capillaries which are made of fused silica and have a length of 25 or 50 m, very good separating results have been achieved.

Of all air pollutants there are mainly the manifold hydrocarbon compounds which are determined gas chromatographically, both emissions as well as pollutants in ambient air. The flame ionization detector is primarily used as gas chromatographic detector of such compounds.

The retention time inside the column is characteristic for each substance. The identification of the substances of interest is carried out by comparing the retention times with those of known pure substances, s. Fig. 5.19. The concentrations of the substances are proportional to the areas A under each peak. These area sections can be converted directly into concentration units by computer controlled integrators.

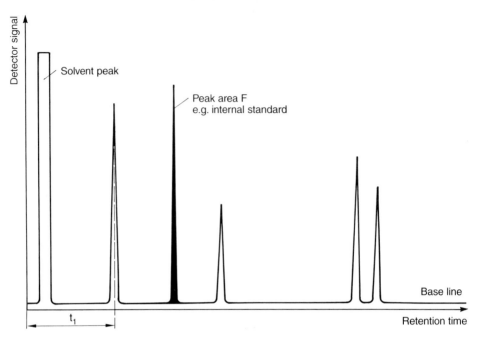

Fig. 5.19. Principle presentation of a gas chromatogram with retention time t_1 and peak area F

The peak area of a specially added substance can act as reference quantity, as a so-called internal standard.

5.2.14.2 Gas Chromatography/Mass Spectrometry

If identification based on retention times is not possible in gas chromatograms with numerous peaks, then a mass spectrometer (MS) is often included in the set-up as a detector after the gas chromatograph (GC). Fig. 5.20 shows the principle of a mass spectrometer: the gas to be investigated is fed into an ion source where charged molecules are generated by bombarding with electrons. By applying high voltage these are formed to an ion beam and focussed with an electrostatic lens. After having passed through a cylinder condenser the different components of the ion beam are separated from each other in a magnetic field and are recorded with a suitable detector. By measuring the strength of the magnetic field, the mass M of the ions can be determined. In Fig. 5.20 mass M_2 of the ions just reaching the detector is greater than M_1, but smaller than M_3.

The analytical possibilities of mass spectrometry have been described in detail in reference publications, e.g., in [77]. Although gas-chromatography-mass-spectrometry (GC/MS) is an expensive analysing technique, it is indispensable today in both trace analysis as also in air quality control.

As gaseous air pollutants do not occur in such high concentrations that they can be injected directly into the gas chromatograph and be analysed there; it is necessary

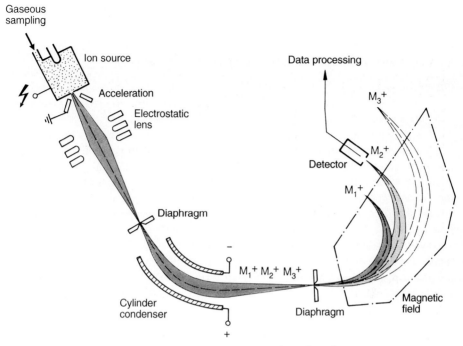

Fig. 5.20. Principle of a magnetic mass spectrometer; M_1^+, M_2^+, M_3^+ are masses of different ions [78]

to enrich the substances most of the time. This, e.g., is carried out on activated carbon or other adsorbants, s. Fig. 5.17d. The substance mixture can then be extracted from the activated carbon, e.g., with a solvent such as carbon disulfide (CS_2) and can then be injected as a fluid mixture into the gas chromatograph. Another method of separating the substances which accumulate on the adsorbent is thermal desorption. For this the sampling tubes are heated and their contents are flushed into the carrier gas flow in a gaseous state.

5.2.14.3 High Performance Liquid and Ion Chromatography

Relatively recent methods for determining pollutant concentrations in solutions such as are shown in Fig. 5.17c are high performance liquid chromatography (HPLC) and ion chromatography as a special version of the HPLC. These methods are particularly suitable for determining several components in solutions. For this, the solutions to be investigated are sent through special separating columns where the species of interest are retained for different lengths of time and accordingly exit at different times. The substances are then recorded with suitable detectors, e.g., with UV absorption or fluorescence detectors in HPLC, with conductivity detectors in ion chromatography. In air and exhaust gas analyses HPLC is used particularly to determine organic pollutants such as aldehydes [79]; ion chromatography today is the common method of determining anions, e.g., sulfate, nitrate, chloride and cations such as NH_4^+, Na^+, K^+, Ca^{2+}, Mg^{2+}, Al^{3+} etc. in aqueous samples [80].

5.2.14.4 Determination of Highly Toxic Organic Compounds

Polycyclic aromatic hydrocarbons (PAH), polychlorinated dibenzodioxins, dibenzofurans and polychlorinated biphenyls are to be mentioned here. More information on structure and occurrence of these substances is given in sect. 2.2.3.

Polycyclic aromatic hydrocarbons and polychlorinated dioxins and furans occur in exhaust gases only in traces. Exhaust gas temperatures in motor vehicles, furnaces and incineration plants range between 60 and 400 °C. Most of these compounds condense when cooling down to temperatures in lower ranges. Apart from condensing, organic compounds are mainly adsorbed by existing soot and particulate matter. At temperatures of approx. 400 °C, however, many compounds occur in a gaseous state [81].

Due to the low concentrations and varying states of aggregation, samplings of these substances are subject to special requirements. When exhaust gas temperatures are low sampling must primarily concentrate on collecting the organic compounds on filters. When exhaust gas temperatures are higher the substances must either be condensed by cooling the exhaust gases and then be separated on filters, or absorption devices (e.g., wash bottles) are included in the set-up after the filters to collect the gaseous components [82, 83].

To guarantee an effective sampling it is the safest to conduct the entire exhaust gas flow over the sampling train, but this is only possible with exhaust gas flows of up to approx. 50 m³/h. Sampling trains operating according to this method are, e.g., used in exhaust gas investigations of motor vehicle engines and in domestic furnaces.

Fig. 5.21 shows the set-up of the sampling system with coolers and filters as it is used for collecting polycyclic aromatic as well as polychlorinated aromatic hydrocarbons [83, 84]. When dealing with large flue gas flows, such as those of industrial plants, only a partial flow can be taken as a sample, but then a grid sampling over the cross-section of the flue gas canal is necessary. As part of the organic substances occur in particulate form, sample taking must take place at the same speed isokinetically (s. sects. 5.3 and 5.4). Measuring trains for such samplings are, e.g., described in [83].

Sample preparations as the step between sample taking and analysis are often complicated and work-intensive. There are three fractions of the sample to be investigated: in the water condensate, in the cooler and on the filters. The filter usually contains the main amount of the PAH. The preparation of the three fractions is shown in a diagram in Fig. 5.22: The PAH of the three fractions are first extracted, the extracts then treated further one by one or together and then separated from the accompanying substances, such as paraffins, by a double liquid-liquid separation. After another pre-cleaning in a silica-gel column the PAH are split into two fractions by *column chromatography* of which one contains the components with 2-3 rings, the other those with 4-7 rings. After this pre-separation the solutions can be analysed gas-chromatographically without having to fear that the PAH peaks of interest in the chromatogram may be obscured by interfering components.

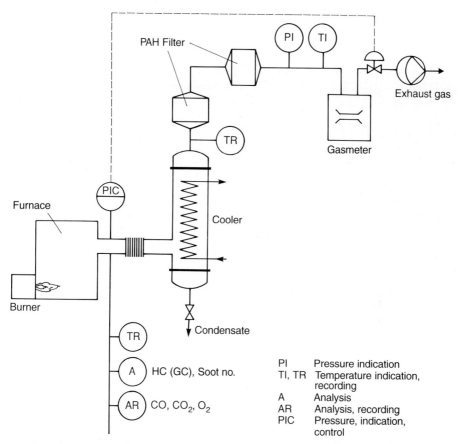

Fig. 5.21. Diagram of a PAH sampling system with the total flow method in a research furnace (acc. to Grimmer [85]); nearly the same apparatus is used for sampling automotive exhaust gases

The appropriate sample preparation for polychlorinated aromatic compounds is also done by extracting the different fractions and by concentrating (vacuum evaporation) the species of interest.

When collecting the wanted components on filters and during preparation of the samples chemical transformations can occur which lead to errors. E.g., Hartung et al. found nitro-derivatives of the PAH as artefacts in samplings of diesel exhaust gas [87]. PAH measurement depends on the reactivity and on the temperature of the PAH, on the concentrations of the interfering components (e.g., oxides of nitrogen) and on the sampling time. In polychlorinated aromatic compounds, also, artefact formation during sample taking on filters must be taken into consideration.

Analysis of the prepared samples is generally carried out gas-chromatographically with separating capillaries and mass-selective detectors or, in chlorinated compounds, with electron-capture detectors [78, 82, 84, 85, 88].

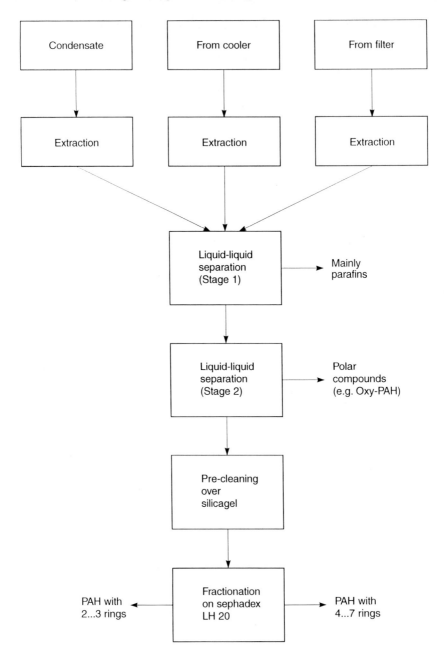

Fig. 5.22. Diagram of PAH sampling preparation and concentration [86]

5.2.15 Method for Determining Odorous Substances – Olfactometry

Odorous substances generally occur in such low concentrations or as such complex combinations of substances that a recording with measuring techniques based on chemical or physical methods is mostly impossible. As odorous substances are often offensive to people a "measuring technique" has been developed which has the human nose itself serve as a detector. For this, equipment is used in which the gas to be investigated can be diluted with pure air to varying extents. This diluted gas is then conducted to a funnel where smelling tests are conducted by a panel of persons. This method is called dynamic olfactometry, s. Fig. 5.23.

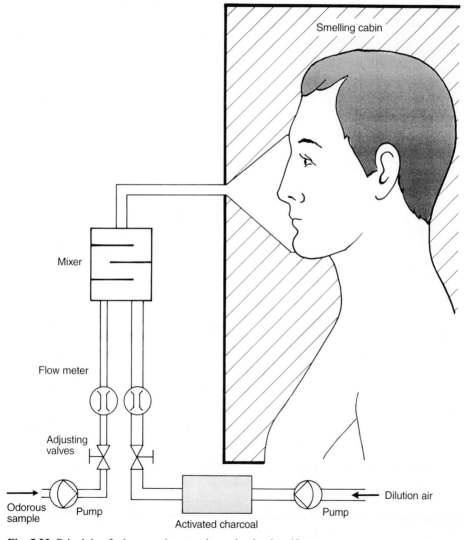

Fig. 5.23. Principle of odorous substance determination by olfactometry

Of the gas to be investigated either a continuous partial flow is conducted into the olfactometer or gas samples in plastic bags which were filled at the source of the odor are entered into the olfactometer [89]. Olfactometry has to fulfill several tasks [90]:
- detecting odor thresholds of individual substances and combinations of substances in the laboratory,
- determination of odor units in unknown combinations of substances at emission sources or in the ambient air,
- assessment of the type of odor (agreeable, disagreeable, terms of association... etc.) = hedonic effect.

The odor threshold in a sense of recognition threshold is the concentration of an odorous substance triggering a just barely perceptible sensation of odor. According to convention it is that concentration where a test person reports a perception of odor in 50 % of the odorous samples. Accordingly, for a panel test (with several test persons) perception of odor is to be indicated 50 % of the persons [90].

To determine the odor threshold the sample gas in the olfactometer is diluted until its odor is clearly below the perceptible range. Then the concentration is raised until the test person indicates just a barely perceptible sensation of odor.

As a relative measure of concentration of undefined odor samples the *odor unit* is used. According to definition, the odor sample with threshold concentration has one odor unit . The number of odor units of an odor sample is identical with the dilution number Z_{50}, which is defined at the point at which the odor threshold is reached in olfactometer measurement [90]:

$$Z = \frac{\dot{V}_P + \dot{V}_R}{\dot{V}_P}, \qquad (5.18)$$

where Z is the dilution number of the dilution of the odor sample, $\dot{V}_P$ is the volume flow of the odor sample, $\dot{V}_R$ is the volume flow of the added pure air, and Z_{50} is the dilution number indicated by 50 % of the panel.

Investigations of the hedonic effects of odors are carried out in the superthreshold range.

Olfactometric measurement must be carried out absolutely painstakingly. Exact instructions are listed in [90]. The deviations of the odor tresholds and of the relative odor concentrations can in some cases be considerable. Results of ring tests also show, however, that the method of olfactometry is essentially suitable for the assessment of odors [91-96].

5.3 Measuring Methods for Particulate Matter

When examining particulate matter in emissions as well as in ambient air the following factors must be taken into account:
- total mass concentration of the particulate matter,
- concentration of fine particles,

- size distribution,
- chemical composition.

The following sections will give an overview of the different ranges of application and some examples of measuring techniques will be decsribed. As a complete representation of all measuring methods is not possible within this book, we recommend that more specialized literature be consulted for this purpose, e.g., [97].

In the air quality range particle sedimentation as well as non-sedimenting suspended particulate matter is of interest, particularly the latter, as it is respirable and can thus carry pollutants into the human body.

5.3.1 Gravimetric Determination of Particulate Matter in Exhaust Gases

The simplest method of determining the particle content of exhaust gases is to suck off a partial flow of the exhaust gas and to conduct it over a filter. Particle concentration is then calculated from the added weight of the filter and the gas volume sucked through. This is a manual and discontinuous method. It is used in acceptance tests, for calibrating automatic particle measuring systems and for the taking of samples of particulate matter whose composition is to be determined subsequently. This measuring technique has been described in detail in the guideline no. 2066 of the German Engineering Association (VDI) [98, 99]. To reduce faulty measurements the partial flow must be sucked off isokinetically, i.e., with the same speed as the exhaust gas flow which makes the whole method more complicated. This is illustrated by Fig. 5.24:

If the sampling rate is too slow (Fig. 5.24a) the exhaust gas moving in stream lines which ought to flow into the sample inlet diverges past it. Due to their inertia the particles do not follow the diversion, but enter the sample inlet instead. The par-

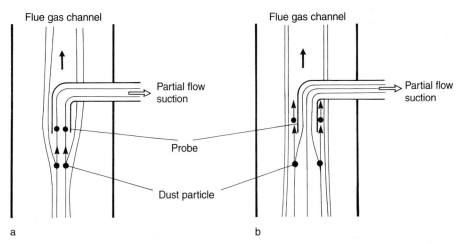

Fig. 5.24. Influence of a non-isokinetic partial flow suction; **a.** suction speed too low, **b.** suction speed too high

ticulate content measured is too high. If the sampling rate is too fast, the exhaust gas from the surrounding area is sucked into the sample inlet, again the particles do not follow the stream line flow, pass by on the outside (Fig. 5.24b) and the particulate content measured is too small. Findings have shown that the errors made when the suction speed is too slow increase much more dramatically than when the suction speed is too high [89, 100]. However, this depends on the grain size and on the density of the particles. The smaller and lighter they are, the better they can follow the divergent flow, and the smaller the likelihood of errors.

For matching sampling rate the speed of the exhaust gas stream must be determined before the actual particulate measurement, distributed over the entire cross-section of the exhaust gas canal. Fig. 5.25 shows a measuring set-up for gravimetric particulate content determination including the set-up for the measurement of the flue gas speed.

The speed of the partial stream is determined with the help of the venturi tube and the manometers and set at the pump. The total gas volume sucked through is read on the gas meter.

When using a pitot tube the speed of the main gas stream can be calculated according to the following equation (Bernoulli):

$$v = \sqrt{\frac{2p_d}{\varrho_f}}, \tag{5.19}$$

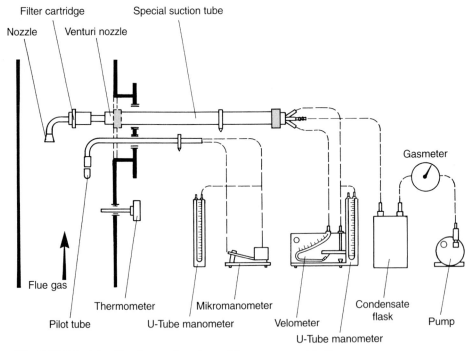

Fig. 5.25. Principle of in-situ gravimetric particle measurement in flue gases

where v is gas speed, p_d is dynamic pressure (= total pressure – static pressure) measured with a pitot tube and ρ_f is the density of the humid flue gas. If the carrier gas is not air but, e.g., furnace flue gas, then for the determination of the flue gas density, the flue gas temperature, static flue gas pressure as well as gas composition have to be determined first, e.g., CO_2, O_2, H_2O and possibly also the SO_2 content. Possible measuring techniques have already been discussed in sect. 5.2.

One possibility to avoid the complex speed measurement is offered by the use of a so-called zero pressure probe [101]. In this method the static pressure is measured on both the inside and the outside of the suction tube, then the pressure difference is set to zero by adjusting the partial flow.

High-grade steel cartridges or glass fiber cartridges stuffed with quartz wool are generally used to collect the particulate matter, s. Fig. 5.26. It is recommended that the particle filter be positioned directly in the flue gas canal. In this way it will be heated by the flue gas; external heating to prevent water vapor condensation in the filter is thus not necessary. In addition, the short suction tube minimizes particulate losses in the sampling system. For special purposes, particularly for the collection of low particulate concentrations, plane filters can also be used [102].

The particulate matter sampled on the filters can be examined for special substances. E.g., to examine for adsorbed sulfuric acid, the particulate matter together with the filter material is dissolved in water and titrated with caustic soda. To determine its combustible content, e.g., soot, the filter is incinerated and the so-called ignition residue is weighed.

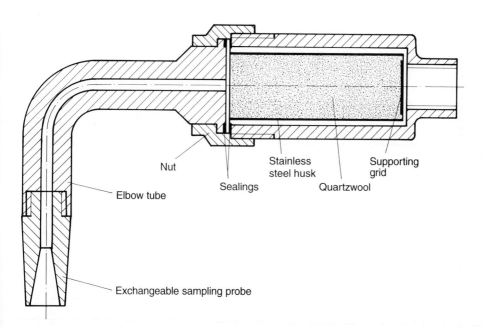

Fig. 5.26. Diagram of an in-situ sampling filter: stainless steel case packed with quartz wool [100]

To analytically determine heavy metals such as barium, cadmium, chromium, nickel, lead, vanadium a.o. the particulate matter is broken down chemically together with the quartz wool filter and the elements are analyzed with the help of atomic absorption spectrometry (AAS) [103]. However, when analyzing metals in exhaust gases it must be considered that, due to the high temperatures, some compounds can also be present as metal vapors. To be able to detect such metal vapors, complicated collecting equipment must be installed after the filter, e.g., gas wash bottles [104].

5.3.2 Continuous Registration of the Flue Gas Particle Concentration

To monitor the effectiveness of particulate removal systems, the correct setting of furnaces and the observance of maximum emission values the concentration of particulate matter in exhaust gases must be measured and recorded continuously. For this, two measuring principles which have passed suitability tests and have been officially approved by the Environment Ministry of the FRG [105]. One category of these are optical (photometric) measuring instruments, and the other is an instrument applying the attenuation of β radiation as measuring effect.

β radiation attenuation

Fig. 5.27 shows the principle of a β dust meter. A partial gas flow is sucked from the flue gas channel isokinetically. The particulate matter contained in the partial stream is collected on a strip of filter paper and the particulate deposit is then penetrated by

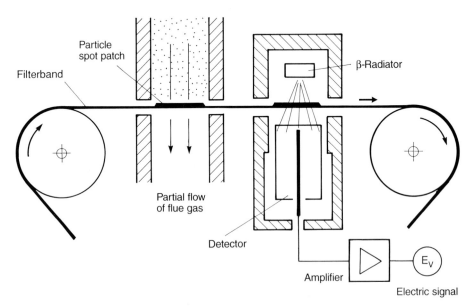

Fig. 5.27. Principle of particle measurement by attenuating β-radiation [106, 107]

radiation from a β emitter. The attenuation in radiation is used as measuring effect; it depends on the particulate mass on the filter.

After a certain period the filter paper is advanced and a new mass of particulate is collected. Particulate concentration is determined by the maximum value achieved. Thus, the method is a quasi-continuous one. The instrument must be calibrated by a gravimetric particle measurement.

Photometric particle measurement – opacimeter

When looking at particle measuring instruments one must differentiate between *simple smoke-density measuring equipment and calibratable particle concentration measuring equipment*. What both types of equipment have in common is that they are positioned directly at the stack or flue gas channel, penetrate it with light and use the light absorption caused by the particulate matter in the flue gas as measuring effect (opacimeter).

Simple smoke density measuring instruments operate with visible light and measure the transmission, which is the ratio of attenuated light intensity to emitted light intensity, s. equation (5.3). Decreasing transmission values are equivalent to increasing smoke density or increasing particle content in the exhaust gas. These instruments are, e.g., used for monitoring the completeness of combustion in furnaces [105, 108, 109].

If an opacimeter is to be used to determine the particle concentration of flue gas, then the extinction is measured (logarithm of the reciprocal value of transmission, s. equation (5.4), (5.5)). According to the Lambert-Beer law, extinction is directly proportional to the concentration of interest in flue gas, s. equation (5.6) [110]. Fig. 5.28 shows a diagram of an opacimeter which can simultaneously determine the particle concentration and the SO_2 concentration of the flue gas via UV radiation absorption.

There are three specific wavelengths which are used for measurement: 313 nm, 436 nm, 546 nm. In each of these wavelengths particles cause light absorption, at 313 nm so does SO_2. The absorption at the wavelength 436 nm is used as auxiliary variable to prevent faulty SO_2 measurements.

Emitter and detector unit are housed together in one casing. On the other side of the stack there is only a mirror. A low-pressure mercury lamp serves as light emitter. In the detector unit the beam of light is diverted by a beam splitter and focused on the photomultiplier. Beforehand, the measuring light beam is spectrally split into the three different wave lengths with the help of a rotating filter disk. The reference light is measured every 15 minutes by inserting mirrors. The intensities of the measuring light and reference light are transformed opto-electronically by the photomultiplier and passed on to the electronic evaluation device. The extinction values for particles and SO_2 are calculated from the sum extinctions measured at the three wave lengths. Every 120 minutes zero and control point are measured by inserting mirrors and filters and recorded on the recording strip. This measuring principle offers the advantage of a so-called in situ measurement, i.e., the measurement takes place directly in the stack without partial stream removal by suction. Problems of a representative

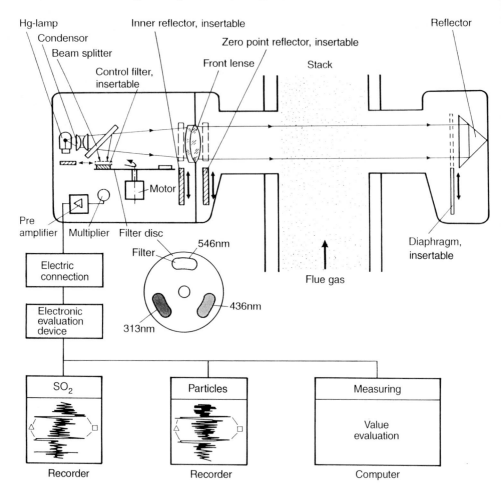

Fig. 5.28. Diagram of an opacimeter for in-situ determination of particle and SO_2 concentration in stack gases (acc. to [111])

sampling point, of cooling the measuring gas, faulty measuring gas pumps, clogged filters etc. (s. sect. 5.4) do not have to be dealt with. However, calibration must be carried out by comparative measurements with other measuring methods; for particles with gravimetric measurement, for SO_2 generally with a wet-chemical reference measuring method. This calibration prodedure is work-intensive as networking measurements at different concentrations are required. The calibration of optical particulate measurement is only valid for one single type of particles and size distribution. If the particles have other optical properties as compared to the particles used in calibration, e.g., finer grain size or a different color, then errors in measurements occur. It is obvious, e.g., that black soot absorbs light quite differently than white gypsum, which reflects part of it.

5.3.3 Determination of the Soot Number of Furnace Flue Gases

To monitor the completeness of combustion in oil furnaces a so-called soot number has to be conformed to [34, 112]. A hand pump into which a white filter paper is inserted serves as "measuring instrument". A defined volume is drawn off from the exhaust gas (generally 10 draws). The soot contained in the flue gas is precipitated on the filter paper and leaves a round, blackened spot. The degree of blackening, also known as the soot number according to Bacherach, is determined with the help of a comparative scale containing 10 comparative degrees of blackness from 0 (white) to 9 (black). This method is standardized with precise requirements having to be fulfilled by pump and filter paper [113]. Furnaces for light fuel oil may not exceed blackening degree 2, furnaces for heavy fuel oil may not exceed degree 3. This method is a qualitative one. Correlations between degree of blackening and particulate mass cannot be easily determined as particles with different optical properties can be emitted depending on furnace and fuel. However, the soot number is a good, easily determinable indicator to monitor the completeness of combustion in oil furnaces and thus limit CO and hydrocarbon emissions.

If *oil derivatives* (unburned fuel oil components) are present on the filter spot, they can be recognized by a yellowish-brown tint, then a so-called *blotting-paper test* is to be carried out to render them visible [114]. This is a simple paper-chromatographic technique: when the filter paper strip is dipped into acetone (formerly pyridine, which is toxic, was used) it becomes saturated beyond the filter spot and flushes out the oil derivatives which become clearly visible around the filter spot.

Furnaces for solid fuels not only emit soot but also tar. Baum and Brocke have developed a scale of comparison, for easy monitoring of emissions in small furnaces. This scale contains 10 blackening degrees and 8 yellowish-brown hues as further parameters, thus extending the entire scale to 80 different hues [115]. A semiautomatic sampling device has also been developed by the authors for this measurement.

5.3.4 Determination of Grain Size Distribution of Particles

Even if collectors for particulate matter show sufficient efficiencies they frequently emit considerable amounts of fine particles, as these do not weigh as heavily as coarser particles when the degree of separation is determined. It is the fine particles, however, which cause the haze in the air because they remain suspended. Fine particles also have the special property of adsorbing gases. Furthermore fine particles are respirable. All this makes the degree of fraction separation of dust collectors all the more interesting today. Particle size analyses are necessary to determine the degree of fraction separation. To determine the size distribution it would be possible to remove particle samples from the flue gas stream, to be then investigated in a pneumatic size analysis or a screen analysis. These analyses, however, require minimum amounts of particulate mass which cannot be obtained easily, when the particulate content of the flue gases is low. To catch larger amounts of particulate matter, the collecting filter would have to be located outside the flue gas channel. In that case the problem arises that probably certain size fractions would tend to adhere to the walls of the sampling pipes. It is therefore preferable to measure or collect the particles with the different grain sizes directly in the flue gas, a so-called in situ measurement.

One possibility for a grain size determination directly in the flue gas is, e.g., offered by cascade impactors [116]. The principle of such impactors consists of perforated plates which are set up in a cascade-like arrangement one above the other with baffleplates below each one. The holes which have the effect of nozzles on the gas stream decrease in size with each level. Thus the speed rises incrementially from the first to the last level. Fig. 5.29 shows the operating principle of a cascade impactor. By means of the nozzles the gas streams are aimed at the baffleplates such that the

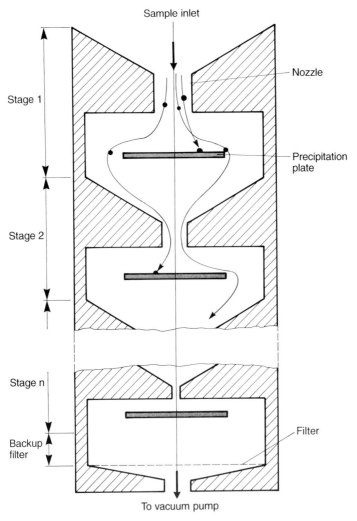

Fig. 5.29. Principle of particle grain size determination by cascade impactor [117]

particles starting from a certain minimum size impact on the plate because of their inertia and are deposited there. Particles below this minimum size are diverted at the baffleplate along with the gas stream and enter the next level where gas speed is higher owing to the smaller hole diameter. Thus, depending on the gas speed, each grain fraction is separated on individual baffleplates of the cascade. The individual levels do not have perfectly separating lines, the actual separating lines are shown in Fig. 5.30 [118]. Each level can be assigned a grain diameter for which the separation probability amounts to 50 %.

The individual baffleplates which are usually made of glass fiber paper are weighed before and after the sampling process, and the relative proportion of the individual grain fractions is determined.

The cascade impactor is available in such a small size that it can be inserted directly into the flue gas channel for in-situ measurements. The precision of impactor measurement depends entirely on the precision of the weighing. Thus, weighing techniques must meet very strict requirements. Among other influences, temperature and humidity can possibly mask the measuring results.

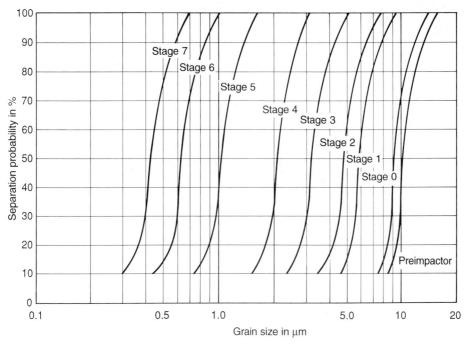

Fig. 5.30. Particle size distribution cascade impactor.
a. Separating lines of the different stages of a cascade impactor [118]

5.3.5 Measurement of Particle Sedimentation in the Ambient Air

If a layer of dirt forms on window sills, car roofs or garden furniture, then this layer is caused by particle deposition. Formerly, this particle deposition used to be considered an archetypal characteristic of air pollution, particularly if the dirt layer was

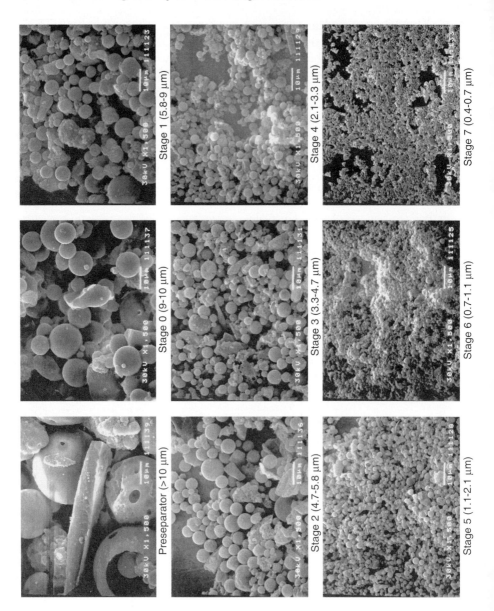

Fig. 5.30. Particle size distribution cascade impactor.
b. Electron mocroscope scan of the particles collected on the different stages of the cascade impactor (sizes in brackets are given by manufacterer) (Photo: Konrad Glaser)

black. Therefore it was relatively early on that horizontal collecting surfaces or collecting containers were set up so that particle deposition per unit of surface and time could be determined.

In the meantime various particle collecting devices have been developed. The Technical Directive on Air Pollution Control (TA Luft) of the FRG [34] mandates the so-called Bergerhoff method. This includes a glass jar which is positioned on a pole in a basket with a bird protection ring, s. Fig. 5.31. Exposure time usually

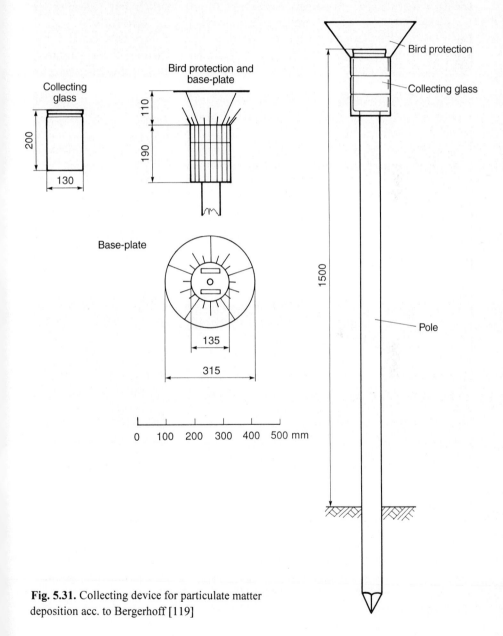

Fig. 5.31. Collecting device for particulate matter deposition acc. to Bergerhoff [119]

amounts to 30 days. Both particulate matter as well as liquid precipitation (rain, snow) are collected. During evaluation total particulate matter is determined gravimetrically and the weight is related to the open area of the collecting jar and exposure time. Particle deposition is then expressed in grams per square meter and day [$g/(m^2 \cdot d)$]. As this is an extremely inexpensive "measuring instrument", particle deposition measurements can be carried out in narrow grids.

Various other collecting devices have been developed, e.g., a bottle with a collecting funnel ("Löbner-Liesegang device") [120] and adhesive foil collecting surfaces which are covered with vaseline as adhesive agent [121]. The collected particulate matter can be viewed under the microscope and even be partially identified, e.g., soot particles, plant pollen etc.. However, with this method particle deposition cannot be completely recorded, as they can be washed away by rain or the adhesive agent can be damaged.

If particle deposition is to be analyzed for its composition, collection on adhesive foils is out of the question. But even collection with the Bergerhoff instrument proves to be problematic: the rain which is collected in the jar can contain dissolved pollutants which crystallize when the rain is vaporized and are then identifiable as particulates. Neither is the Bergerhoff method well-suited for the analysis of liquid precipitation as, conversely, particulates can dissolve in water and cause errors. Thus, Rohbock and Georgii developed a so-called dry-and-wet sampler for research purposes [122] which closes the rain collecting vessel with a lid when the weather is dry and opens it when there is precipitation. A particle collecting vessel, on the other hand, opens in dry weather and closes in wet. In the meantime this dry-and-wet sampler is also commercially available.

5.3.6 Measurement of Particle Concentration in the Air

5.3.6.1 Filter Sampling Instruments

As is the case with gravimetric particle emission measurements the air to be investigated in ambient air measurements is also sucked through a filter by a pump, the mass increase is measured and the air volume sucked through is determined. In ambient air measurements an isokinetic suction is irrelevant, as samples must, e.g., also be taken from air without any flow. Depending on the measuring operation the instruments used vary in their rate of air flow, filter size and intake port.

High-volume particulate sampler

If the components of the particles sampled are to be analyzed by several techniques, then so-called high volume samplers are used to collect enough material for these analyses. These samplers achieve rates of air flow up to 1.5 m^3/min. (corresponding to 90 m^3/h). Generally, the air is sucked in through a glass fiber or other filters having a large diameter (e.g., 257 mm) on which almost all particles with an aerody-

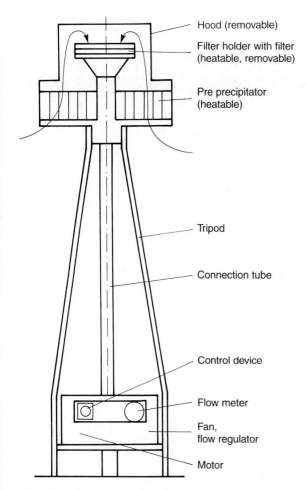

Fig. 5.32. Diagram of a high-volume sampler [123]

namic diameter of > 0.3 µm settle. Exposition time per filter usually amounts to 24 hours. Fig. 5.32 shows a diagram of such a high volume sampler [123]. Filter and preseparator are heatable, as the pressure loss at the filter would increase during the collection of, e.g., fog droplets. Owing to the heater the droplets quickly evaporate. Besides, when collecting fog droplets in winter there would also be the danger of the filter freezing over.

Low-volume particulate samplers

Apart from high volume samplers there are also low-volume samplers with smaller filters, e.g., 120 or 50 mm ⌀ [124-127]. These instruments usually have a rate of air flow of between 2.5 and 16 m³/h and are set up for sampling times between 1 and 24 hrs. Sometimes there is so much dust on the filter, that in addition to determining the

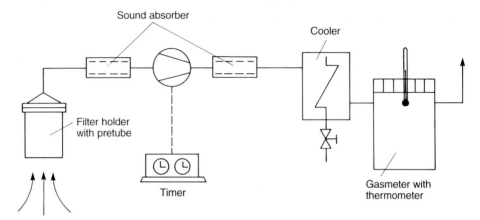

Fig. 5.33. Set-up of a particulate matter sampling device [124]

weight, identification of the individual particles collected can also be carried out. Fig. 5.33 shows the diagram of such a dust sampler.

Minimization of chemical reactions on filters

As mentioned earlier, particle collection filters should be heated to avoid problems when collecting fog droplets. Damp filters also have another disadvantage – gaseous SO_2 dissolves in the humidity, reacting with certain kinds of dust particles, e.g., with calcium, to become sulfate [128]. Thus sulfate and sulfur determinations of the particles collected would lead to errors, so-called artefacts, as part of the sulfate was formed on the filter. Artefact formations have not only been observed with sulfates but also with other substances, e.g., with nitrates and organic substances [117]. Heating the filters solves this problem for the most part. In this context it must be noted, that some filter materials are particularly prone to absorb humidity due to its hygroscopic properties. In the presence of acidically reacting gases such as SO_2 and NO_2, glass fiber filters may not be used, as they are usually slightly alkaline and, combined with the residual humidity, cause the gases to remain on the filter [117].

Heating the filters, however, can also entail dangers: highly volatile particles can vaporize, particularly when working with high volumetric flow rates.

When identifying dust components one must also pay attention to the fact that the particles separated on the filter can interreact chemically with each other. The longer the filter exposure time, the more transformations can take place. In a particulate sampling optimized for these factors the following must be observed:
– careful choice of filter material,
– gentle heating,
– the lowest possible volumetric flow rates,
– short sampling times.

Naturally, compromises must be made, as small volumetric flow rates and short sampling times only lead to small volumes of particulate matter being sampled which in turn magnifies errors in subsequent analyses.

5.3 Measuring Methods for Particulate Matter

Particle fractionation during sampling and determination of particle size distribution

Generally, particle sampling instruments have filter covers to keep rain out and to prevent particle depositions or separation of too large fragments, e.g., leaves, from being sucked in. Depending on the design of the cover on the filter holder different

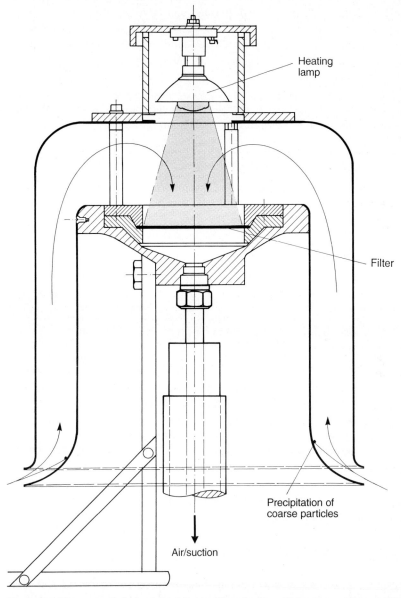

Fig. 5.34. Diagram of a heated particle sampling device with coarse particle precipitator (PM 10 µm)

sized particles are separated. E.g., at a vertical flow rate of approx. 0.55 m/s, particles in a diameter range below 80 µm are collected in the suction tube shown in Fig. 5.33 [124]. In particulate measurements mainly for the measurement of respirable particles a specially shaped preimpactor holding back all particles with a diameter larger than 10 µm [129] from the filter is necessary. In air pollution measurements in the forests of Southern Germany suction tubes working according to this principle are used to preseparate coarse particles, s. Fig. 5.34 [130, 131]. Particles with an aerodynamic diameter > 20 µm as well as water and large fog drops are impacted during gas diversion in the suction slit. A filter heater is also included in this set; it consists of an electric bulb with a reflector radiating onto the filter. This type of heating guarantees an even heat distribution and direct contact of the heating filaments with the filter is avoided.

In ambient air measurements *particle size determinations* of the particles to be collected and analyzed are carried out almost exclusively with impactors of the most varying types of construction. The principle of particle fractionation with a cascade impactor has already been explained in the context of Fig. 5.29.

5.3.6.2 Automatically Recording Particle Concentration Measuring Instruments

Many cases require a periodic continuous particle concentration measurement in the air. Instruments with automatic filter changers or long strips of filters are suitable for this purpose only to a certain extent as the collecting times for each filter do not become shorter and often amount to at least 12 hours.

For continuously recording particulate content determinations, as is the case with particle emission measurements, either measuring instruments are used which use as measuring effect the reduction of β-radiation on filters which are loaded with particles or optical instruments are employed. Both principles only permit determinations of total dust content but not identification of component particles which are often of greater interest.

β particle meters

The principle of ambient particulate measurement according to β-radiation absorption is the same as in emission measurement, s. Fig. 5.27. In the Federal Republic of Germany two measuring instruments operating on this principle are approved for ambient particulate measurements [132, 133]. The sampling systems for these instruments have also been standardized.

Scattered-light photometer [134]

If a light beam is sent through a gas volume laden with particles the light is both attenuated and at the same time scattered by the particles. In optical smoke density and dust content measurements for emissions light attenuation is measured as a

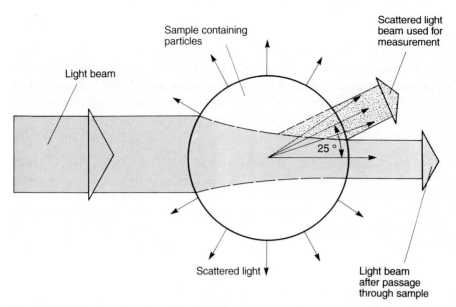

Fig. 5.35. Light scattering of particles [134]

measure for the particle content. In ambient air measurements, however, the instruments used evaluate scattered light which is scattered forward at a 25 ° angle to the incident light beam and is thus deflected from the main light beam, s. Fig. 5.35.

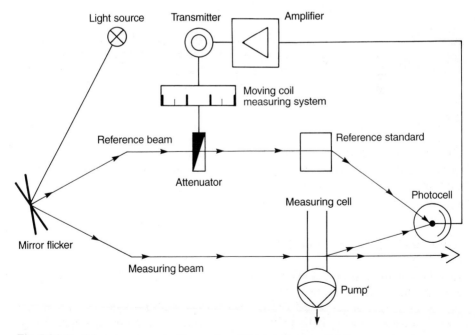

Fig. 5.36. Particle measurement with a scattered light photometer [134]

The principle of the measuring instrument based on this scattered light process can be seen in Fig. 5.36 [134]:

A beam of light is sent by a mains-powered light source to the *flicker mirror* which alternately generates a *measuring beam* and a *reference beam*. The *measuring beam* reaches the *measuring cell* and is scattered by the particles there. The scattered light is received by the *photocell*. The reference beam is passed through an attenuator and the *reference standard*. If the beams of light alternately reaching the photocell all have the same light intensity, a direct current results, the optical bridge is balanced. More particles in the measuring cell generate more scattered light resulting in an alternating current adjusting the attenuator (making it penetrable) for a period of time until the alteration ceases owing to the same light intensities of both beams. Obstruction of the attenuator until bridge balance is achieved again is shown on the *moving coil measuring system* and is a direct, electrical measure for the amount of scattered light. The amount of scattered light depends on the number and size of the particles in the air volume to be measured.

5.3.7 Determination of Chemical Composition of Particulates

A wide variety of methods are applied to determine components of particulate sedimentation and suspended particles.

The following paragraphs introduce some examples of analytical methods without laying any claim to completeness. Further information can be obtained from specialized literature, e.g., [117, 135].

Microscopy

The particles collected on, e.g., adhesive foils, are examined under a light microscope. Soot particles, tire abrasions, plant pollen etc. can be identified easily [136].

Ashing and residue determination

Organic suspended particulate content can be determined by ashing in an annealing furnace and by determining the loss of weight. This can, e.g, be of significance if the organic substance content consists mainly of soot, e.g., in the heating season during the winter months.

Atomic absorption spectrometry (AAS)

AAS is a standard laboratory method for identifying metals. The metals are extracted from the filters by means of a solution and then vaporized in a flame. A beam of light with a wavelength which can be absorbed by the species of interest is sent through the vaporized sample. Light absorption is measured and the metal mass is determined according to the Lambert-Beer law by comparison with standards.

X-ray fluorescence

The sample on the filter is excited with the help of monochromatic X-rays. When excited in this manner, each element emits characteristic X-rays (X-ray fluorescence) whose wave-length or energy is used for identification and whose intensity is a measure of the element mass on the filter surface being examined [137].

Particle-induced X-ray Emission (PIXE)

With this method particles heavier than sodium can be analyzed. The sample is bombarded with a particle beam, usually protons. During this, the elements in the sample are excited, leading to an emission of X-rays whose wavelengths or energies are characteristic of each element [138].

Neutron activation analysis (NA)

The sample is bombarded with neutrons and radioactivity thus induced in the particles, particularly the characteristic gamma rays released, are measured [117, 135].

Mass spectrometry (MS) and laser microprobe mass spectrometry (LMMS)

Mass spectrometry is a general method for the identification of organic substances (s. also sect. 5.2.13), but some inorganic components can also be analyzed with it. To analyze individual particles a special version, laser microprobe mass spectrometry (LMMS), is applied. The particles settling on a thin, organic carrier foil are vaporized individually with laser impulses under microscopical control. The conditions present in the laser focus lead to the formation of element, molecule and molecule fragment ions. The masses of the ions thus formed are analyzed in the mass spectrometer (s. also Fig. 5.20) via time of flight [139].

Colorimetry

Ions adhering to different particles or ions of which the particles themselves consist, e.g., Cl^-, NH_4^+, SO_4^{2-}, NO_3^- a.o. can be analyzed with colorimetric methods. For this, the particle sample is put into a chemical reaction solution where the ions of interest cause color to form. Color intensity is determined photometrically (s. also sect. 5.2.2).

Ion chromatography (IC)

The ion content of particles, e.g., sulfate, nitrate, chloride etc. is frequently determined with ion chromatography. The sample is put into an aqueous, possibly acidified solution, the particles are filtered off and then the solution is subjected to ion chromatographic separation and analysis [140].

The precondition for the application of each method of substance identification is that the *filter materials* do not interfere with the analysis. This is particularly the case when the filter itself contains elements which are to be analyzed or when the filters, e.g., in physical analyses (AAS, X-ray fluorescence etc.) generate an excessive noise. Materials therefore need to be chosen accordingly. In the VDI guideline 2463, no. 1, filters suitable for the measurement of particles are listed [141].

5.4 Setting up of Measuring Equipment, Sampling Techniques and their Influences on the Accuracy of the Measurements

When measuring air pollution, be it in the form of emissions or in the ambient air, the method of sampling has a decisive influence on accuracy. Applying the most expensive measuring instrument is of no use if, e.g., the pollutant to be investigated is influenced during its sampling to such a degree that it either cannot reach the measuring instrument at all or not in its original state. Requirements which must be met in sampling when carrying out emission and ambient air measurements vary to some extent. For this reason both topics will be dealt with separately here.

5.4.1 Emission Measurements in Furnaces and Process Equipment

5.4.1.1 Location of Sampling

The choice of a certain sampling location for emission measurements depends on the task to be accomplished. If emissions reaching the environment are to be monitored in a large furnace, then the measuring instruments must be installed at a section of the exhaust gas system where the exhaust gases are not further transformed, i.e., in the stack or in the exhaust gas channel just before the stack, Fig. 5.37.

If, however, burning behavior is to be monitored with a CO or soot measurement, then the measuring instruments must be installed as close as possible to the end of the furnace, s. Fig. 5.37. To control the efficiency of exhaust gas cleaning devices, as in the example for the reduction of NO_x or SO_2 shown in Fig. 5.37, measuring instruments must be installed at both ends of the cleaning system.

When flue gas channel cross sections are large it must be taken into consideration that due to possible streaks in the flue gas the concentration distribution may not be even. As an example, Fig. 5.38 shows the distribution of NO_x and O_2 concentrations in the flue gas channel of a pulverized coal furnace. Gas samples for continuously operating measuring instruments must be collected at a representative spot, i.e., where concentrations are as close to the mean value as possible. However, concentration distribution can also shift when the load changes. If mass flows of pollutants are to be determined from the concentrations, then it must be taken into account that

5.4 Setting up of Measuring Equipment

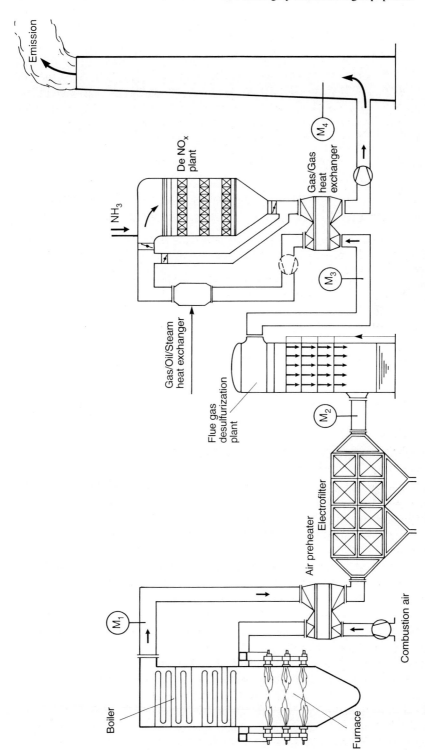

Fig. 5.37. Set-up of the measuring points (M1-M4) in a power plant furnace with flue gas cleaning system

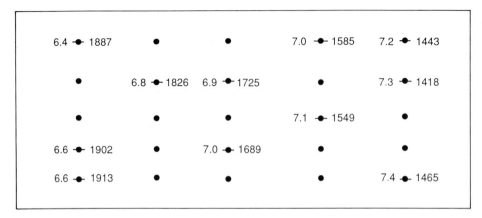

Fig. 5.38. Different NO_x and O_2 concentrations in the flue gas channel of a pulverized-coal furnace (acc. to [142]). Numbers on the left indicate the O_2 concentration in vol. %, numbers on the right indicate the NO_x mass concentration as NO_2 in mg/m³ related to 0 °C and 1013 mbar, dry and 5 vol. % O_2. Mean value calculated from all points of net measurement: NO_x as NO_2=1 690 mg/m³

differing gas velocity distributions can also occur. In emission measurements of the particulate content of flue gases grid measurements in the flue gas channels are prescribed [98] so that representative results will be achieved. For this, the flue gas channel cross section is divided into rectangles or rings with the same area, sample gas is sucked off from the center of each area, and the mean value of the particulate emission for the total channel cross section is determined.

5.4.1.2 Setting up of the Site of Measurement – Sampling System

The sampling system must collect the gas to be measured from the flue gas channel, remove interfering components and feed the gas into the measuring instruments. The most important interfering factors influencing the measurement of gaseous pollutants in furnaces are the particle and water vapor content of the flue gases as well as leaks in the measuring system. Besides, certain measuring instruments or measuring techniques can be highly sensitive towards other gaseous components in the flue gas. Ambient influences, above all temperature and vibrations, can also have a disturbing influence on the measurements. The two latter interfering factors of extreme sensitivity and ambient influences, can be kept at a minimum by choosing the right measuring instrument. The first-mentioned interfering factors, however, have to be overcome by the sampling system.

If dust particles enter the measuring system they can cause clogging of the devices and faulty readings. The exhaust gases of most furnaces contain a lot of water vapor which condenses when cooling. Condensed water in the measuring system inevitably leads to malfunctioning. Moreover, acid exhaust gas components, such as SO_3, dissolve in the condensed water, so that apart from the interfering influence of the water itself corrosive effects have to be coped with.

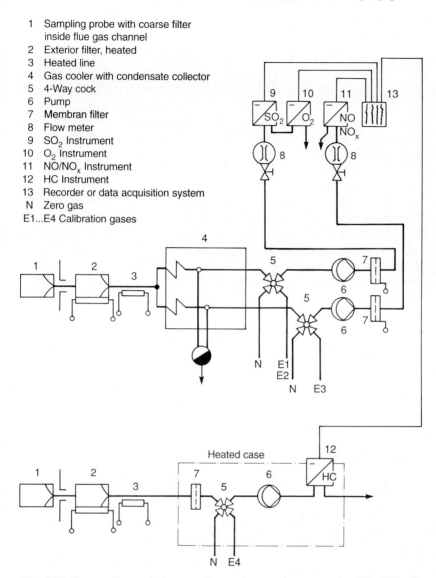

1 Sampling probe with coarse filter inside flue gas channel
2 Exterior filter, heated
3 Heated line
4 Gas cooler with condensate collector
5 4-Way cock
6 Pump
7 Membran filter
8 Flow meter
9 SO_2 Instrument
10 O_2 Instrument
11 NO/NO_x Instrument
12 HC Instrument
13 Recorder or data acquisition system
N Zero gas
E1...E4 Calibration gases

Fig. 5.39. Set-up of an emission sampling and measuring system for SO_2, NO_x, O_2 and HC concentrations

Fig. 5.39 shows the set-up of a sampling system in which interfering influences ought to be minimized as much far as possible.

The particle filters which are located before the water condensator must be heated. Coarse particle filtering is best carried out in the sampling probe directly in the exhaust gas channel (1). The heat of the exhaust gas serves as heating. The connecting tube between sampling probe and measuring gas cooler should be kept as short as possible. To be consistent, it should be heated, too (3), although this is not always the case in practice. Possible errors caused by this will be dealt with further in the next section.

Gas coolers with condensate separators (4) are now standard for drying the measuring gas in measurements of the classic pollutants SO_2, NO_x, CO and O_2. Errors which can appear due to this type of measuring gas drying will also be dealt with in the next section. In the measurement of most gas components gas drying with absorbants is not possible, as these drying substances frequently do not only absorb water vapor but also the components of interest. Residual drying can also be carried out with so-called permeation dryers [143]. However, drop separation cannot be replaced by this. Since the water is separated from the exhaust gas, the measuring results finally refer to dry exhaust gas.

The fine filter is important (7). If the fine particulates are not filtered out carefully, the cuvette windows in, e.g., infrared or UV gas analyzers gradually become covered, thus leading to a reduction of the instruments' sensitivity. In chemiluminescence measuring instruments (for NO_x) fine particulates gradually clog the measuring gas capillaries which also leads to malfunctioning. To be able to detect in continuous operation whether condensate escapes from the cooler before it reaches the measuring instruments, fine dust filters are often fitted with condensate sensors (conductivity measurement).

If the sample gas contains gas components which would themselves condense on cooling, the gas cannot be conducted through coolers. In this case the entire measuring system including the measuring instrument must be heated to avoid condensation. This is, e.g., general practice in hydrocarbon measurements with flame ionisation detectors, s. Fig. 5.39. In this case, water vapor condensation takes place after the measuring instrument.

5.4.1.3 Possible Faults during Sampling

Leakages

The greatest source of faults during sampling are leakages. If the particle filters in the measuring gas probe become clogged, a vacuum can build up to the measuring gas pump. Under these circumstances even the slightest of leakages will lead to the penetration of ambient air and thus to the dilution of the measuring gas and to faulty measurements. The most effective leakage control can be exercised via O_2 measurements by introducing pure nitrogen into the measuring system. If infiltrated air penetrates into the system at any point, then the O_2 measuring instrument indicates the oxygen in the air. By introducing nitrogen into the system at different points the leak can be pin-pointed.

Measuring gas losses in the condensate

Faults can also occur when the measuring components of interest dissolve in the condensed water of the gas cooler or at other points of the sampling system and thus

Table 5.7. Solubilities (Henry coefficients H)[1] of some air polluting gases in water at 25 °c without chemical reactions acc. to [117]

Gas	H (mol l^{-1} bar^{-1})
O_2	$1.3 \cdot 10^{-3}$
NO	$1.9 \cdot 10^{-3}$
C_2H_4	$4.9 \cdot 10^{-3}$
NO_2	$1 \cdot 10^{-2}$
O_3	$1.3 \cdot 10^{-2}$
N_2O	$2.5 \cdot 10^{-2}$
CO_2	$3.4 \cdot 10^{-2}$
SO_2	1.24
HONO	49
NH_3	62
H_2CO	$6.3 \cdot 10^{3}$
H_2O_2	$(0.7 \cdots 1.0) \cdot 10^{5}$
HNO_3	$2.1 \cdot 10^{5}$
HO_2	$(1 \cdots 3) \cdot 10^{3}$
PAN	5
CH_3SCH_3	0.56

[1] In German literature these Henry Coefficients are also called solubility L: $L=c/H_d$, where c=concentration in the fluid in mol per l and H_d=Henry coeffcient acc. to German definition in bar.

do not reach the measuring system completely. Particularly with highly water soluble and condensable gases losses are to be expected. The more water-soluble the gases are, the more problematic the sampling becomes. Table 5.7 shows the water solubilities (Henry coefficient) of some pollutant gases.

As can be seen, sulfur dioxide (SO_2) is 100 x more soluble than nitrogen monoxide (NO); ammonia (NH_3) is still 50 x more soluble than SO_2. For this reason, it is very difficult to measure NH_3. This has recently become evident, especially when measuring NH_3 slippage behind catalytic converters for the reduction of nitrogen oxides in power plant emissions. Continuous NH_3 trace measurement is difficult enough, but problems of the sampling side are even greater. Apart from its high water solubility this gas reacts with SO_3 below approx. 280 °C to become ammonia sulfite and ammonia sulfate. To avoid losses in this way it is necessary to heat the entire system up to 300 °C. This is not an easy task to accomplish as, the optical systems integrated in the measuring instruments can warp due to the different expansion coefficients of the components of the instruments.

SO_3 measurement, e.g., is just as problematic as NH_3 measurement. Apart from a highly maintenance-intensive instrument operating on wet-chemical principles there are no automatic measuring instruments for this exhaust gas component. Results from manually measured analytical methods decisively depend on how the sample was collected. To avoid SO_3 losses sampling lines are not used at all, and the measuring device is fitted directly at the nozzle of the sampling probe. The particle filter used must also be submitted to an analytical SO_3 determination as SO_3 can be adsorbed to the dust particles filtered out.

When dealing with critical measuring components such as hydrocarbons, NH_3, SO_3 and other gases, drying of the gas to be measured by cooling and condensation is to be avoided. However, when measuring the classic gas components SO_2, NO, NO_2, CO, CO_2 and O_2 *gas coolers* are used without any reservations. Hardly any losses of CO, NO and also O_2 and CO_2 which have low water solubility are to be expected, the reason being, among others, that the wetted condensation distance in the gas cooler and with it the residence time of the gases over the condensation water are short. With SO_2, however, whose water solubility is relatively high, losses in the measuring gas cooler must be expected. Experimental investigations of furnace exhaust gases with an SO_2 content of 1.6 g/m^3 and 10 % water vapor revealed SO_2 losses in the cooler of upto 4.5 % [144]. Losses depend on cooler temperature and on the water and SO_2 content of the exhaust gases. Theoretical calculations [145] agree quite well with the experimentally gained results [144]. The higher the water content of the exhaust gases and the lower the SO_2 content, the higher the losses are (the larger amount of condensation water can dissolve more SO_2). Cooler temperature has the greatest influence. The warmer the condensation water, the lower the losses. If the cooler is warmer, however, the efficiency of water condensation declines. Due to the increased water vapor content of the gas, this in turn leads to an increased interference of the measuring instruments, particularly in infrared gas analyzers. If the SO_2 content is low and cooler temperatures are at 2-5 °C losses can amount to over 10 %.

For the calibration of SO_2 measuring instruments the Hartmann & Braun Company [146] recommends wetting the calibration gases which are then to be conducted over the gas cooler just like the measuring gas. In the cooler the calibration gas then acquires a water vapor content corresponding to the one in the measuring gas. In this way faulty responses of the infrared gas analyzers sensitive to water vapor are compensated for. However, SO_2 losses in the cooler do not correspond to those gas conditions of the sampled gas as the cold calibration gas cannot absorb as much water vapor as the hot exhaust gas. However, a certain fault compensation is to be expected by this measure, so that SO_2 losses in the cooler are not as critical.

As *unheated measuring gas lines* are sometimes used in practice it was tried to determine the ensuing SO_2 losses experimentally [144]. However, it was not possible to achieve a definite result as SO_2 losses basically occur whenever there are condensation phenomena in unheated measuring gas lines. On the other hand, liquid or dried condensate in the tube has a certain storage effect so that behind an unheated measuring gas tube more SO_2 can be found after heating than was actually present in the exhaust gas. Thus, unheated gas lines lead to undefined conditions in the sampling system when sampling water-soluble and reactive gases.

5.4.2 Emission Measurements for Motor Vehicles

One must distinguish between pollutant measurements done on engines running on test blocks and those on motor vehicles directly. E.g., measurements of engines serve the purpose of determining the dependency of pollutant emissions on certain engine factors such as performance and speed. The measurement set-up is similar to that used for furnaces.

To control emission threshold values measurements must be carried out for the motor vehicle itself and not just for the engine alone. For this it must be taken into account that pollutant emissions depend to a high degree on the driving behavior of the vehicle. Driving behavior is simulated on roller dynamometers by various driving programs and the pollutants emitted are measured.

5.4.2.1 Exhaust Gas Sampling and Measurement According to the CVS Method

Though the prescribed driving programs (driving cycles) and emission threshold values vary from country to country, the method of exhaust gas sampling and meas-

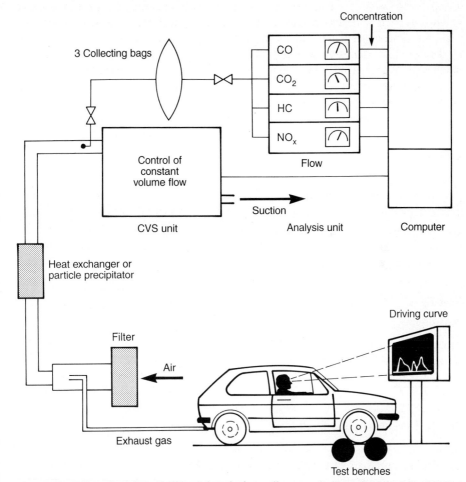

Fig. 5.40. CVS method for sampling and analyzing pollutant emissions of automobiles [149] (internal combustion system; for diesel vehicles an additional sampling system for particulate matter is included)

urement has in the meantime been standardized in the USA, Japan and Europe. The so-called CVS method (constant volume sampling) is applied. The principle underlying this type of sampling of motor vehicle exhaust gases is shown in Fig. 5.40. For this, the exhaust gas produced by the motor vehicle during the test cycle is diluted with purified ambient air in such a way that a constant volume flow of exhaust gas and air is being sucked through. A small partial flow is continuously taken from this air-exhaust gas flow, sampled in bags and analyzed after the test with continuously operating measuring instruments. Pollutant concentrations in the sampling bags correspond to the mean values of the exhaust gas-air mixture suctioned off during the test. The pollutant quantities emitted during the driving cycle can then be calculated on the basis of the known volume flow and the individual gas concentrations [147, 148].

The advantage of this dilution method is that the dew point of the exhaust gas is lowered and water condensation is thus avoided. By keeping the air-exhaust gas volume flow at a constant level the problem of continuously recording continually changing exhaust gas volume flows is conveniently solved.

The limited exhaust gas components and the measuring methods which have in the meantime been standardized in most countries are listed in Table 5.8.

To determine evaporation losses (total hydrocarbons, HC) in the fuel system the vehicle to be tested is placed in a gas-tight chamber. After going through a one-hour test the increase of the C_nH_m concentration in the chamber is measured [147].

Table 5.8. Limited compounds in motor vehicle exhaust gases and measuring methods applied

Component		Measuring Method
Carbon monoxide	CO	Non-dispersive infrared absorption (NDIR)
Nitrogen oxides	NO_x ($NO+NO_2$)	Chemiluminescence instrument
Hydrocarbons (HC)	Total HC	Flame ionization detector (FID)
Particulate matter	Soot	Not standardized yet; collection on filters or opacity of exhaust gas
Vaporization losses	Total C_nH_m (HC)	FID
in addition		
Carbon dioxide (for fuel consumption)	CO_2	NDIR
Oxygen (for controlling of fuel/air ratio)	O_2	Zirconium dioxide sensor or paramagnetism

5.4.2.2 Various Driving Cycles

The driving programs carried out on the roller dynamometers for the measurement of exhaust gas vary greatly.

In the USA a driving program was introduced which approximately simulates the driving conditions existing on a Los Angeles highway in the morning [148]. The

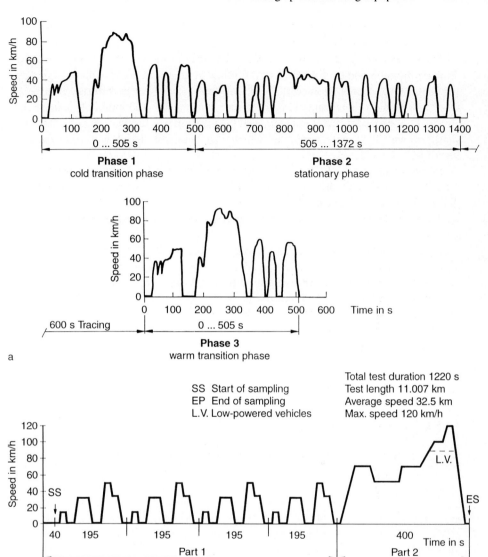

Fig. 5.41. Different driving cycles. **a.** US-Test 75 (FTP-75: Federal Test Procedure) **b.** Part 1: conventional ECE-driving cycle (city); Part 2: Extra-Urban Driving Cycle (EUDC); Part1+Part2: MVEG (Motor Vehicle Emission Group)

driving curve which must be simulated on the roller dynamometer is shown in Fig. 5.41a. This test is characterized by numerous acceleration and braking operations and has a highway share of 90 km/h in the cold and in the warm phase respectively. The USA have the most stringent regulations, i.e.:
- HC (C_nH_m): 0.41 g/mile,
- CO: 3.40 g/mile (California 7.0),
- NOx: 1.00 g/mile (California 0.4).

With the exception of the additional warm phase the US test 75 has been adopted by Switzerland and Sweden.

The EU countries apply the so-called ECE test, s. Fig. 5.41b. This test was to simulate the stop-and-go traffic of city driving during rush hours with low speeds and long idling phases. However, the actual driving behavior is reflected in a very simplified fashion. Sampling starts 40 s after starting the cold engine (point BP in Fig. 5.41). It proved difficult not to exceed the maximum values during this cold start phase, above all in motor vehicles which had been retrofitted with exhaust gas cleaning equipment.

In the meantime the ECE test has been supplemented by a second part, so that extra urban traffic with driving speeds of 80 to 120 km/h is taken into consideration (Extra Urban Driving Cycle, EUDC) [148].

The ECE threshold values are staggered according to the engine size of vehicles, s. [147].

In Japan, on the other hand, other tests than those in the USA and Europe are prescribed. It is therefore a difficult undertaking to compare pollutant emissions of motor vehicles which were determined with different testing programs.

As even the US test, which covers many driving conditions, is merely a simulation, and actual driving behavior can be quite different in each individual case, measurements in actually running motor vehicles under various real traffic conditions are now carried out in addition to the tests on roller dynamometers [150].

5.4.3 Ambient Air Measurements

5.4.3.1 Significance of the Locations of the Measuring Sites

When measuring pollutant gases in the air a difference must be made between mobile measurements with measuring vehicles and stationary measurements at measuring stations. Mobile measurements are carried out at random at regularly changing locations to determine the spatial distribution of the air pollutants, whereas stationary measurements continuously record the temporal distribution at only a few points of a certain area, e.g., in a city. Particle and rain precipitation measurements, similar to the distribution of the mobile measurements of the gases, are frequently carried out at numerous locations with collecting vessels which have a temporally integrating effect (s. sect. 5.3.5).

Stationary measurements must be carried out at a representative point of the area to be investigated. For this purpose it would be a good idea to know the spatial distribution of the ambient air pollutants which is determined with mobile measurements at the corner points of a 1 x 1 km grid over the area to be investigated.

In the early and middle seventies isolated measuring stations in the industrial and urban centres of the FRG developed into measuring networks of the German states which have been continuously expanded. When the recent forest damage became known, forest measuring stations were built by the German states in the early eighties. Moreover, the German Federal Environmental Agency (Umweltbundesamt) runs a measuring network in the FRG with several measuring stations outside the urban and industrial centres. In the eighties, the long-range transport processes of air pollutants were determined, based on the results of these "clean air stations". These processes were the reason for high pollutant concentrations being found even in rural areas during certain weather situations, in particular high SO_2 values. This lead to the fact that the existing ambient air measurement networks were recently expanded by further stations, above all in the smaller towns.

Overviews of the measuring networks in the FRG are given in [152, 153].

Selection of the right sampling site can be of great importance for stationary as well as for mobile measurements, on which, e.g., the calculations for the mean values of 1 x 1 km areas are based.

In the FRG guidelines have been drawn up so that in the measuring networks the site locations for automized measuring stations are chosen according to standardized criteria. These guidelines are to serve as a basis when planning telemetric air quality measuring networks, i.e., with telemetric data transfer to a central station [151]. When selecting site locations for ambient air quality measuring stations the following factors must be taken into consideration:
- goal: ambient air sample representative of the area,
- measuring grid with constant mesh size: multiple of 1 km, in a north-south/east-west direction,
- distance of measuring site to the closest flow obstacle of twofold height or width of the obstacle; sample suction below half of the mean building height,
- keep influence of local emittents low: keep distance to sources (industry, domestic furnaces, streets with heavy traffic) ≥ 20 m, special investigation when industrial sources are situated nearby,
- free flow accessability of sampling systems: within a radius of ≤ 10 m no flow obstacles such as trees or buildings,
- excluding influence on measuring site by topographically caused local circulation.

The guidelines also indicate that for the investigation of special pollution and depending on the specific issue the appropriate principles are to be applied when selecting the site location.

The measuring stations of the FRG were generally set up according to the above mentioned guidelines. Since 1985, an EU guideline requires that the pollutant gas NO_2 must be measured continuously, particularly "in locations with the suspected highest load risk", i.e., also near roads with heavy traffic, in street canyons etc. [154].

When carrying out air quality measurements it is important to define the problem precisely (submission of a measuring plan) and to choose the site locations or measuring stations accordingly. The significance of spatial and temporal distribution of air pollutants and their identification will be dealt with in greater detail in sect. 6.2.

5.4.3.2 Sampling Systems in Measuring Stations

The following sections will deal with the air suction in measuring stations for pollutants which can be recorded with continuous measurements. Nowadays mobile measurements are primarily carried out with measuring vehicles which are equipped with continuously operating measuring instruments. Thus, the measuring set-up corresponds to the most part to the one in stationary measuring stations.

Sampling system in automized measuring stations

It is the task of the sampling system to feed the ambient air into the measuring instruments in the most as unadulterated form possible. The guidelines for the automized measuring stations in the measuring networks of the FRG also provide standards concerning the design of the sampling lines for gaseous and particulate pollutants. Accordingly, the sampling line for gaseous pollutants consists of a sampling nozzle, a guide tube, the central sampling tube, sampling lines leading from the central sampling tube to the individual measuring instruments and a fan or a pump. The sampling nozzle should be constructed as a preseparator for particles and precipitation.

The sampling line should extend 1 m beyond the station roof.

In the standard measuring stations particle filtering in the sampling system is not provided for to preclude possible reactions of the gases on the filters. To protect the measuring instruments particle filters are only installed at the input of the instruments themselves. A special sampling nozzle is prescribed for the collection of particulates [151].

Sampling for special measuring tasks, e.g., the measurement of concentration profiles

In air quality investigations pollutant concentration profiles must frequently be measured. Examples for this are the determinations of pollutant concentrations in and over forest stands at different altitudes [131, 155] or the registration of the concentration decline in the vicinity of streets with heavy traffic [156, 157]. To be able to simultaneously or quasi-simultaneously measure the gas concentrations at several locations there is the possibility of sucking the air in via tubes and to conduct it to a set of measuring instruments with the help of a measuring gas change-over in short periods of alternation.

When measuring gas is sucked in through long tubes the influencing of gaseous substances is generally to be expected even if the tube wall material is inert, e.g., teflon. This depends on the reactivity of the gases to be investigated. Examinations with long tube lines have shown that the component ozone is most prone to losses in the tube system whereby a clear dependency on the rate of flow through the tube is to be observed. Ozone losses in measuring gas tubes can be minimized with short periods of contact which can be achieved by sufficiently high suction speeds [158, 159].

With the gas components NO, NO_2 and SO_2 no losses have been observed in the tube even during low volume flows [158, 159].

Apart from measuring gas losses caused by wall effects, faulty readings of the measuring instruments caused by low pressure in the suction system must be taken into consideration and avoided [160].

Diverging from the above mentioned guideline [151] is deviated from, the air sucked in must be filtered when entering the tube to avoid dust depositions on the relatively large inner surface of the long tubes. A tube soiled on the inside surface would interfere with concentration measurements by uncontrolled ad- and absorption to an undefined extent.

The influence of the amount of particles precipitated on the filter on trace gas measurement was examined: when filters were dry no influencing of the gases O_3 and SO_2 could be measured. However, great losses occur with the water-soluble gas SO_2 when the filters are very damp or saturated with water. For O_3 there were no measurable influences even in this case [159].

As has already been indicated in sect. 5.3.6.1, gases react particularly with particulates on damp filters. Apart from interfering with the separated particle components there is also the danger of adulterating the measured gases. In the suction hood shown in Fig. 5.34 the losses of trace gases were minimized by the filter heater.

5.4.3.3 Set-up of Air Quality Measuring Stations – Example of a Forest Measuring Station

Based on the findings described in the section above, two measuring stations were set up in the Black Forest [131, 164]. Within the framework of forest damage research similar measuring stations were set up in Switzerland [161] and and by other institutions in Germany [162, 163]. Fig. 5.42 shows a diagram of a set-up of such a forest measuring station with measuring gas suctions. The measuring gas suction hoods (corresponding to the one shown in Fig. 5.34) and different meteorological measuring instruments are installed on a 42 m high lattice tower at different altitudes.

The gas sampling system equipped with 2 suction inlets as well as the set-up of the pollutant measuring instruments are shown in a diagram in Fig. 5.43. Apart from the instruments for the measurement of pollutants there are numerous other instruments for the recording of meteorological parameters such as wind direction and wind speed, temperatures, global radiation, duration of rain and bedewing, amount of rain etc..

Measured value registration is carried out by a computer controlled electronic data acquisition system. The computer also controls the valve timing and correctly stores the measured values according to the given valve position. Half-hourly mean values are calculated which are are stored on disks and printed out. Further evaluation of the data is then carried out on a large computing system. Some measuring stations work with direct data transmission to a central computing station. For reasons of safety the measured values are additionally recorded by multichannel continuous-line recorders or multipoint recorders independently of the computer controlled measured value recording system. If the computer breaks down these recorder graphs can be evaluated if necessary. Besides, the computer's recordings can be verified once again with the help of these strip-charts.

The measuring principles of the pollutant measuring instruments are listed in Table 5.9.

To record meteorological parameters the measuring instruments customarily applied by the meteorological services are used.

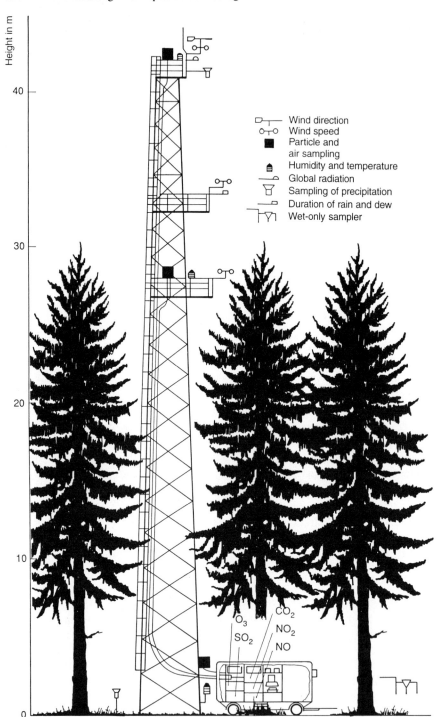

Fig. 5.42. Diagram of a forest measuring station with air sampling at different heights above the canopy and within the stand

5.4 Setting up of Measuring Equipment

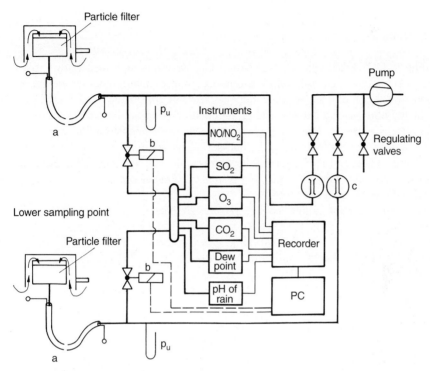

Fig. 5.43. Set-up of the sampling system of a forest measuring station with air suction through long tubes and with sampling gas switching; *a.* Teflon tube, heated, 45-26 m long; *b.* switch-over valve; p_u low-pressure measurement; *c.* flow control

Table 5.9. Principles of measuring instruments for air quality control in a forest monitoring station

Compound to be measured	Measuring Principle
NO, NO$_2$	Chemiluminescence
SO$_2$	UV fluorescence
O$_3$	Chemiluminescence or UV absorption
CO$_2$	IR absorption
Dew point or humidity	Dew point level
pH of rain	Precipitation monitor including flow-through cell with pH electrode

With the measuring technology described here it is possible to continuously determine the pollutant load in forests [131, 163, 164].

5.5 Calibration for Pollutant Measurements

5.5.1 Definitions

According to DIN 1319 *adjusting* signifies setting or trimming a measuring instrument as accurately as possible, while calibrating signifies determining the deviation measured as compared to the accurate value or the value considered as such [165].

In air pollution measurements calibration conditions are set up with the help of *calibration gas mixtures*. These are gas mixtures whose composition has been predetermined with sufficient certainty by measuring basic parameters such as mass, volume, time, amount of substance (molar number) or with the help of independent analysis methods and which influence the system to be calibrated in the same way as the air samples to be analyzed [166].

During calibration the calibration gas mixtures are entered into the measuring set-up (measuring instrument or measuring process), the values indicated are read and compared with the values of the calibration gases assumed correct, then the deviations are recorded. Frequently, calibrating a measuring instrument is accompanied by other measures such as maintenance, function control, trimming and others.

Hartkamp et al. [166] divide the total of all calibration processes into three types of calibrations:
- basic calibration
- routine calibration
- and control calibration.

Basic calibration

The basic calibration provides the fundamental relationship between given calibration gas concentrations and recorded signals, and is called calibration function. In many instruments linear calibration functions are indicated. During the basic calibration not only gradient and zero point of this calibration function are checked but also the linearity itself.

The basic calibration can basically also be carried out by comparative measurements with reference measuring methods, e.g., directly in the exhaust gases to be measured [167, 168]. As reference methods mostly wet-chemical measuring methods are used in which the measurement is based on determining basic parameters such as mass, volume etc..

Routine and control calibrations

According to Hartkamp routine and control calibrations ensure the validity of the calibration data of the basic calibration. This control verifies whether the data obtained by the basic calibration carried out last are still valid. The results of routine and control calibrations are yes-no-decisions. The many aspects which must be taken into consideration during calibration and which depend on the task at hand also the different calibration methods for air pollution measurements have been described in great detail by Hartkamp et al. [166].

5.5.2 Calibration Gases

The principle of calibration gas production is based on the procedure of adding a known amount of the gas of interest to a known volume or volume flow of carrier gas (mostly N_2 or air) in certain doses.

This dosing can be carried out via mass determination (gravimetrically), volume determination (volumetrically) or via a partial pressure determination. When converting the concentration values thus obtained into other units (e.g., volume concentration in mass concentration) it must be taken into consideration that there are possible deviations from the ideal law.

Whether the calibration gas is produced with static or dynamic (i.e., dosing as substance flow) processes depends on the required amount of calibration gas and on other frame conditions, e.g., on the desired concentration and on the required exactness and stability of the calibration gases.

The following sections list the different processes of calibration gas production. Most of the methods are described in detail in ISO standards VDI guidelines [76] or federal regulations [176].

5.5.2.1 Static Methods for the Production of Calibration Gases

Gravimetric methods. The measuring and the carrier gas are filled into the pressure gas cylinder one after the other and the mass increase is determined.

Volumetric methods. The individual volumes of the gases to be mixed are determined close to atmospheric pressure and are then filled into the space of a known volume, e.g., into a glass container.

Manometric methods. Changes in pressure occurring while adding each mixture component are measured (e.g., during the filling of pressure gas cylinders).

5.5.2.2 Dynamic Methods – the Mixing of Volume Flows

Gas mixing pumps. Carrier gas and measuring gas are conveyed through a pump system with two separate, simply operating pistons in defined flows and subsequently mixed (suitable for relatively high concentrations).

Periodic injection with plugs or loops. Small, constant and defined volumes of the measuring gas are periodically flushed into a continuously flowing, constant volume flow of the carrier gas by means of plugs or loops and are subsequently mixed well.

Continuous injection with capillaries or critical orifice. The measuring gas is injected continuously into a continuously flowing, constant basic gas flow with a capillary or a critical orifice.

Permeation through membranes. The measuring gas is added to a constant, continuously flowing volume flow of the carrier gas from a liquid or gaseous stock via permeation through a membrane. The rate of permeation is, e.g., obtained by determining the weight loss of the stock.

Saturation methods. The measuring gas is added to a constant, continuously flowing carrier gas flow via gas saturator, e.g., by dropping below the dew point or by a vaporization method.

Ozone calibration gas by UV radiation. By exposing a continuous air flow to UV light part of the oxygen molecules can be excited, whereby ozone is formed via atomic oxygen.

In all methods the aim is being able to determine the final gas concentration from basic quantities such as mass or volume determinations. To make sure, however, an analytic determination with a reference measuring methods is frequently carried out.

5.5.2.3 Example of a Calibration Gas Production

The method of periodic injection with plugs will be described here as an example of calibration gas production [36, 169-171]:

The measuring gas is discontinuously added to a certain amount of carrier gas (nitrogen or even air) in very small dosages. By passing the gases through mixing vessels a homogenous mixing of the gases is achieved.

The set-up of a system operating according to this principle is shown in Fig. 5.44. The carrier gas quantity taken from a gas bottle is set by a valve. The carrier gas flow rate must be determined with a gas meter; the rotameters indicated merely serve as a flow rate control.

A certain amount of measuring gas is added to a part of the carrier gas with the rotating plug. The measuring gas or carrier gas flows over the overflow capillaries O if the measuring gas cannot pass through the plug because the latter is not set accordingly. The carrier gas, having flowed by, is conducted back to the mixture whereas the overflowing measuring gas is drained off as excess gas. The gas mixture concentration for one depends on the volume of the plug and the rotational speed of the plug, for another on the carrier gas rate of flow.

The gases mix in the mixing cylinder. A relief pressure vessel behind it has the task of preventing excessively high or low pressures. After the relief pressure vessel the gas reaches the outlet. In the first degree of dilution the concentration set approximately corresponds to that of emission concentrations. Mixtures with ambient air concentrations can also be produced with this unit. For this, the gas has to be once again diluted in the same way.

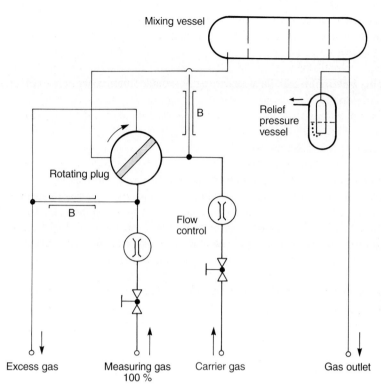

Fig. 5.44. Dynamic calibration gas production by periodic injection with rotating plugs. $\ddot{U}$ bypass capillary

5.5.2.4 Difficulties in the Production of Calibration Gas

The most convenient way of calibrating is with commercially available calibration gases in pressure gas cylinders (produced with static methods). In some cases, however, it is of no use to produce calibration gases in pressure cylinders as they are unstable.

Particularly when dealing with reactive gases in the trace range one is frequently forced to produce suitable calibration gases oneself, something which is usually done with dynamic methods.

The difficulties of calibration gas production for the ambient air are summarized by Hartkamp et al. [166]:
- extreme ratios of pollutant and carrier gas, usually in the ranges of $1:10^6$ and $1:10^{10}$,
- stringent requirements for the purity of the basic gas,
- highly unfavorable proportion of substance amount to vessel, line and instrument volume which lead to considerable sorptive substance losses as well as to memory effects,
- occasionally considerable reactivity of the measuring components towards resid-

ual impurities (e.g., water) which are unavoidable even in highly pure carrier gases, and towards the vessel and line material,
- stringent requirements concerning the stability of the calibration gas properties,
- stringent requirements concerning the compatability of the calibration equipment with changing environmental influences (e.g., climatic fluctuations, changes in pressure, vibrations).

5.5.3 Significance of the Calibration

The accuracy of air pollution measurements stands and falls with the precision of the calibration methods used, particularly of the calibration gases used. In this the measuring task has a decisive influence on the requirements to be made on the calibration. If, e.g., compliance with emission threshold or alarm threshold values (e.g., in smog alarm situations) is to be monitored, calibration must be carried out with particular precision. If, however, characteristic profiles of pollutant concentrations are to be recorded, then it is sometimes more important to achieve a good reproducibility of the measurements rather than to be absolutely precise. For this, the equipment may not drift. These drifts again must be checked by control or routine calibrations.

In principle, the comparability of different measurements depends decisively on the precision of the calibration. In this respect it is abolutely essential that it is carried out both conscientiously and meticulously.

5.6 Accuracy of Measuring Methods and Measuring Instruments

5.6.1 Overview of Characteristic Values

The accuracy of measuring methods and measuring instruments is characterized by performance characteristics. There is a general description of the performance characteristics customary in air analysis techniques in VDI guideline 2449, part 1 [172] and in DIN 1319, part 3 [173] as well as in DIN ISO 5725 [174].

Performance characteristics are listed for each of the measuring methods and measuring instruments described in the VDI guidelines [76]. In the FRG instruments used for monitoring limited emissions or for official air quality measurements must be approved by the Environmental Protection Agency. To obtain this license the measuring instruments must meet certain minimum requirements [167, 175]. When applied for by the producer the instruments will be submitted to so-called qualification tests which are carried out by recognized measuring institutions to establish whether the instruments meet these minimum requirements.

Table 5.10 lists an overview of the most important performance characteristics of the measuring techniques for air pollution. The minimum requirements listed are valid for instruments which are used in the field of air quality control regulations. Different requirements concerning performance characteristics may need to be met for other measuring tasks, e.g., in research.

5.6 Accuracy of Measuring Methods and Measuring Instruments

Table 5.10. Important performance criteria of measurement methods for air pollutants and minimum requirements for the certification of instruments for exhaust gas and air quality measurements in Germany

Performance Characteristics [172-174,176]	Minimum Requirements	
	Emission Measurement Instruments [167]	Air Quality Monitoring Instruments 175]
Calibration function $x = f(A)$ Taking into account blind values: $x - \bar{x}_0 = f(A)$ In case of linearity: $x - \bar{x}_0 = K \cdot A$ Verification of linearity through graphic depiction and regression calculation		
Analytical function $A = f(x)$ (Reversal of the calibration function)	To be determined by a reference method with regression calculation	To be determined by calibration gases with regression calculation
Sensitivity Quotient resulting from the changes of the output and the input signals (slope of the calibration function, factor K)		
Standard deviation s (empirical) (most important parameter for the determination of the random variation of n single values of a measuring series around their mean value $\bar{x}$) $$s = \sqrt{\frac{1}{n-1} \sum_{i=1}^{n}(x_1 - \bar{x})^2}$$ Variance s^2 or σ^2 (square of standard deviation)		
Variation coefficient (empirical) (also called relative standard deviation) $$v = \frac{s}{|\bar{x}|}$$ v is frequently expressed in %		
Measuring range		Terminal value at the end of range $\geq$ threshold value,
Zero point	"Life zero point", to be able to recognize drifts	
Detection limit $\underline{x}$ [172, 177] Smallest quantity which can be distinguished from zero with a statistical certainty of 95%	$\leq 2\%$ of the most sensitive range	$\underline{x} \leq 10\%$ of the longterm air quality threshold value

Table 5.10. (continued)

Performance Characteristics [172-174,176]	Minimum Requirements	
	Emission Measurement Instruments [167]	Air Quality Monitoring Instruments 175]

$\underline{x} = 3 \cdot S_{xo}$ or $\underline{x} = t \cdot S_{xo}$

Threefold (t-fold) standard deviation of the blind values x

$$S_{x0} = \sqrt{\frac{1}{n-1}\sum_{n=1}^{n}(x_{oi}-\bar{x}_0)^2}$$

Response threshold [176]
  value of a required minimal change of the quantity to be measured with a recognizable change on the signal

Uncertainty range U
Difference between upper and lower limit of the single values' interval and the mean value at a confidence level of 95%

$U = \pm s \cdot t$

Student factor t
  see Table 5.11.

The uncertainty range is frequently related to the mean value and is indicated in % ($c \pm \Delta c$)

Accuracy: The closeness of agreement between the true value and the measured value.

Precision: The closeness of agreement between the results obtained by applying the experimental procedure several times under prescribed conditions.

Repeatability standard deviation σ_r
Repeatability r.
Quantitatively, the value below which the absolute difference between two single test results measured under the same conditions (same operator, same apparatus, same laboratory and short intervals of time) may be expected with a probability of 95%.

$r = 1.96\sqrt{2} \cdot \sigma_r = 2.77 \cdot \sigma_r$
  (in DIN ISO 5725 the value 2.83 (=$2 \cdot \sqrt{2}$) is used instead of 2.77)

5.6 Accuracy of Measuring Methods and Measuring Instruments

Table 5.10. (continued)

Performance Characteristics [172-174,176]	Minimum Requirements	
	Emission Measurement Instruments [167]	Air Quality Monitoring Instruments 175]
Application, e.g., in repeated measurements of calibration gases Systematic errors are not recognizable		
Comparison standard deviation σ_R Reproducibility R. Quantitatively, the value below which the absolute difference between two single test results on identical material obtained by operators in different laboratories, using the standardized test method, may be expected with a probability of 95%.		
$R = 1.96\sqrt{2} \cdot \sigma_R = 2.77 \cdot \sigma_R$ (in DIN ISO 5725 the value 2.83 (=$2 \cdot \sqrt{2}$) is used instead of 2.77) Reproducibility is, e.g., used in double determinations and in measurements of several laboratories at the same measuring object (ring tests), s. especially [174] and [172]		
Dead time Time from sudden change of gas concentration to the moment at which the indication reaches a value conventionally fixed at 10 % of the final change in indication		
Rise time time from 10% to 90% of the expected value		
Response time total of dead time+rise time (90%-time)	≤ 200 seconds including sampling system	≤ 180 seconds
Maintenance interval	must be determined and indicated; e.g., 14 days	
Temperature dependency of the indication at zero point	when ambient temperature changes by 10 K: ≤ ±2% of the most sensitive range	≤ ±2% of the measuring signal at the threshold value (is also valid for gas temperature)

Table 5.10. (continued)

Performance Characteristics [172-174,176]	Minimum Requirements	
	Emission Measurement Instruments [167]	Air Quality Monitoring Instruments 175]
Zero point drift Change of the signal when measuring signal at the operating with zero gas over a length of time without adjusting	Within maintenance interval ≤±2% of the most sensitive range	Within 24 hours ≤±2% of the measuring signal at the threshold value within the maintenance interval ≤±10% of the measuring signal at threshold value
Temperature dependency of the sensitivity	When ambient temperature changes by 10 K: ≤±3% of the slope of the analytical function	≤±2% of the slope of the analytical function (is also valid for gas temperature)
Sensitivity drift	Within maintenance interval ≤±3% of the slope of the analytical function	Within 24 hours ≤±2%; within maintenance interval ≤±10% of the slope of the analytical function
Interference	≤±4% of the most sensitive range	≤±6% of threshold value when CO_2, H_2O, SO_2, NO, NO_2, H_2S, NH_3, some HC interfere
Availability	In a three-month long-term test ≥ 90%, aim: 95%	≥ 80%, aim: 90%
Other operating conditions – power-supply voltage – ambient temperature – relative humidity – liquid water content in the air – resistant to vibrations	Defined, a.o. acc. to DIN 43 745	

Explanation of symbols of Table 5.10.
- x measured value
- x_i single value
- x_{oi} single blind value
- $\bar{x}_0$ mean blank value (blind value)
- A given quantity of object to be measured
- K slope factor
- $\underline{x}$ detection limit
- s, σ standard deviation
- s_{x0} standard deviation of blind values
- n number of single values
- v variation coefficient
- U uncertainty range
- t numerical value of t distribution acc. to Student
- r repeatability
- R reproducibility

Table 5.11. Values for Student factor t with a confidence level of 95 % (acc. to [173]) (in pollutant analyses confidence levels of 95 % are commonly used)

Number n of individual values	t
2	12.71
3	4.30
4	3.18
5	2.78
6	2.57
8	2.37
10	2.26
13	2.18
20	2.09
30	2.05
32	2.04
50	2.01
80	1.99
100	1.98
125	1.98
200	1.97
more than 200 ($t=t_\infty$)	1.96

The student factors required for the calculation of the detection limit and the uncertainty of measurements are listed in Table 5.11 for the confidence interval of 95 %.

Some performance characteristics and their determination will be described in the following section in more detail.

5.6.2 Linearity of the Calibration Function and Sensitivity

The calibrating functions of infrared (IR) gas analyzers are normally not linear. The measuring outputs of these instruments have been linearized by the producers so that users of this equipment do not have to determine the result via the calibration function each time a measurement is carried out (which is very inconvenient) but it can be read off directly. The simplest way of linearizing is to install a special scale in the indicating instrument. As the measured values are usually recorded electrically, the electric outputs are generally linearized also. In older equipment this type of linearization did not always function flawlessly. Fig. 5.45 compares as examples the calibration functions of an old and a new IR CO gas analyzer recorded with CO calibration gases [39]. The old CO instrument shows a non-linearity leading to a deviation from the ideal linear calibration function of approximately 10 % in the medium stretch of the measuring range. If regression is linear, the fault is somewhat smaller in the medium stretch of the measuring range; in the lower stretch, however, there are considerable deviations from the actual values. As can be seen, this non-linearity does not become visible in a two-point calibration (zero point and one span value).

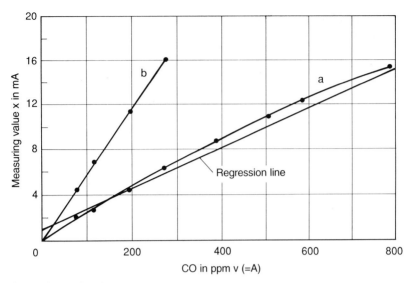

Fig. 5.45. Calibration functions of two infrared gas analyzers for CO detection. x measured value, A given quantity of the object to be measured, in this case CO test gas concentration in ppm v, K slope factor of the calibration function. **a.** old version of an IR instrument, linear regression: $x - 0.55 = 0.02\,A$ (but non-linear !), sensitivity: $K = 0.02$ mA/ppm CO; **b.** newer version of an IR instrument, linear regression: $x - 0.026 = 0.059\,A$; sensitivity: $K = 0.059$ mA/ppm CO

In the newer CO measuring instrument the calibration function is strictly linear. The zero point deviations resulting in the regression calculation are so small that they cannot be recorded.

The new instrument is approximately three times as sensitive as the old one.

5.6.3 Interference

The disturbing influence of accompanying substances in the gas to be investigated is called interference. Interferences can have a linear or a non-linear effect on the measuring signal.

Fig. 5.46 shows as examples the possible interferences in both an automatic, physical measuring method and in a wet-chemical analysis. In the IR gas analysis technique the danger of interference is due to the fact that the IR absorption bands partially overlap (s. Fig. 5.3). This type of interfering factor can exist particularly when measuring low CO concentrations in the presence of high CO_2 contents, as is usually the case in furnaces. Producers of measuring instruments try to minimize this interference by technical means, particularly by installing selective filters. Fig. 5.46a shows that this concept has been carried out relatively successfully in the new IR analyzers. With a CO_2 volume concentration of 15 %, a customary value in furnaces, the interference amounts to <5 ppm (CO indication). In the old IR instrument I a faulty indication of 30-40 ppm must be taken into consideration; the second old IR

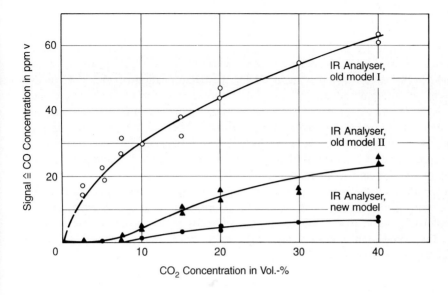
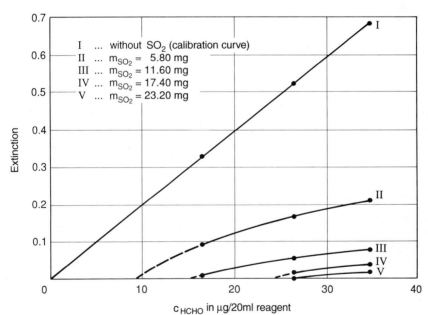

Fig. 5.46. Representation of interferences in an automatic, physical measuring instrument and a wet-chemical measuring method. **a.** interferences of CO infrared measurement with CO_2 [39]. **b.** sulfite-pararosaniline method for the detection of formaldehyde; dependency of color intensity on HCHO concentration with different SO_2 contents (interferes with the calibration function) [179]

analyzer showed a fault of approx. 10 ppm. Fundamentally, in IR analyzers a high sensitivy to water vapor and corresponding interference must be taken note of.

As an example fig. 5.46b shows the influence of SO_2 on the development of color intensity in formaldehyde determination according to the sulfite-pararosaniline method [178]. This method is intended for the determination of formaldehyde in ambient air. Attempts to use this method in furnaces revealed a high interference from SO_2. The development of color intensity is highly impaired by SO_2, i.e., the sensitivity of the method is reduced. In contrast to the interference shown in Fig. 5.46a, where an increase in the measured value is wrongly indicated, a decrease in the measured result is shown in this case. Its influence varies depending on the the different SO_2 contents (non-linear interference).

Depending on the measuring task one must be aware of possible interferences when using measuring methods, and, if necessary, additional tests must be carried out to clarify the situation. VDI guidelines generally point out interferences which were discovered in the course of protracted investigations and which were, if possible, eliminated before the guidelines were passed.

5.6.4 Determination of the Efficiency of Measuring Methods by Ring Tests

In ring tests either a calibration gas (e.g., in a gas cylinder) is sent from laboratory to laboratory and the results of the analyses are compared, or samples are simultaneously taken from a test gas or exhaust gas line and analyzed, or simultaneous measurements are carried out in these lines with continuously operating instruments. In the past years numerous ring tests have been carried out to make measurements reprocuceable. Most of these experiments are carried out at the State Institution for Ambient Air Quality Protection (Landesanstalt für Umweltschutz, LAU) in Essen, where a ring line was installed specifically for this purpose [180, 182], recently also in the EU research center in Ispra, Italy. There, the most varied test gases or test gas mixtures are produced and fed into the line in specific doses. However, ring tests with actual exhaust gases have also been carried out [39].

Ring tests offer the following possibilities:
- to check the feasibility of measuring regulations,
- to improve the efficiency of individual laboratories, when their measuring results deviate excessively,
- to determine the repeatability and reproducibility of measuring methods,
- to simultaneously and reproduceably determine the interferences of different measuring instruments and measuring methods,
- to gain information through mutual exchange of experiences of the ring test participants.

Fig. 5.47 shows as an example the result of a ring test in the test gas line of the LAU for the testing of techniques for the measurement of nitrogen oxide in ambient air [182]. Both wet-chemical measuring methods as well as automatic chemiluminescence instruments were used.

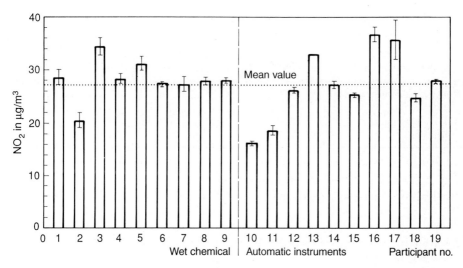

Fig. 5.47. Comparison of wet-chemical methods and automatic instruments for NO_2 ambient air measurement, results of a ring test [182]. m mean value, $m = 27.3$ µg/m³, r repeatability, $r = 7.53$ µg/m³, s_r repeatability standard deviation, $s_r = 2.66$ µg/m³, v_r repeatability variation coefficient, $v_r = 9.7$ %; R reproducibility, $R = 34.8$ µg/m³, s_R reproducibility standard deviation, $s_R = 12.3$ µg/m³, v_R reproducibility variation coefficient, $v_R = 45$ %

The deviations of the individual participants can be regarded as an indication of the quality of their measurement. Their mean value with a repeatability coefficient of variation of v_r=9.7 % lies just within an acceptable range of repeatability. However, the measuring results of the different participants differ strongly from each other: The reproducibility with a coefficient of variation of v_R=45 % is very poor. It is to be seen here that with relatively high precision (small repeatability standard deviation, e.g., with participant 10 and 11) measurements can be carried out very wrongly from an absolute point of view. This is a result of bias during calibration, e.g., by using different calibration gas standards. This does not mean that the mean value is the true one under all circumstances.

The results of ring tests, however, can turn out considerably better. Wet-chemical measurements, e.g., of CO calibration gases in steel cylinders (emissions) achieved reproducibility variances of <2 % (with 510 ppm v/v) and of <10 % (with 40 ppm v/v) [39].

The reproducibility of the results of ring test participants improves when the circle of participants discusses problems and draws the appropriate conclusions from the deviations. To harmonize the measuring techniques and standards of the measuring networks of the FRG a first nitrogen oxide ring test was carried out in 1982. After the results has shown relatively high deviations the problems were discussed and the ring test repeated. Table 5.12 shows how the reproducibility of the results could be improved in this way.

As experience has shown, variances in the reproducibility up to 5 % must be regarded as a good result in ring tests, variances between 5-10 % are still acceptable.

Table 5.12. Relative variation coefficients (mean ranges) of participant mean values of the NO/NO$_2$ ring tests in May 1982 and October 1982 in % [181]

	May 1982	October 1982
NO (automatic instruments)	9.0...12.0	3.5...6.1
2NO$_2$ (automatic instruments)	6.9...13.7	4.1...5.1
NO$_2$ (wet-chemical, Saltzman)	2.6...12.2	2.1 (1.8)...5.1

With variances in excess of 10 % measuring methods must be improved and the standards used must be checked and corrected if necessary.

5.6.5 Fault Consideration using the Example of a Complete Emission Measurement

Apart from pollutant concentrations, auxiliary quantities such as the exhaust gas volume flow and the O$_2$ content of the exhaust gases must be measured to determine emission flows. By this, the inaccuracy of the original measurement is magnified. As an example of the determination of nitrogen oxide emissions from a furnace the possible fault will be roughly estimated here:

The variance of a function $z(x_1,...,x_k)$ can be estimated from the variances of the individual variables according to the law of error propagation:

$$\sigma_z^2 = \sum_{i=1}^{k} \left(\frac{\partial z}{\partial x_i}\right)^2 \sigma_i^2 . \tag{5.20}$$

The percentage faults which are possible in the individual measuring phases and which can be estimated from experience are shown in Fig. 5.48.

Simplified, the possible total error (empiric variation coefficient v_G) is for this purpose estimated as sum of the single errors as shown in Fig. 5.48 (empiric variations s_i^2) for the most favorable and for the most unfavorable case. There may be varying dependencies in individual cases, e.g., a concentration dependency of the error:

$$v_G = \sqrt{s_1^2 + s_2^2 + s_3^2 + s_4^2 + s_5^2 + s_6^2 + s_7^2 + s_8^2 + s_9^2} , \tag{5.21}$$

most favorable case:

$$v_G = \sqrt{4+25+25+4+4+4+25+4+25} = \sqrt{120}$$

$$v_G = 11 \% ,$$

most unfavorable case:

$$v_G = \sqrt{25+100+25+25+4+4+100+25+100} = \sqrt{408}$$

$$v_G = 20 \% .$$

5.6 Accuracy of Measuring Methods and Measuring Instruments

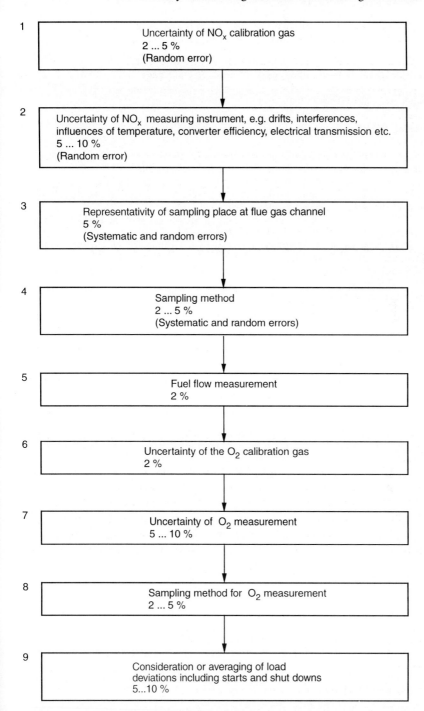

Fig. 5.48. Possible relative errors (variation coefficients), estimates based on experience in the individual steps of emission rate determination. Example of NO_x measurement of a furnace

According to this, the determination of emission flows from concentration measurements includes an error of 10-20 %. This, however, implies that the accuracy (reproducibility) of emission measurements (fault 1+2) lies between 5 and 10 %, which is not necessarily guaranteed, as different ring test results have shown.

5.7 Bibliography

1 Christen, H.-R.: Grundlagen der allgemeinen und anorganischen Chemie. Frankfurt, Berlin, München: Salle und Aarau, Frankfurt, Salzburg: Sauerländer 1985
2 Brügel, W.: Einführung in die Ultrarot-Spektroskopie, Darmstadt: Steinkopf 1969
3 Zeller, M.V.; Juszli, P.P.: Vergleichsspektren von Gasen. Angewandte Infrarotspektroskopie, Heft 11, Bodenseewerk Perkin-Elmer, 1974
4 Birkle, M.: Meßtechnik für den Immissionsschutz – Messen der gas- und partikelförmigen Luftverunreinigungen. München Wien: Oldenbourg 1979
5 Karthaus, H.; Engelhardt, H.: Physikalische Gasanalyse, Grundlagen. Druckschrift L 3410, Fa. Hartmann & Braun, Frankfurt
6 Luft, K.F.; Kesseler; Zörner, K.H.: Nichtdispersive Ultrarot-Gasanalyse mit dem UNOR. Chem.-Ing.-Techn. 39 (1967) 937–945
7 Siemens: Eignungsgeprüfte Emissionsmeßeinrichtungen. Prospekt Bestell Nr. E 681, Siemens AG, Bereich Meß- und Prozeßtechnik, Karlsruhe 1986
8 Leybold-Heraeus: BINOS-IR, Prinzip der Infrarot-Gasanalysatoren. Druckschrift 43-200.1/2, Leybold-Heraeus GmbH, Hanau 1982
9 Measurex Rauchgas Analysator 2225. Firmenprospekt, measurex GmbH, Eschborn 1980
10 Wilks Scientific Lim.: MIRAN-Gasanalysator, verschiedene Versionen. Firmenprospekte und Bedienungsanleitung, Antechnika GmbH, Ettlingen bei Karlsruhe 1982
11 Grisar, R.; Preier, H.: Laserspektroskopische Analyse von Gasen und Flüssigkeiten. Fraunhofer-Gesellschaft, FhG-Berichte 3–81, S. 20–25
12 Stephan, K.; Hurdelbrink, W.: Laserstrahl analysiert Gasgemische. VDI-Nachrichten Nr. 26, 26.6.1981, S. 8
13 Staab, J.; Klingenberg, H.; Schürmann, D.: Strategie of development of a new multicomponent exhaust emissions measurement technique. SAE Tech. Pap. 830437, (Soc. Automot. Eng.) 1983
14 Hartmann & Braun AG. NDUV-Betriebsphotometer Radas 1G. Gebrauchsanweisung 42/20-22-1 bzw. Listenblatt 20-1.53, Frankfurt 1982
15 DFVLR, Institut für Reaktionskinetik: Meßgerät zum Nachweis von Stickstoffoxiden. Exponatbeschreibung, Stuttgart 1979
16 Dasibi Environmental Corp.: 1008 Ozone Analyzer. Firmenprospekt und Bedienungsanleitung, Antechnika GmbH, Ettlingen bei Karlsruhe, 1983
17 Monitor Labs, Inc.: Ozon-Analysator 8810. UV-Absorption, Firmenprospekt, Monitor Labs Inc. Allershausen bei München, 1988
19 Kohn, D.: Entwicklung eines spektroskopischen Meßverfahrens für Formaldehyd-Emissionen. Diplomarbeit Nr. 2106 am Institut für Verfahrenstechnik und Dampfkesselwesen der Universität Stuttgart, 1982
20 Krauss, L.; DFVLR: Formaldehydnachweis II. Patentschrift P 3537 482.9-52, Deutsches Patentamt, München 1986
21 Brinkmann, A.: Entwicklung eines UV-spektroskopischen Meßverfahrens für Formaldehyd. Studienarbeit Nr. 2249 am Institut für Verfahrenstechnik und Dampfkesselwesen der Universität Stuttgart, 1987
22 Berner-Moundrea, V.: Schwankungen der NO_2-Konzentration in der bodennahen Atmosphäre von Wien. Staub-Reinhalt. Luft 32 (1972) Nr. 5, S. 210–212
23 Platt, U.; Perner, D.: Direct Measurements of Atmospheric CH_2O, HNO_2, O_3, NO_2 and SO_2 by Differential Absorption in the Near UV. J. Geoph. Res. 85, C 12 (1980) 7453–7458 und in: Killinger, D.K.; Mooradian, A. (eds.): Optical and laser Remote Sensing. Berlin: Springer 1983

24 Opsis AB: Analysis of Gases with Opsis Technologie. Prinzipbeschreibung und Firmenprospekte, Lund/Schweden 1989
25 Graber, W.K.; Taubenberger, R.: Differential Optical Absorption Spectroscopy of Atmospheric Trace Gases. TM-52-89-01, Paul Scherrer Institute, Villingen, Switzerland, 1988
26 Weitkamp, C.: Luftüberwachung durch ortsaufgelöste Fernmessung von SO_2 und NO_2. Bericht über die Tagung der Arbeitsgemeinschaft der Großforschungseinrichtungen (AGF), 3./4.11.1983, Bonn-Bad Godesberg
27 Zolner, W.J.; Cieplinski, E.W.; Dunlap, D.V.: Measurement of Ambient Air SO_2 Concentration using a pulsed Fluorescent Analyzer. Themo Electron Corporation, Waltham USA; deutsche Vertretung, Duisburg, Firmenprospekt 1984
28 SO_2-Analysator 8850, UV-Fluoreszens-Detektor, Firmenprospekt, Monitor Labs, Inc. – Allershausen bei München, 1989
29 Paffrath, D.: DFVLR-Meßsystem zur Erfassung der räumlichen Verteilung von Umweltparametern in der Atmosphäre mit mobilen Meßträgern. Forschungsbericht 85-09, DFVLR Institut für Physik der Atmosphäre, Oberpfaffenhofen 1985
30 Kaiser, R.: Chromatographie in der Gasphase, Teil 4: Quantitative Auswertung. Mannheim: Bibliographisches Institut 1965
31 Schomburg, G.: Gas-Chromatographie. Weinheim: Chemie 1977
32 Leithe, W.: Die Analyse der Luft und ihrer Verunreinigungen. Stuttgart: Wissenschaftliche Verlagsgesellschaft 1974
33 Staab, J.; Baronick, J.D.; Kroneisen, A.: Improving the Method of Hydrocarbon Analysis. SAE Tech. Pap. 810427 (Soc. Automot. Eng.) 1981
34 Erste Allgemeine Verwaltungsvorschrift zum Bundes-Immissionsschutzgesetz (Technische Anleitung zur Reinhaltung der Luft – TA Luft) vom 27.2.1986, GmBl. S. 95 f
35 VDI: Richtlinie 3481, Bl. 6, Vorentwurf, Messen gasförmiger Emissionen, Auswahl und Anwendung von C-Summenverfahren. Düsseldorf: VDI 1986
36 Gans, W.; Baumbach, G.: Kalibrierverfahren zur quantitativen Bestimmung flüchtiger organischer Substanzen in Abluft und Abgasen mit dem Flammenionisationsdetektor. VDI-Fortschritt-Berichte Reihe 15, Nr. 32, Düsseldorf: VDI 1986
37 Winkelbauer, W.; Paul, H.; Baumbach, G.: Abscheidung von Ölnebeln aus der Abluft von Vergüteanlagen. Staub-Reinhalt. Luft 46 (1986) Nr. 6, S. 300–302
38 Gasanalysen-Meßanlage Typ Ultragas U4S. Gerätebeschreibung, Fa. Wösthoff oHG, Bochum
39 Weller, L.; Baumbach, G.: Entwicklung eines Referenzmeßverfahrens für Kohlenmonoxid. Umweltforschungsplan des Bundesministeriums des Innern, Luftreinhaltung, Forschungsbericht 104 02 116, Umweltbundesamt Berlin 1982
40 Gasanalysen-Meßgerät Mikrogas TE-SO_2. Bedienungsanleitung und Firmenprospekt, Fa. Wösthoff oHG, Bochum 1988
41 VDI: Richtlinie 2462, Bl. 5: Messen gasförmiger Emissionen; Messen der Schwefeldioxid-Konzentration; Leitfähigkeitsmeßgerät Mikrogas-MSK-SO_2-E1. Berlin: Beuth 1979
42 Ultragas-SO_2-Analysatoren. Datenblätter der verschiedenen Typen, Fa. Wösthoff oHG, Bochum
43 VDI: Richtlinie 2451, Bl. 5E: Messung gasförmiger Immissionen; Messung der SO_2-Konzentration; Leitfähigkeitsverfahren (Ultragas U3ES). Berlin: Beuth 1977
44 Picoflux 2T, Chemisch-physikalischer Gasspurenanalysator. Listenblatt und Bedienungsanleitung, Fa. Hartmann & Braun, Frankfurt
45 VDI: Richtlinie VDI 2451, Bl. 4: Messung gasförmiger Immissionen; Messung der SO_2-Konzentration; Leitfähigkeitsverfahren (Picoflux). Berlin: Beuth 1986
46 Konrad, G.: Luftverunreinigungen in einem Schwarzwaldtal bei Inversionswetterlagen, Teil 1: Meßtechnik. Studienarbeit Nr. 2202 am IVD der Universität Stuttgart, 1986
47 Schütz, H.: Aufbau moderner Meßstationen zur Luftreinhaltung. Einzelbericht zu H & B-Meßwerte, Hartmann & Braun AG, Frankfurt, 1969
48 Pfeffer, H.-U.: Das telemetrische Echtzeit-Mehrkomponenten-Erfassungssystem TEMES zur Immissionsüberwachung in Nordrhein-Westfalen. Landesanstalt für Immissionsschutz des Landes Nordrhein-Westfalen, LIS-Bericht Nr. 19, Essen 1982
49 Breuer, W.: Nachr. Chem. Tech. 18, Nr. 14 (1970) 287 f

50 Picos, Elektro-chemischer Gasspurenanalysator. Bedienungsanleitung, Hartmann & Braun, Frankfurt 1971
51 Oxycom 25D, Sauerstoff-Meß- und Warngerät. Prospekt Nr. 4561, Drägerwerk AG Lübeck, 1980
52 Comowarn, der tragbare CO-Wächter. Prospekt 4111, Drägerwerk AG Lübeck, 1979
53 Fabian, L.: Elektrochemische Sensoren für die Rauchgasanalyse. cav 1986, Dezember, S. 95–102 und Rauchgasanalyse-Computer MSI 2000 P. Firmenprospekt, Fa. MSI Measuring Systems Industrial, Schwerte-Geisecke 1987
54 Rauchgasanalyse-Computer IMR 3000P/IMR 30109P. Bedienungsanleitung und Firmenprospekt, Fa. IMR Industrielle Mess- und Regelsysteme für Umwelttechnologie, Leingarten
55 Philips Elektronik Industrie GmbH: Schwefeldioxid-Meßgerät PW 9700, Firmenprospekt, Hamburg 1973
56 Rumpel, K.-J., Umweltbundesamt: Meßstation Deuselbach, persönliche Mitteilung, 1989
57 LFE Laboratorium für industrielle Forschung und Entwicklung: COSO2-SO_2-Gas-Analysator. Firmenprospekt und Betriebsanleitung, Maintal 1987
58 Bühler, H.: Grundlagen und Probleme der pH-Messung. Dr. W. Ingold AG, Frankfurt 1980
59 Schwabe, K.: pH-Meßtechnik, Dresden 1976
60 DIN 19620: pH-Messung: Allgemeine Begriffe für das Messen in wässrigen Lösungen. Berlin: Beuth 1971
61 DIN 19261: pH-Messung: Begriffe für Meßverfahren mit Verwendung galvanischer Zellen. Berlin: Beuth 1971
62 DIN 19262: Steckbuchse und Stecker geschirmt für pH-Elektroden. Berlin: Beuth 1959
63 DIN 19263: pH-Messung: Meßfertige Gaselektroden. Berlin: Beuth 1961
64 DIN 19264 E: pH-Messung: Meßfertige Bezugselektroden. Berlin: Beuth 1983
65 DIN 19265: pH-Messung: pH-Meßzusatz, Anforderungen. Berlin: Beuth 1967
66 DIN 19266: pH-Messung: Standardpufferlösungen. Berlin: Beuth 1979
67 DIN 19267: pH-Messung: Technische Pufferlösungen (Eichung). Berlin: Beuth 1978
68 DIN 19268 E: pH-Messung von klaren, wässrigen Lösungen. Berlin: Beuth 1982
69 Fa. Eigenbrodt, G.K.W.: Niederschlagsmonitor für elektrische Leitfähigkeit, pH-Wert und Niederschlagsmenge nach Dr. Winkler. Firmenprospekt, Königsmoor bei Harburg 1986
70 Fa. Thies, Clima: pH-Meßanlage, Listen-Nr. 2945/79. Gebrauchsanleitung und Firmenprospekt, Göttingen 1984
71 Jurkschat, R.: Entwicklung und Bau eines kontinuierlich arbeitenden Regen-pH-Meßgerätes. Diplomarbeit am Institut für Verfahrenstechnik und Dampfkesselwesen der Universität Stuttgart, 1984
72 Bran & Luebbe: On-line Analysensysteme, Meßprinzipien und Applikationen. Katalog Nr. 1.1 und 3.4, Bran & Luebbe GmbH, Norderstedt bei Hamburg 1986
73 Fischer, W.; Rohr, F.J.: Beispiele für die Anwendung von Festelektrolyten. Chem.-Ing. Techn. 50 (1978) Nr. 4, S. 303–305
74 Eisenhardt, H.: Untersuchung des ZrO_2-Meßverfahrens zur Bestimmung des O_2-Gehaltes in Abgasen. Studienarbeit Nr. 2021 am Institut für Verfahrenstechnik und Dampfkesselwesen der Universität Stuttgart 1979
75 Bundeseinheitliche Praxis bei der Überwachung der Emissionen nach §§ 2a, 4 der Ersten Verordnung zur Durchführung des Bundes-Immissionsschutzgesetzes; hier: Bekanntmachung geeigneter Meßgeräte für die Bestimmung der Rußzahl, des Kohlendioxidgehaltes sowie geeigneter Rußzahl-Vergleichsskalen. RdSchr. d. BMI vom 19.9.1979–UII8-556 134/2, GMBl. S. 539
76 VDI: VDI-Handbuch Reinhaltung der Luft, Bd. 5. Berlin: Beuth wird ständig ergänzt.
77 Budzikiewicsz, H.: Massenspektrometrie. Taschentexte 5, Weinheim: Chemie 1972
78 Gebefügi, I.: Dioxine – eine Herausforderung an die moderne Analytik. gsf mensch + umwelt, Gesellschaft für Strahlen und Umweltforschung, München, August 1985
79 Kuwata, K.; Uebori, M.; Yamasaki, Y.: Determination of aliphatic and aromatic aldehyds in polluted airs as their 2, 4-dinitrophenylhydrazones by high performance liquid chromatography. J. of Chromatogr. Sci. 17 (1979) 264–268
80 Gjerde, D.T.; Fritz, J.S.: Ion Chromatography, 2. ed.. Heidelberg: Hüthig 1987
81 Grimmer, D. u.a.: Luftqualitätskriterien für ausgewählte polyzyklische aromatische Kohlenwasserstoffe. Umweltbundesamt Berichte 1/79, Berlin: Erich Schmidt 1979

82 Hagenmaier, H.; Brunner, H.; Kraft, M.: Emissionsmessungen von polychlorierten Dibenzodioxinen und Dibenzofuranen an Abfallverbrennungsanlagen. In: Schriftenreihe Aktuelle Probleme der Luftreinhaltung, Band 2, S. 199–224, Düsseldorf: VDI Kommission Reinhaltung der Luft 1986
83 Burk, H.D.: Probenahmetechniken zur Erfassung gas- und partikelförmiger Emissionen von polychlorierten Dibenzodioxinen und Dibenzofuranen. VDI Ber. 608 (1987) 191–208
84 VDI: Richtlinie 3872, Bl. 1E: Messen von Emissionen. Messen von polycyclischen aromatischen Kohlenwasserstoffen (PAH) – Messen von PAH in Abgasen von PKW-Otto- und Diesel-Motoren, Gas-Chromatographische Bestimmung. Berlin: Beuth 1987
85 Grimmer, G.; Hildebrandt, A.; Böhnke, H.: Probenahme und Analytik von PAK in Kfz-Abgasen, Erdöl und Kohle, Erdgas. Petrochemie 25 (1972) 531–536
86 Nicht limitierte Automobil-Abgaskomponenten. Druckschrift, Volkswagen AG, Forschung und Entwicklung, Wolfsburg 1988
87 Hartung, A.; Schulze, J.; Kieß, H.; Lies, K.-H.: Nitroderivate der PAK als Artefakte bei der Probenahme aus dem Dieselabgas. Staub-Reinhalt. Luft 46 (1986) Nr. 3, S. 132–135
88 Ballschmiter, K.; Zell, M.: Analysis of polychlorinated biphenyls (PCB) by glas capillary gas chromatography. Fresenius Z. Anal. Chemie 302 (180) 20–31
89 VDI: Richtlinie VDI 3881, Bl. 2: Olfaktometrie – Geruchsschwellenbestimmung, Probenahme. Berlin: Beuth 1987
90 VDI: Richtlinie VDI 3881, Bl. 1E: Olfaktometrische Technik der Geruchsschwellen–Bestimmung – Grundlagen. Berlin: Beuth 1983
91 Thiele, V.: Olfaktometrie von H_2S – Ergebnisse eines VDI-Ringvergleichs. Staub-Reinhalt. Luft 42 (1982) Nr. 1, S. 11–15
92 Bahnmüller, H.: Olfaktometrie von Dibutylamin, Acrylsäuremethylester, Isoamylalkohol und eines Spritzverdünners für Autolacke – Ergebnisse eines VDI-Ringvergleichs, Staub-Reinhalt. Luft 44 (1984) Nr. 7/8, S. 352–358
93 Thiele, V.: Olfaktometrie an einer Emissionsquelle – Ergebnisse des VDI-Ringvergleichs, Staub-Reinhalt. Luft 44 (1984) Nr. 7/8, S. 348–351
94 König, A.; Rohlfing, H.: Geruchsbelästigung durch Kfz-Abgase, Staub-Reinhalt. Luft 44 (1984) Nr. 9, S. 396–398
95 Dollnick, H.W.O.; Thiele, V.; Drawert, F.: Olfaktometrie von Schwefelwasserstoff, n-Butanol, Isoamylalkohol, Propionsäure und Dibutylamin, Staub-Reinhalt. Luft 48 (1988) 325–331
96 Winneke, G.; Berresheim, H.-W.; Kotalik, J.; Kabat, A.: Vergleichende olfaktometrische Untersuchungen zu Formaldehyd und Schwefelwasserstoff. Staub-Reinhalt. Luft 48 (1988) 319–324
97 Baum, F.: Luftreinhaltung in der Praxis. München: Oldenbourg 1988
98 VDI: Richtlinie 2066, Bl. 1: Messen von Partikeln; Staubmessungen in strömenden Gasen, gravimetrische Bestimmung der Staubbeladung, Übersicht. Berlin: Beuth 1975
99 VDI: Richtlinie 2066, Bl. 2: Messen von Partikeln; manuelle Staubmessung in strömenden Gasen; gravimetrische Bestimmung der Staubbeladung, Filterkopfmeßgerät. Berlin: Beuth 1981
100 Krebs, R.: Staubgehaltsmessungen in strömenden Gasen. Lurgi Apparatebau Gesellschaft mbH, Forschungslaboratorium, Frankfurt 1967
101 Düwel, U.; Dannecker, W.: Neuartige Probenahmeeinrichtung zur Staubkonzentrationsmessung in Reingasen von Großfeuerungsanlagen zum Zwecke der Bestimmung anorganischer und organischer Staubinhaltsstoffe. Staub-Reinhalt. Luft 43 (1983) Nr. 7, S. 277–284
102 Bühne, K.-W.; Jockel, W.: Das Planfilterkopfgerät. Staub-Reinhalt. Luft 49 (1989) Nr. 3, S. 93–98
103 VDI: Richtlinie 2268E, Bl. 1: Stoffbestimmung an Partikeln; Bestimmung der Elemente Ba, Be, Cd, Co, Cr, Cu, Ni, Pb, Sr, V, Zn in emittierten Stäuben mittels atomspektrometrischer Methoden. Berlin: Beuth 1984
104 Dannecker, W.; Redmann, W.A.; Düwel, U.: Erfassung filtergängiger Metalle und Metalloide, Staub-Reinhalt. Luft 45 (1985) Nr. 7/8, S. 331–338
105 Bundeseinheitliche Praxis bei der Überwachung der Emissionen und Immissionen, Eignung von Meßgeräten zur laufenden Aufzeichnung von Emissionen. RdSchr. d. BMI vom 21.7.1980, Gemeinsames Ministerialblatt, Ausgabe A, 31 (1980) Nr. 21, S. 342–355

106 Fa. Verewa Meß- und Regeltechnik Spohr: Staub-Emissionsmeßgerät „Beta-Staubmeter" F 50 (stationäre Ausführung) bzw. F 60 (transportable Ausführung). Firmenprospekt, Mühlheim/Ruhr 1978
107 Borho, K.: Staubmeßverfahren. Messen, Steuern und Regeln in der chemischen Technik. Band II, S. 216–223; Hrsg.: Hengstenberg, J.; Sturm, B.; Winkler, O. Berlin: Springer 1980
108 Fa. E. Sick GmbH: RM 61-03-Rauchdichtemeßgerät. Firmenprospekt, Waldkirch 1988
109 Fa. Durag Elektronik: Rauchdichte-Meßsystem D-R216. Firmenprospekt, Hamburg 1982
110 VDI: Richtlinie 2066, Bl. 4E: Messen von Partikeln; Staubmessung in strömenden Gasen; Bestimmung der Staubbeladung durch kontinuierliches Messen der optischen Transmission. Berlin: Beuth 1980
111 Fa. E. Sick GmbH: GM 21-Schwefeldioxid- und Staubgehalt-Meßgerät. Firmenprospekt, Waldkirch 1985
112 Erste Verordnung zur Durchführung des Bundes-Immissionsschutzgesetzes (Verordnung über Kleinfeuerungsanlagen – 1. BImSchV) vom 15.7.1988. Bundesgesetzblatt I, S. 1059
113 DIN 51402, Teil 1: Prüfung der Abgase von Ölfeuerungen; Bestimmung der Rußzahl. Berlin: Beuth 1978
114 DIN 51402, Teil 2: Prüfung der Abgase von Ölfeuerungen; Fließmittelverfahren zum Nachweis von Ölderivaten. Berlin: Beuth 1979
115 Baum, F.; Brocke, W.: Entwicklung einer einfachen Methode zur Messung der Staub-, Ruß- und Teeremissionen aus Hausbrandeinzelöfen für feste Brennstoffe. Schriftenreihe der Landesanstalt für Immissions- und Bodennutzungsschutz des Landes Nordrhein-Westfalen; H. 17, S. 7–17. Essen: Giradet 1969
116 Laskus, L.; Bake, D.: Erfahrungen bei der Korngrößenanalyse von Luftstäuben mit dem Andersen-Kaskadenimpaktor, Staub-Reinhalt. Luft 36 (1976) Nr. 3, S. 102–106
117 Finlayson-Pitts, B.J.; Pitts Jr., J.N.: Atmospheric Chemistry. New York: John Wiley & Sons 1986, S. 829–832
118 Kaskaden-Impaktor Andersen-Mark II. Gebrauchsanleitung, Vertrieb: Fa. Schäfer, Neu-Isenburg, 1979
119 VDI: Richtlinie 2119, Bl. 2, Messung partikelförmiger Niederschläge; Bestimmung des partikelförmigen Niederschlags mit dem Bergerhoff-Gerät (Standardverfahren). Berlin: Beuth 1972
120 VDI: Richtlinie 2119, Bl. 3, Messung partikelförmiger Niederschläge; Bestimmung des partikelförmigen Niederschlags mit dem Hibernia- und Löbner-Liesegang-Gerät. Berlin: Beuth 1972
121 VDI: Richtlinie 2119, Bl. 4, Messung partikelförmiger Niederschläge; Bestimmung des partikelförmigen Niederschlags mit Haftfolien. Berlin: Beuth 1972
122 Rohbock, E.; Georgii, H.W.: Ein Depositionssammelgerät zur getrennten Erfassung der trockenen und feuchten Deposition atmosphärischer Schadstoffe. Bericht im Auftrag des Umweltbundesamtes Forschungsprojekt 104 02 600, Frankfurt: Eigenverl. des Universitätsinst. für Meteorologie und Geophysik 1982
123 VDI: Richtlinie VDI 2463, Bl. 2E: Messen von Partikeln, Messen der Massenkonzentration von Partikeln in der Außenluft – High Volume Sampler – HV 100. Berlin: Beuth 1977
124 VDI: Richtlinie VDI 2463, Bl. 4: Messen von Partikeln, Messen von Partikeln in der Außenluft – LIB-Verfahren. Berlin: Beuth 1976
125 VDI: Richtlinie VDI 2463, Bl. 7: Messen von Partikeln, Messen der Massenkonzentration (Immission), Filterverfahren. Kleinfiltergerät GS 050. Berlin: Beuth 1982
126 VDI: Richtlinie VDI 2463, Bl. 9: Messen von Partikeln, Messen von Massenkonzentration (Immission), Filterverfahren. LIS/P – Filtergerät. Berlin: Beuth 1987
127 Satorius-Membranfilter GmbH: Staubsammelgerät Gravikon, Firmenprospekt SM 702.50.673.GU, Göttingen
128 Marfels, H.; König, R.: Zum „Gipsfaser-Problem" bei der Messung faserförmiger Stäube in der Außenluft. Staub-Reinhalt. Luft 45 (1985) Nr. 10, S. 441–444
129 Wedding, J.B.; McFarland, A.R.; Cermack, J.E.: Large particle collection characteristics of ambient aerosol samplers. Environmental Science & Technology 11, Nr. 4, 1977, pp. 387–390
130 Baumann, K.: Aufbau und Erprobung einer Meßstation zur Erfassung der Schadstoffausbreitung an einer Autobahn. Studienarbeit Nr. 2193 am Institut für Verfahrenstechnik und Dampfkesselwesen der Universität Stuttgart, 1985

131 Baumbach, G.; Baumann, K.; Dröscher, F.: Luftverunreinigungen in Wäldern, Institut für Verfahrenstechnik und Dampfkesselwesen der Universität Stuttgart, Abteilung Reinhalt. Luft, Bericht Nr. 5 — 1987
132 VDI: Richtlinie VDI 2463, Bl. 5: Messen von Partikeln, Messen der Massenkonzentration (Immission), Filterverfahren. Automatisiertes Filtergerät FH 62I, Berlin: Beuth 1987
133 Bundeseinheitliche Praxis bei der Überwachung der Emissionen und Immissionen, Rdschr. des BMI vom 2.2.1983, GMBl. 34 (1983), Nr. 4, S. 76—82
134 Sigrist Photometer AG: Staubgehaltsmessungen zur Emissionskontrolle an Abgaskanälen, hinter filternden Abscheidern oder zur Immissionskontrolle, Firmenschrift, Ennetbürgen/Schweiz, 1985
135 Malissa, H.: Analysis of Airborne Particles by Physical Methods, CRC Press, Oca Rotan, Fl. USA, 1978
136 Manuntschehri, A.; Candala, J.P.: Gewinnung und mikroskopische Untersuchung von Staubproben aus der Atmosphäre, Studienarbeiten am Institut für Verfahrenstechnik und Dampfkesselwesen der Universität Stuttgart, 1959
137 Schreiber, H.: Energiedispersive Röntgenfluoreszensanalyse, KfK-AFR 006, 103—110, Kernforschungszentrum Karlsruhe 1983
138 Traxel, K.; Wätjen, U.: Particle-induced X-Ray Emission Analysis (PIXE) of Aerosols; in Physical and Chemical Characterization of Individual Airborne Particles (Ed. K.R. Spurny), Ellis Horwood (J. Wiley & Sons), 298—330, Chichester UK, 1986
139 Wieser, P.; Schreiber, H.; Greiner, W.: Quellenspezifische Merkmale partikelgebundener atmosphärischer Spurenstoffe der bodennahen Luft. Vergleichende Untersuchungen mit dem Lasermikrosonden-Massenanalysator LAMMA 500 und der Röntgenfluoreszensanalyse. Projekt Europäisches Forschungszentrum für Maßnahmen zur Luftreinhaltung (PEF) am Kernforschungszentrum Karlsruhe (KfK), KfK-PEF-Bericht 20. April 1987
140 VDI: Richtlinie VDI 3497, Bl. 3: Messen partikelgebundener Anionen in der Außenluft. Analyse von Chlorid, Nitrat und Sulfat mittels Ionenchromatographie mit Suppressortechnik nach Aerosolabscheidung auf PTFE-Filtern, Berlin: Beuth 1988
141 VDI: Richtlinie VDI 2463, Bl. 1: Messen von Partikeln in der Außenluft — Übersicht, Berlin: Beuth 1974
142 Thoenes, H.W.; Lützke, K.; Guse, W.: Stand der Meßtechnik und Probleme bei der Emissionsmessung von Staub, Stickstoffoxiden und Schwefeloxiden aus Feuerungsanlagen, in VDI-Berichte Nr. 495, Düsseldorf 1984
143 KNF Neuberger GmbH: KNF Perma Pure Gastrockner, Firmenprospekte, Freiburg 1985
144 Knoß, M.; Kopka, M.: Beeinflussung der Meßgenauigkeit bei SO_2-Emissionsmessungen durch das Probenahmesystem, Studienarbeiten Nr. 2184 und 2185 am Institut für Verfahrenstechnik und Dampfkesselwesen der Universität Stuttgart, 1985
145 Divisek, J.; Fürst, L.; Nürnberg, H.W.: Kontinuierliches SO_2-Meßgerät für die zuverlässige Überwachung von Feuerungsabgasen, Staub-Reinhalt. Luft 45 (1985) Nr. 9, S. 414—418
146 Hartmann & Braun AG: Rauchgasanalyse und Emissionsüberwachung, Technische Information 30 PY103, Frankfurt 1981
147 Robert Bosch GmbH (Hrsg.): Abgase von Verbrennungsmotoren. In: Kraftfahrtechnisches Taschenbuch, 20. Aufl. Düsseldorf: VDI 1987
148 Mercedes-Benz: Abgas-Emissionen. Grenzwerte, Vorschriften und Messung: Druckschrift Nr. 13 EP/EB, Stuttgart März 1991
149 Volkswagen AG: Druckschrift „Abgasmessungen", Wolfsburg 1986
150 Staab, J.; Pflüger, H.; Schröter, D.; Schürmann, D.: Ein kompaktes Abgasmeßsystem zum Einbau in Personenkraftwagen für Messungen bei Straßenfahrten. DK 621.43.019.9, Volkswagen AG, Wolfsburg 1988
151 RdSchr. des Bundesministers des Innern betreffend Bundeseinheitliche Praxis bei der Überwachung der Emissionen und Immissionen — UI8 — 556134/4 —, II. Richtlinien für die Wahl der Standorte und die Bauausführung automatischer Meßstationen in telemetrischen Immissionsmeßnetzen, vom 2.2.1983, GMBl. S. 76—78
152 Immissionsmeßnetze in der Bundesrepublik Deutschland — Stand 1987. In: Monatsberichte aus dem Meßnetz 8/87, Umweltbundesamt Berlin
153 Projektgruppe Bayern zur Erforschung der Wirkung von Umweltschadstoffen (PBWU): Atlas der Immissionsmeßstationen Europas, 2. Aufl. Gesellschaft für Strahlen- und Umweltforschung, München, GSF-Bericht 25/87

154 Richtlinie des Rates vom 7. März 1985 über Luftqualitätsnormen für Stickstoffdioxid. Amtsblatt der EG, L 87/1
155 Kost, W.-J.; Baumbach, G.: Messung der vertikalen Konzentrationsgradienten der Spurengase NO, NO_2, SO_2 und O_3 in Waldbeständen im Schönbuch und Schwarzwald. VDI Ber. 560, S. 205–239, Düsseldorf 1985
156 Esser, J.: Schadstoffkonzentration im Nahbereich von Autobahnen in Abhängigkeit von Verkehr und Meteorologie. In: Kolloquiumsbericht „Abgasbelastungen durch den Kraftfahrzeugverkehr im Nahbereich verkehrsreicher Straßen". Köln: TÜV Rheinland 1982 und Umwelt 3/82, S. 158–164
157 Baumann, K.: Schadstoffausbreitung im Nahbereich einer Autobahn. Institut für Verfahrenstechnik und Dampfkesselwesen der Universität Stuttgart, Abt. Reinhalt. Luft, Bericht Nr. 8–1987
158 Käß, M.: Untersuchungen zur Erfassung des Schadstoffeintrages in Waldbestände. Diplomarbeit Nr. 2139 am Institut für Verfahrenstechnik und Dampfkesselwesen der Universität Stuttgart 1984
159 Baumbach, G.; Käß, M.: Ermittlung von vertikalen Schadstoffkonzentrations-Profilen in Waldbeständen in Baden-Württemberg – Meßtechnik und erste Ergebnisse. Staub-Reinhalt. Luft 45 (1985) Nr. 6, S. 274–278
160 Gehrig, EMPA, Dübendorf/Schweiz: Persönliche Mitteilung, 1987
161 Nater, W.: Das atmosphärenphysikalische Meßsystem des Projektes „Luftschadstoffe" (Waldschäden) an der Lägeren. Eidgenössisches Institut für Reaktorforschung, Würenlingen/Schweiz, EIR-Bericht Nr. 616, Juni 1987
162 Sattler, T.; Jaeschke, W.: Automatisierte Bestimmung von Immissionsprofilen anorganischer Luftschadstoffe in verkehrsreichen Waldschneisen; Staub-Reinhalt. Luft 47 (1987) Nr. 11–12, S. 261–266
163 Michaelis, W.; Schönburg, M.; Stößel, R.-P.: Trocken- und Naßdeposition von Schwermetallen und Gasen. In: Bauch, J.; Michaelis, W. (Hrsg.): Das Forschungsprogramm Waldschäden am Standort „Postturm" Forstamt Farchau/Ratzeburg, GKSS-Forschungszentrum Geesthacht GmbH, GKSS 88/E/55, 1988
164 Baumbach, G.; Baumann, K.; Dröscher, F.: Schadstoffbelastung im Nordschwarzwald und Schönbuch. In: Verteilung und Wirkung von Photooxidantien im Alpenraum, Gesellschaft für Strahlen- und Umweltforschung, München, GSF-Bericht 17/88, S. 609–616
165 DIN 1319, Teil 1: Grundbegriffe der Meßtechnik – Allgemeine Grundbegriffe. Berlin: Beuth 6/1985
166 Hartkamp, H.; Buchholz, N.; Klukas, F.; Münch, J.: Ermittlung und Erprobung von Kalibrierverfahren für Immissionsmeßnetze. Umweltforschungsplan des Bundesministers des Innern, Forschungsbericht 104 02 216, UBA-FB 83–034, Materialien 3/83, Berlin: Erich Schmidt 1983
167 Richtlinien für die Eignungsprüfung, den Einbau und die Wartung kontinuierlich arbeitender Emissionsmeßgeräte. RdSchr. des BMI vom 21.7.1980 – UII8-556 134/4-, GMBl. S. 343
168 Güdelhöfer, P.; Hönig, H.-J.: Erprobung von Referenzmeßverfahren zur Feststellung gasförmiger Emissionen. Umweltforschungsplan des Bundesministers des Innern, Forschungsbericht 104 02 100, Köln: TÜV Rheinland 1979
169 Fa. Wösthoff oHG: Gasmischvorrichtungen. Firmendruckschrift 511–2a, Ausgabe 1074, Bochum 1977
170 Baumbach, G.: Messung von Stickstoffoxiden in der freien Atmosphäre. Diplomarbeit am Institut für Verfahrenstechnik und Dampfkesselwesen der Universität Stuttgart, 1973
171 VDI: Richtlinie 3490, Bl. 7: Messen von Gasen – Prüfgase, Dynamische Herstellung durch periodische Injektion. Berlin: Beuth 12/1980
172 VDI: Richtlinie 2449, Bl. 1: Prüfkriterien von Meßverfahren – Datenblatt zur Kennzeichnung von Analysenverfahren für Gas-Immissionsmessungen. Berlin: Beuth 10/1970
173 DIN 1319, Teil 3: Grundbegriffe der Meßtechnik – Begriffe für die Meßunsicherheit und für die Beurteilung von Meßgeräten und Meßeinrichtungen. Berlin: Beuth 8/1983
174 DIN ISO 5725: Präzision von Meßverfahren – Ermittlung der Wiederhol- und Vergleichspräzision von festgelegten Meßverfahren durch Ringversuche. Berlin: Beuth April 1988
175 Bundeseinheitliche Praxis bei der Überwachung der Immissionen. Richtlinien für die Bauausführung und Eignungsprüfung von Meßeinrichtungen zur kontinuierlichen Überwa-

chung der Immissionen. — RdSchr. d. BMI vom 19.8.1981 — UII8-556 134/4 —, GMBl. S. 355
176 DIN 1319, Teil 2: Grundbegriffe der Meßtechnik — Begriffe für die Anwendung von Meßgeräten. Berlin: Beuth 1/1980
177 VDI: Richtinie 2449, Bl. 2E: Grundlagen zur Kennzeichnung vollständiger Meßverfahren, Begriffsbestimmung. Berlin: Beuth 11/1979
178 VDI: Richtlinie 3483, Bl. 1: Messen gasförmiger Immissionen, Messen von Aldehyden — Bestimmen der Formaldehyd-Konzentration nach dem Sulfit-Pararosanilin-Verfahren. Berlin: Beuth 1/1979
179 Baumbach, G.: Emissionen organischer Schadstoffe von Ölfeuerungen. Dissertation, Universität Stuttgart 1977
180 VDI-Kommission Reinhaltung der Luft: Ringversuch „Immissionsmessungen von Stickstoffoxiden" am 11./13. Mai 1981 bei der LIS in Essen. Ergebnisbericht, Düsseldorf April 1983
181 Pfeffer, H.U.: Qualitätssicherung in automatischen Immissionsmeßnetzen, Teil 3: Ringversuche der staatlichen Immissions-, Meß- und Erhebungsstellen in der Bundesrepublik Deutschland, Ergebnisse für die Komponenten SO_2, NO_x, O_3 und CO. LIS-Bericht Nr. 52, Landesanstalt für Immissionsschutz, Essen 1984
182 Landesanstalt für Immissionsschutz (LIS): Ringversuch für NO, NO_2, SO_2 und O_3, Essen 1985

6 Evaluation of Air Pollution Measurements

In the field of emissions as well as in air quality control air pollutants are generally recorded by measuring instruments in concentrations. This determination of concentrations is sufficient for the purpose of comparing them with threshold values; they must be mass concentrations, however, which requires a transformation of the measured results, as many measuring techniques determine volume concentrations.

To assess air pollution it is not sufficient to merely indicate concentration values, rather it is the mass flows which are frequently of interest. Basically, mass flows are calculated from pollutant concentrations multiplied with the volume flows in which they are contained. Determination of these volume flows is mostly very time-intensive and involves inaccuracies.

Mass flows are described by the following definitions:
- *Emission flow* is the mass of an air polluting substance escaping into the atmosphere per unit of time.
- *Pollutant flux* is the mass of an air polluting substance entering the acceptor (human, animal, plant, soil, materials) per unit of time.
- According to [1] the integral of the pollutant flux for the time during which the acceptor was exposed to the polluting substance is called pollutant dose .

The term pollutant dose is sometimes simply used as the product of concentration and time [2].

The following sections will deal with the evaluation of air pollution measurements in the fields of emission and air quality control with regard to the different possibilities of assessment and quantification.

6.1 Determination of Pollutant Emissions

6.1.1 Detection of Pollutant Emissions from Concentration Measurements of Exhaust Gases

6.1.1.1 Emission Flows and Emission Factors

The means of determining emission flows and so-called emission factors from concentration measurements in exhaust gases is illustrated in Fig. 6.1 with the example of exhaust gases from furnaces. The determination of emission flows and emission factors in industrial production plants is, in principle, carried out in much the same

6.1 Determination of Pollutant Emissions

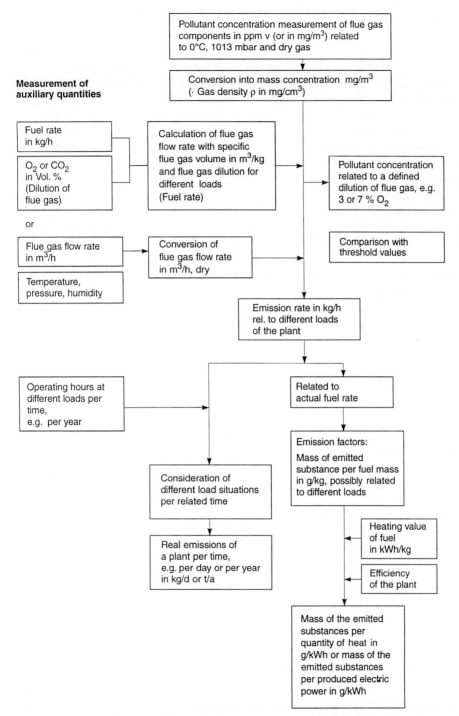

Fig. 6.1. Determination of emission rates and emission factors from concentration measurements of flue gases – diagrammatic example for furnaces

way. Instead of dealing with fuel flows, production flows are to be dealt with. When trying to determine emissions in motor vehicles the matter is somewhat more complicated, as measurements are rarely carried out on vehicles in traffic and the different driving conditions must be simulated on roller dynamometers instead, s. [3-6].

The emission flows can be determined within the error limits described in sect. 5.6.5. Greater errors are to be expected when the emission factors are applied to other plants, as the emissions of different components, e.g., NO_x, depend heavily on individual combustion conditions.

When determining pollutant emissions (CO, C_nH_m, soot) stemming from instationary combustion conditions, which are formed, e.g., during incomplete combustion, the error is magnified. These substances develop, e.g., when starting or closing down furnaces and during break-downs. For these cases the individual determination of the flue gas volume flow is highly inaccurate – in the majority of cases even impossible. In domestic furnaces with solid fuels the individual degree of charging and burn out enter into the calculation. If even determining the emission flows of the components CO, C_nH_m and soot gives rise to problems, it is obvious that the determination of emission factors which are transferable to other plants can only represent a rough estimation.

Considering these points of view the emission factors mentioned in literature [7, 8] should be assessed and used carefully.

6.1.2 Calculation of Pollutant Emissions from Fuel Properties

6.1.2.1 Sulfur Dioxide

As was shown in chap. 2 the sulfur bound in fuels burns almost completely to sulfur dioxide. The latter is emitted or, in some coal combustion processes, bound into the ashes (s. Table 2.4). It is advisable to calculate the SO_2 emission flows from fuel flows or fuel consumption. Precondition for this is, that the sulfur content of the fuels is known and does not change in the time relevant. When carrying out the calculation the following molecule balances must be considered:

$$S + O_2 = SO_2 \tag{6.1}$$

$$1 \text{ Mol} + 1 \text{ Mol} = 1 \text{ Mol} \tag{6.2}$$

$$32 \text{ g} + 32 \text{ g} = 64 \text{ g} \tag{6.3}$$

$$1 \text{ g} + 1 \text{ g} = 2 \text{ g}. \tag{6.4}$$

Thus, 2 g SO_2 are formed from 1 g bound sulfur.

Example: 1 kg of light fuel oil with a sulfur mass content of 0.2 % is burnt→ 4 g SO_2 are formed from 2 g of sulfur.

From the consumptions of individual fuels the SO_2 emissions of an entire area can be calculated relatively accurately for a defined period of time. The emission values by the German Environmental Protection Agency are based on such calculations [9].

These calculations, however, become inaccurate if the sulfur content of the fuels are not known exactly. For light fuel oil, for instance, a maximum sulfur content of 0.2 mass proportion in % was prescribed [10], and calculations were usually carried out using this amount. The mineral oil industry, however, stated that the sulfur content of some light fuel oils was lower and that the average for the oil commercially available supposedly amounted to 0.24 % in 1984 [11].

The accuracy of the SO_2 emission calculations also decreases the more flue-gas desulfurization equipment is used. As these desulfurization units are at varying technological levels and have different removal efficiencies, measurements must be carried out directly in the flue gas when determining SO_2 emission flows. For this, the considerations on methods of determination and accuracy made in the previous chapter apply.

6.1.2.2 Nitrogen Oxides

As was shown in sect. 2.1.3 it is not possible to calculate nitrogen oxide emissions from the nitrogen content of the fuel. For one thing, not all of the fuel nitrogen is transformed into nitrogen oxide during combustion, for another, more or less atmospheric nitrogen is oxidized into nitrogen oxide depending on the combustion temperature.

When calculating flames and furnace dimensions it is advisable to calculate NO_x emissions beforehand. However, this is only feasible for furnaces with defined dimensions and with known fuel properties [12, 13]. These models are at a stage where results calculated with them must be compared and verified with measured results. When applying the results of one furnace to another, furnace dimensions, fuel properties and load conditions would have to be included in the calculation process. The purpose of such computation models is not to collect emission data for NO_x but rather to calculate in advance NO_x emissions of furnaces at the planning stage so as to be able to design the construction of the burner and the furnace chamber such that NO_x emissions are kept at a minimum.

When determining NO_x emission flows from the many different plants one is therefore as dependant now as before on measurements in flue gases.

6.1.2.3 Products of Incomplete Combustion (PIC)

It is not possible to determine products of incomplete combustion (CO, C_nH_m and soot) from fuel properties as the emissions of these substances are influenced almost exclusively by the combustion process. Only a rough estimation of emission is possible by using emission factors determined at similar furnaces.

6.1.2.4 Heavy Metal Emissions in Oil Furnaces

While soot and total dust emissions in oil furnaces cannot be calculated from fuel properties, the metals contained in the oil can be determined in the flue gases as oxides. It has, however, not yet been checked whether this determination is quantitatively possible [14]. Particularly when burning heavy fuel oils metal emissions have a special significance. *Nickel and vanadium oxides*, for instance, are emitted in the process. When tracing atmospheric pollutants these metals can serve as indicators of flue gas components from heavy oil furnaces [15].

In light fuel oils metals occur only in the slightest of concentrations so that the emissions of the relevant oxides have as yet had no significance. Up until now *lead* was added to gasoline in the form of lead tetraethyl as an antiknock agent. The lead emissions of the motor vehicles correspond to the lead content of the gasoline used and can be extrapolated from them. However, that does not tell us in which form the lead is emitted. E.g., halogenous compounds are added so that lead compounds volatilize easier.

When burning coal there is a high degree of ash components in the flue dust. However, the fact that part of the ash is already retained in the furnace chamber and the largest part of the flue ash is kept back in the flue gas particle precipitators makes calculating particle emissions from fuel composition impossible. For coal furnaces particle emissions can only be determined by flue gas measurements.

6.1.3 Registering Pollutant Emissions of an Area in Emission Inventories

6.1.3.1 Spatial Pollutant Distribution

To be able to recognize relationships between emissions and pollutants in a certain area it is important to begin by compiling an overview of the dispersion of pollutant emissions of the area concerned. In these emission inventories the extent of the emissions from the different main source groups is listed:
– motor vehicle traffic,
– domestic furnaces and those of small-scale industries,
– industry: furnaces including power plant furnaces and industrial processes.

For these inventories the area to be assessed is usually divided into subsections of 1 x 1 km and pollutant emissions released in these areas are calculated per annum. Emission conditions are depicted graphically by coloring these subsections in topographical maps. Owing to this type of presentation one can, e.g., infer from these maps in which sections of a town high amounts of pollutants are released and where emissions are low. In combination with air quality inventories areas can be identified where air pollution control measures should be preferably taken. Therefore emission and air quality inventories are vital components of air quality control concepts.

The temporal dispersion of the pollutant emissions cannot immediately be inferred from the inventories. It must be calculated separately and according to [8] it is to be shown also for the period of one calendar year with the proportions of the individual emittents visibly marked.

Pollutants

According to [8] the following atmospheric pollutants, in particular, are to be included in the emission inventories:
- particulate,
- fine particles <0.01 mm (aerodynamic diameter up to 85 % <0.01 mm),
- lead and lead compounds – listed as Pb,
- sulfur dioxide (SO_2),
- nitrogen dioxide – listed as NO_2,
- carbon monoxide (CO),
- gaseous and vaporous organic compounds,
- chlorine and gaseous anorganic chlorine compounds – listed as Cl^-,
- fluorine and gaseous anorganic fluorine compounds – listed as F^-.

Frequently, there are no measuring data available on the individual pollutants emitted, particularly for the not so common pollutants (e.g., fine particles, organic compounds, Cl^- and F^-). Therefore, the values used are usually based on estimates or emission factors based on them. Thus, great accuracy may not be expected here.

6.1.3.2 Determination of Pollutant Emissions of Motor Vehicle Traffic

The path of computation for the determination of pollutant emissions of motor vehicle traffic is shown in a diagram in Fig. 6.2. The traffic inventories of cities where the daily traffic load for the most important roads is given can serve as an important basis for computing emissions. If these types of data are not available, then traffic surveys must be carried out separately, where according to [8] traffic load must be determined depending on the time of day by counting motor vehicles in separate categories of passenger cars and trucks. Roads which are not registered as so-called line sources must be considered plane sources where the traffic load must be determined by random surveys. Driving behavior or driving speeds must be determined for the individual streets, if necessary by random in-traffic test drives. The Technical Control Board Rhineland in Germany have introduced so-called "driving modi" to characterize driving behavior. The emission factors of these modi were measured on roller dynamometers [3, 4]. The emission factors listed in [8] are based on the investigations of the Technical Control Board Rhineland.

The annual loads of SO_2, NO_x and CO per street in t/(km · a) are obtained via traffic load, driving behavior and the applicable emission factors. With the pollutant emissions thus calculated the streets in the city map can be color-coded according to their loads, e.g., the emission map for the source group traffic of the cities of Stuttgart, Mannheim and Karlsruhe [16-18]. For these latter maps the total sum of hydrocarbons, lead and soot was also determined and depicted, in addition to the common pollutants SO_2, NO_x and CO.

The streets with their specific pollutant emissions can then be allocated to the 1 x 1 km grid squares of the area under investigation and the lengths of the individual streets be measured in the different squares. The specific emissions multiplied by the

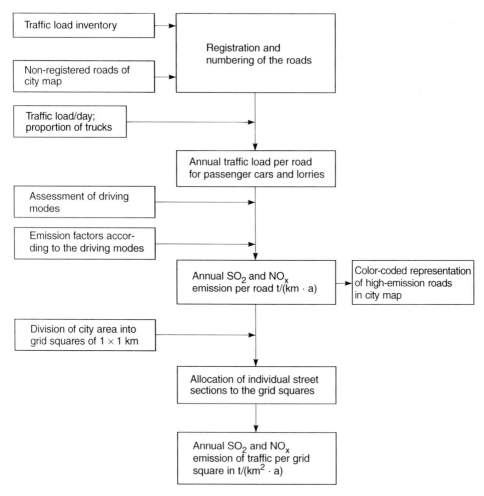

Fig. 6.2. Calculation mode for determining the SO_2 and NO_x emissions from traffic per road and per square kilometer in a city (also valid for other compounds)

relevant street length give the annual emissions of each street per determined square kilometer. The sums of all streets of a grid square then represent the pollutant emissions in $t/(km^2 \cdot a)$ caused by motor vehicle traffic [16-18].

6.1.3.3 Pollutant Emissions of Domestic Furnaces and Small-Scale Industries

The 5th Administrative Regulation for the Implementation of the Federal Air Pollution Control Act [8] stipulates that the number and type of furnaces, the fuels used, the nominal capacities of the furnaces, the source heights and the geographic position are to be determined. For this, the data secured by the chimneysweeps on their annual check of domestic furnaces is to be used as far as possible. Determining the number of furnaces, however, is relatively complex. The accuracy thus achieved is partially undone by the fact that there are no data available on the fuel consumptions of the furnaces and that estimates must be made for this as the furnace capacity does not necessarily say anything about its degree of utilization.

Occasionally, studies on district heating also include data on heat requirement densities of individual town sections. If the annual heat supplies of gas and piped heat are known (these can be obtained relatively easily), then the consumption of light fuel oil can be estimated by calculation. Pollutant emissions can thus be determined for the inventories via the individual heat consumptions and with the help of emission factors. E.g., an air quality control concept for the city of Heilbronn, Germany, was carried out in this way [19].

Fig. 6.3 shows the computational method applied for this. When drawing up the official administrative emission inventories in Germany surveys were not carried out with the utmost consistency, the final energy consumptions being determined via computational methods similar to the one shown in Fig. 6.3 instead [20, 21].

In his "Manual on the Setting up of Clean Air Plans" [22] Dreyhaupt gives directions on how to carry out statistical surveys in large urban areas without interviewing all households (a so-called microcensus).

As an example an emission inventory was drawn up for a small town for which approximately 70 % of all households were individually interviewed to collect data on type of fuel and annual fuel consumption [23]. An emission inventory compiled in this manner naturally surpasses all other survey methods in accuracy. However, it cannot be implemented in larger towns as it is very work-intensive.

Pollutant emissions from *small-scale industries* depend on the materials used, on the type of construction and the mode of operation of the furnaces. The following businesses shall be listed as examples: gasoline stations, printing shops, dry cleaners, paint shops, woodworking, smokehouses, light fuel oil depots and plants for the treatment of parts with chlorinated hydrocarbons. Dreyhaupt [22] gives detailed instructions on how to collect data for this purpose, too. To do this, questionnaires must cover all types of plants. The calculation of the emissions is carried out, e.g., with the help of specific emission factors and the amount of produced or consumed materials.

6.1.3.4 Pollutant Emissions from Industries

When compiling an emission inventory for Industry one cannot avoid individually interviewing all plant operators. In the "polluted areas" for which the authorities must set up an air quality control concept the operators of "plants requiring official approval" [24] must make so-called emission statements [25]. The emission statements are requested by the approving authorities, e.g., the industrial inspection boards. Apart from the information covering the plants' operating hours, the types and amounts of materials used (e.g., fuel consumptions), direct statements pertaining to the emission flows must be made. The emission statement is to be up-dated every two years.

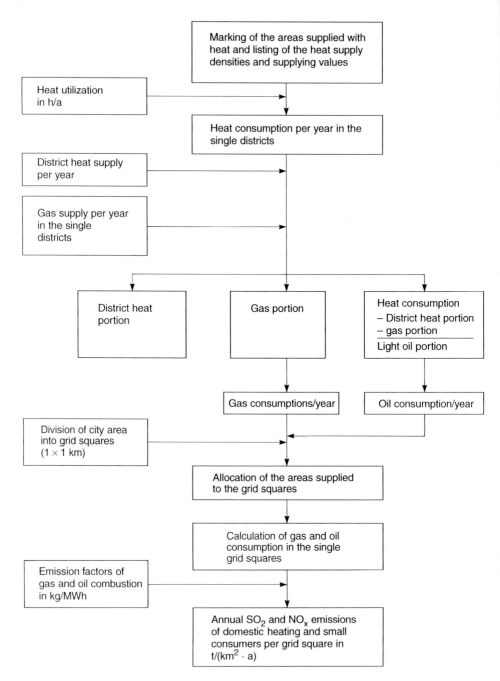

Fig. 6.3. Calculation mode for determining SO_2 and NO_x emissions from domestic heating systems per square kilometer [19]

The annual emissions of the plants are either determined from the individual emission flows by multiplying them with the operating hours, or they are calculated, e.g., in industrial furnaces, from the fuel consumption with the help of specific emission factors. Depending on the location of the individual plants the pollutant emissions thus determined are allocated to the grid squares of the survey area and are color-coded in the emission inventory maps [26, 27].

6.1.3.5 Summary of Annual Emissions

To determine their spatial dispersion the pollutant emissions of the individual source groups are combined for each square kilometer of the surveyed area (based on the 1 x 1 km Gauss-Krueger grid[1]) and color-coded according to their concentrations. In this way the emission inventory maps were set up for the polluted areas as being essential to the air quality control concepts [28-30]. As an example the emission dispersion of the pollutant NO_x ($NO+NO_2$) determined and depicted in the manner described is shown for the town of Heilbronn, Germany, s. Fig. 6.4 [19].

As can be seen, the highest NO_x emissions come from a power plant, even if emitted via high stacks. Besides these power plant emissions the highest emissions occur in the inner city area on the one hand and in the northern part of the town near a highway on the other. This NO_x emission dispersion which is mainly caused by motor vehicle traffic is typical of urban areas. In areas with high traffic densities which unfortunately occur frequently in the city centers along with high residential density traffic-related pollutant emissions (NO_x, CO, hydrocarbons) are also at their highest level. Apart from these, highways with heavy traffic represent large nitrogen oxide sources.

The emissions of the different source groups do not become effective as pollutants to the same extent due to their varying source heights (motor vehicle exhausts directly in the respiratorial space, industrial stacks up to a height of 250 m). It is therefore attempted to weight the emissions according to their individual contribution of pollution to the ambient air. For this, apart from the source heights, the different weather conditions which influence the dispersion of the pollutants are included. After comparing dispersion calculations and pollution measurements Kutzner [31] has defined so-called pollution assessment factors for the different emission sources for the example of Berlin. Assessments of this type [19, 29] can occasionally be found in air quality control concepts. Such an assessment for the pollution component NO_x is shown graphically in Fig. 6.5. It must however be admitted, that the values indicated for the polluted areas are rough estimates and rough mean values for an entire urban area (there are definite local differences). Such considerations, however, can be of help in deciding on pollution control measures.

---

[1] In topographic maps the 1 x 1 km Gauss-Krueger grid is plotted with the coordinates of the x- and y-axis.

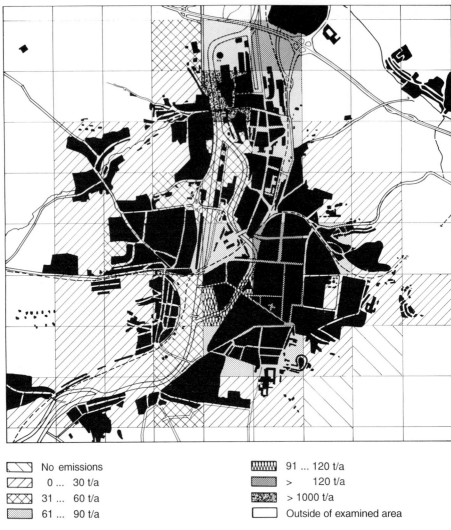

Fig. 6.4. Distribution of NO_x emissions in the city of Heilbronn, FRG [19]

6.1.3.6 Temporal Pollutant Distribution

To determine relationships between emissions and pollution a fine temporal resolution is often useful. For pollution in urban areas, e.g., characteristic diurnal courses can be determined which are on the one hand created by meteorological dispersion conditions such as height of mixed layer, but on the other hand the emission diurnal course has a considerable influence on the pollution profile. Thus, at least for research which aims at tracing atmospheric pollution from emission to deposition it is important that diurnal courses of emissions are known with an hourly or half-hourly resolution.

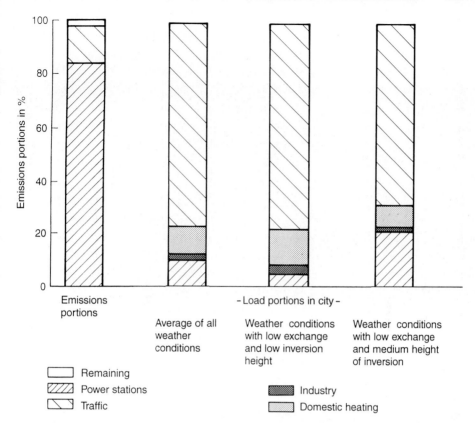

Fig. 6.5. Emissions of source groups traffic, domestic heating, industrial furnaces and power plants with their contribution to air pollution in an urban area (example for the component NO_x) [19, 32]

To determine diurnal courses of emissions for the source group traffic the diurnal courses of the traffic load of the most important streets in the area surveyed must be determined by counts. Based on the relevant driving behavior (driving modi) the profiles of pollutant emissions can be calculated. Fig. 6.6a shows, for instance, the hourly traffic volume on a street on a weekday. Weekdays have characteristic diurnal variations with peaks of rush-hour traffic in the morning and in the afternoon [16]. In this case there is an additional maximum on the street concerned at noon. Saturdays with shopping trips and Sundays with excursion trips have other diurnal courses. The diurnal courses determined for the pollutants NO_x and SO_2 resulting from motor vehicle movement and the heating operations are shown in Fig. 6.6b. The NO_x diurnal course virtually reflects the traffic load, as domestic furnaces emit considerably less NO_x. Concerning the component SO_2 traffic has practically no influence on emissions; the SO_2 is caused exclusively by furnaces.

In March 1985 a large-scale experiment called TULLA was carried out in Baden-Wuerttemberg, Germany, to compile an SO_2 mass balance for the area of Baden-Wurttemberg. As part of of this experiment a fine-meshed 1 x 1 km map of the SO_2

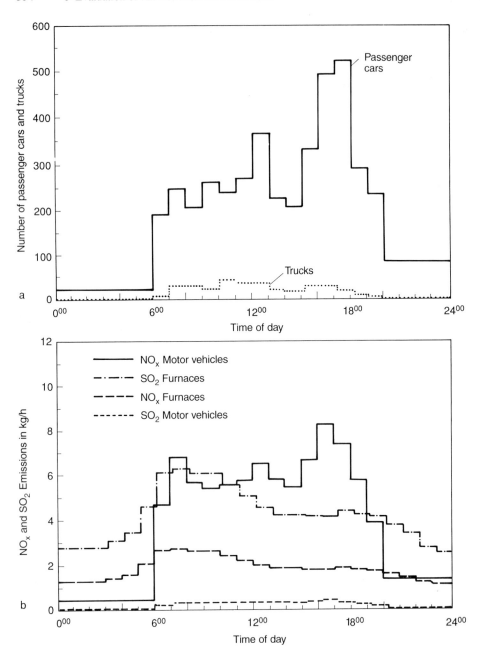

Fig. 6.6. Diurnal variation of vehicle number (**a**) NO_x and SO_2 emissions (**b**) during a working day in a small town [23]

and NO_x emissions with an hourly resolution was drawn up [33, 34]. The emissions of small furnaces were computed with models based on the evaluation of data of the ambient conditions. For the larger industrial plants emission rates were determined by interviews.

The profiles of the NO_x emissions in Baden-Wuerttemberg during the two weeks of this field experiment are shown in Fig. 2.18 as an example of this evaluation. The large share of motor vehicle emissions with completely specific diurnal courses are to be seen: peaks in the mornings and in the afternoons, at night always a drop to lowest values. The two lower peaks in the middle of the diagram indicate the lower week-end traffic. The biggest part of these motor vehicle-related NO_x emissions comes from the urban areas and from the country's highly frequented highways.

Knowing the spatial and temporal structure of the pollutant emissions can in many cases explain why ambient air pollution occurs.

6.2 Evaluation and Graphic Representation of Air Quality Measurements (with K. Baumann)

Concentrations of the different air pollutants can be subject to marked fluctuations as far as time and space are concerned. It is the task of measuring technology to obtain as much information as possible on this.

Depending on the problem at hand different measuring plans can be set up and differentiated evaluations and graphic representations of the pollution measurements can be carried out. Instructions on setting up measuring plans can, e.g., be found in the 4th General Administrative Regulation for the Implementation of the Federal Air Pollution Control Act "Determination of Air Quality in Polluted Areas" [35]. It should be noted that the layout of the measuring plan and the representation of the results can even imply an evaluation of the results. The following sections will provide possibilities of evaluating and graphically representing measured data.

6.2.1 Temporal Resolution and Mean Value Formation

The temporal profile of pollutant concentrations can be recorded with the help of continuously operating measuring instruments. Fig. 6.7a shows the profile of the SO_2 concentration of one day; it was recorded by a measuring instrument and continuously plotted with a line recorder. The measuring instrument used shows a delayed reading (generally 180 s), so that the profile shown in Fig. 6.7a is somewhat smoother than the actual course.

To further evaluate and assess the pollution measurements in Germany half-hourly mean values are generally formed from the measured data as smallest unit of reference. In the international field also one-hour mean values are in use. In the simplest case the mean values can be formed by evaluating the recordings; most of the time, however, data acquisition systems are connected parallel to the line recorders which automatically calculate half-hourly or hourly mean values. Fig. 6.7b shows

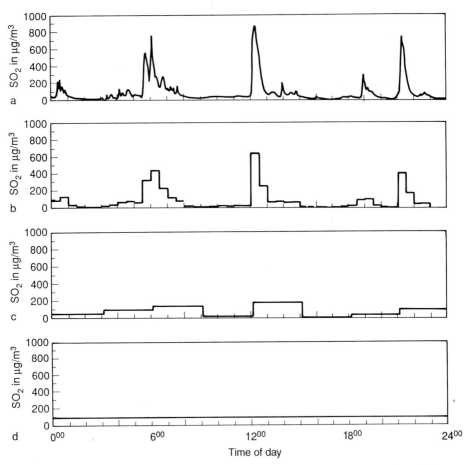

Fig. 6.7. Example of SO$_2$ measurement over 24 hours with different averaging times [36]. **a.** continuous measurement, **b.** half-hour, **c.** 3-hour, **d.** 24-hour averaging

the concentration profile of Fig. 6.7a, averaged with half-hourly values. If three hourly mean values are formed from the profile which, e.g., are the time basis in smog alarm plans, then there is a concentration profile as shown in Fig. 6.7c. The daily mean value (24 hour mean value) is shown in Fig. 6.7d. As can be seen, averaging the values causes a loss of information on the structure of the temporal profile.

The temporal averagings can also be the result of the measuring technique used. Very low concentrations in outside air, for instance, can sometimes only be measured by enriching them through longer sampling times. In this way particle concentrations, e.g., are sometimes recorded as daily mean values [37].

The extent of the temporal resolution of air quality measurements depends on their biological effect, or on any other effect which is to be investigated or which is to be cautioned against via this measurement.

Table 6.1 shows the different response times of different effects of atmospheric pollutants.

6.2 Evaluation and Graphic Representation of Air Quality Measurements

Table 6.1. Examples of different reaction times of individual effects of air pollutants (acc. to [38])

Objects of effect	Characteristic reaction times		
	Short-term effects (seconds)	Medium-term effects (days and possibly constantly repeated short-term effects)	Long-term effects (months to years)
Humans	Perception of odors, visibility, irritation of mucous membranes and eyes	Acute diseases of the respiratory system cancer	chronic diseases of the respiratory system, lung
Vegetation	Reduced yield of field crops, damage on ornamental plants (acute necroses)	Reduced yield of field crops, damage on ornamental plants	Soil acidification, reduced yield of field and forest crops
Materials	Acid-caused pit corrosion destruction of nylon fibers	Brittleness of rubber and synthetics, tarnishing of silver, darkening of colors	Corrosion, decomposition of art monuments, surface soiling
Measuring methods	Continuously operating measuring instruments	Continuously operating measuring instruments or constantly repeated sampling	Random sampling measurements with mean value formation or integrating measurement (exposure of collecting tanks, exposure of test pieces/objects)

Even today it has not been clarified whether the recent forest damage is the result of long-term effects caused, e.g., by pollutant intrusion into the soil or whether short- or medium-term effects of high pollutant concentrations have triggered off the damage. The influence of pollutants occurring for short or medium length periods of time has hardly been investigated up to this point, as mostly random samples were taken in forest areas, or medium- to long-term sampling was carried out which merely permits an assessment of long-term effects. Only of late have more frequent continuous pollutant measurements been carried out which can throw light upon the actual course of events. It is the task of measuring technology to reflect the informational content of the results of the continuous measurements through suitable means of representation and not simply to reduce the value of the measuring results to an assessment of long-term effects via mean value calculations over shorter or longer periods of time.

6.2.2 Summary and Graphic Representation of Measuring Data from Continuous Measurements

6.2.2.1 Unsmoothed Monthly Profiles

For the representation of its measuring results the Environmental Protection Agency of Germany has chosen as smallest unit of reference mean values of 1 hour [37]. A complete graphic representation of all 1-hour mean values for one month is shown in Fig. 6.8. Every 1-hour mean value is plotted as a vertical line, so that all peaks, all low values and all failures of measurement are recognizable. Such a compact representation saves considerably more space than the plottings of a continuous-line recorder and can be grasped at a glance, which is not the case with tables. Fig. 6.8 compares the SO_2 concentration recordings for two months of the German measuring site Waldhof. In both months a monthly mean value of 11 µg/m^3 was calculated from the individual values. These graphic representations show how by solely indicating mean values actual events can become blurred: in August 1983 there were two peak concentrations of 130-150 µg/m^3, whereas in September very brief peaks amounted to only 70 µg/m^3. The different monthly profiles are with certainty different in their relevance of effects.

6.2.2.2 Mean Value Calculation for Smog Warnings

To assess smog hazards sliding 3-hourly mean values are formed from the half-hourly values measured as usual. In this case sliding means, that after every newly measured half-hourly value the previously calculated 3-hour mean value is updated. In the smog regulations of Germany [39] 3-hour threshhold values are fixed for different alarm phases; when these are exceeded the appropriate smog warning is to be issued. E.g., a smog warning must be given when two measuring stations in a town record values exceeding 0.6 mg/m^3 of SO_2 or 0.6 mg/m^3 of NO_2.

Moreover, the smog regulations also provide threshold values for a 24-hour averaging time. In this, the measured air-borne particulate concentration is included in the assessment.

6.2.2.3 Diurnal Courses

If specific situations are to be investigated in greater detail, then pollutant profiles are usually shown as diurnal courses based on half-hourly values (s. Fig. 6.7b). As an example of special situations Fig. 6.9 compares the concentration profiles of a day with an inversion and a rainy day. The diurnal courses of wind speed are also included. As can be seen, there are only low wind speeds on the inversion day. When the wind speed picks up some in the early afternoon pollutant concentrations immediately decrease. On the rainy day there is a strong wind throughout the whole day and the pollutants provide only low values. The concentration increase in the morning hours is not a result of unfavorable exchange conditions but of the rush-hour

6.2 Evaluation and Graphic Representation of Air Quality Measurements

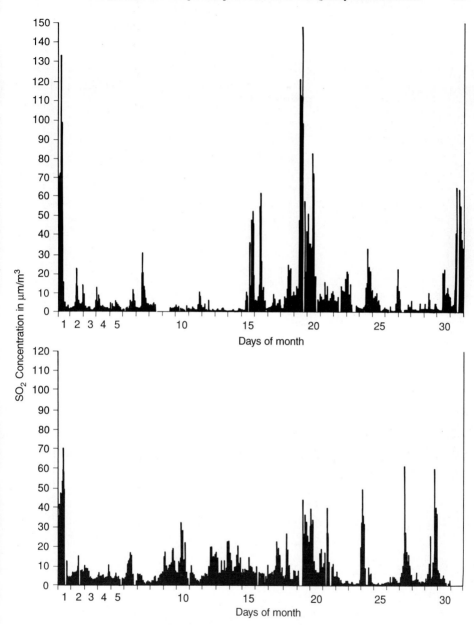

Fig. 6.8. Comparison of SO_2 concentrations (in $\mu g/m^3$) in August (**a**) and in September (**b**) at a measuring site of the German Umweltbundesamt (Federal Environmental Agency). Monthly average of both months: 11 $\mu g/m^3$ [4]

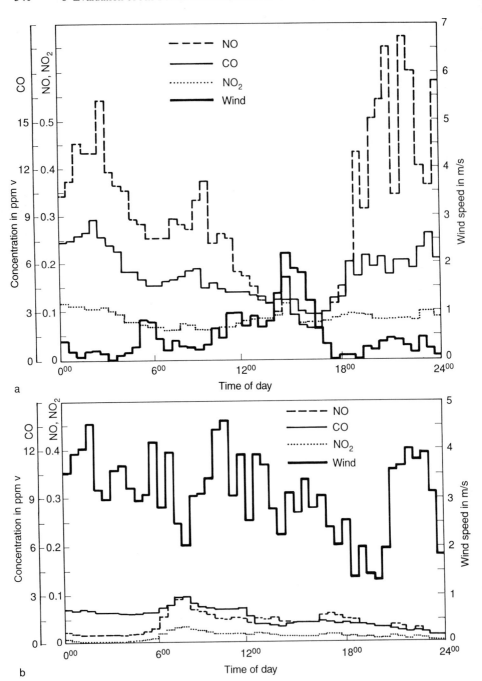

Fig. 6.9. Comparison of diurnal courses of the pollutants NO, NO_2, CO and wind speed on a day with inversion (**a**) and on a rainy day (**b**) – measurements were taken at a high traffic road [40]

6.2 Evaluation and Graphic Representation of Air Quality Measurements   341

traffic peak during this time of day. This fact can be confirmed by a comparison with the diurnal course of the traffic load [40].

6.2.2.4 Long-Term Annual Profiles

In the simplest method of representing the pollutant profiles of several years the annual mean values are included graphically, s. [41]. However, one does not obtain a lot information from this as concentrations can fluctuate considerably within one year.

One possibility of representing the development of pollutant concentrations over several years and in different towns as a comparison is shown in Table 6.2. Here, too, the fluctuations occurring within the individual half years cannot be recognized. Further details, however, would go beyond the scope of this tabular representation and would make it less clear.

If one takes a look at the response times for the atmospheric pollutants listed in Table 6.1 to take effect it becomes clear that such long-term mean values (semi-annual and annual values) cannot be directly used for assessing whether there is a hazard for acute respiratory tract diseases or even mucuous membrane irritations. If anything, these values could be regarded in connection with chronic respiratory tract diseases.

To include the maximum information possible on the fluctuations and peaks different possibilities of representation are available, s. Fig. 6.10.

In Fig. 6.10a all 24-hour mean values for the period from June 1, 1981 to February 28, 1985 are marked as vertical lines. Here one can clearly recognize the structure of the concentrations in this relatively long period of time [43]. In Fig. 6.10b the monthly mean values (I1) over the same period of time are entered as horizontal lines. In addition, the monthly 98 %-values of the frequency distribution (I2) ar shown [43] (for further information on frequency distribution s. sect. 6.2.3). In a complete measurement of one month the 98 %-value (I2) represents the concentration that is exceeded by the 29 highest half-hourly values of the month. Thus, the 98 %-value of a month is always higher than a daily mean value (24 hours = 48 half

Table 6.2. Half-annual mean values of SO_2 concentrations (in mg/m^3) of different measuring sites of the State Authority for Environmental Protection in Baden-Wuerttemberg; Germany (acc. to [42])

Measuring station	Sum. 79	Win. 79/80	Sum. 80	Win. 80/81	Sum. 81	Win. 81/82	Sum. 82	Win. 82/83
Mannheim – south	0.11	0.13	0.09	0.11	0.04	0.09	0.03	0.05
Karlsruhe – downtown	0.04	0.08	0.03	0.07	0.04	0.08	0.03	0.06
Freiburg – west	–	–	0.02	0.04	0.02	0.06	0.03	0.03
Heilbronn	–	–	0.04	0.07	0.04	0.07	0.04	0.06
Stuttgart – Marktplatz (downtown)	0.02	0.06	0.03	0.07	0.04	0.13	–	0.03
Ulm	0.02	0.04	0.02	0.06	0.03	0.06	0.04	0.03

6 Evaluation of Air Pollution Measurements

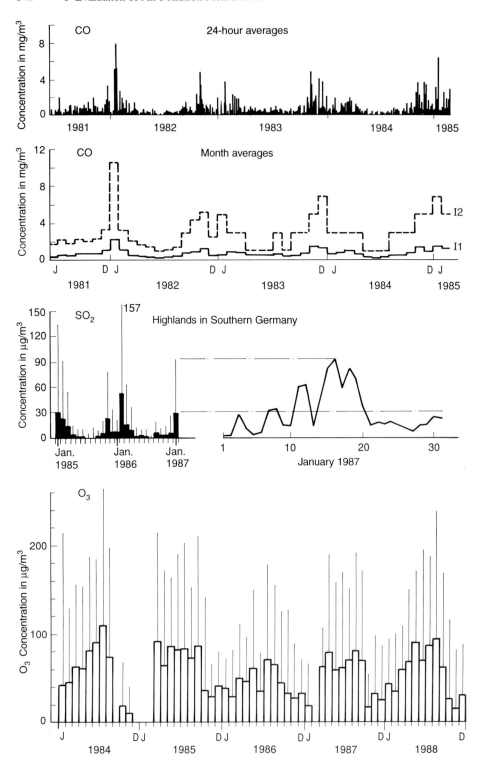

hours). The representation in Fig. 6.10c is similar to the one in Fig. 6.10b, except that instead of the 98 %-values (I2) the highest daily mean values (vertical lines on the left, polygonal course on the right) are plotted together with the monthly mean values (bars). In addition to the monthly mean values (bars) the highest half hourly values (vertical lines) are included with every month in Fig. 6.10d [45].

6.2.3 Frequency Distribution

Graphic representation of frequency distribution serves to investigate how frequently certain pollutant concentrations occurred in a measuring period or measuring series. For this, the pollutant concentration is divided into the required number of classes on the abscissa, and the frequency of the measured values in the individual classes, either in their absolute number or in percent in reference to all measuring values is shown on the ordinate. Fig. 6.11 shows a frequency distribution of this type.

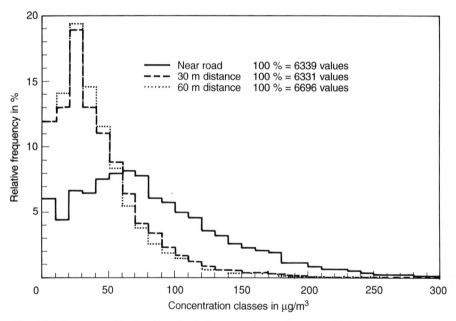

Fig. 6.11. Frequency distribution of an NO_2 measuring series near a highway, measuring cycle from April to November 1985

Fig. 6.10. Different possibilities of representing annual courses including information on maximum values; **a.** 24-hour mean values; **b.** monthly mean values (I1) and monthly 98 % values (I2); **c.** monthly mean values and highest daily mean value as well as polygonal curve of the daily mean values of a month; **d.** monthly mean values and highest half-hourly values [43-45]

Atmospheric pollutants generally have an asymmetrical distribution, i.e., there are many low measuring values and only a few high ones. The latter are, e.g., caused by low exchange weather conditions which are not very persistent in typical Central European weather conditions and do not occur very frequently. In this asymmetrical distribution the mean value of the measurement is generally not identical with the most frequent value (highest point of the envelope curve).

The frequency distribution is different for every measuring site; in sites close to a source it will be shifted to the right to higher concentrations as compared to sites far from sources. One can also see from the diagram how much the measuring values fluctuate. A slender envelope curve signifies slight, a wide envelope curve considerable fluctuations.

Another method of representing frequency distribution is the *cumulative frequency curve*. This means that for every concentration value (on the abscissa) the number of measuring values which are below this value are added up. Generally the share in percent of the total number of the measuring values is indicated on the ordinate. Fig. 6.12 shows a cumulative frequency curve of a measuring value collective. The condition for exceeding or keeping a certain threshold value is in many cases coupled with a frequency distribution. E.g., the EU threshold value for nitrogen dioxide (NO_2) of 200 µg/m³ [47] is exceeded if 2 % of all values for one year are in excess of this concentration. The German regulations, e.g., the Technical Directive on Air Pollution Control (TA Luft) [46] provides maximum values (IW2) which are to be compared with the 98 percentile of a measuring series (I2). The 98 percentile (c_{98})

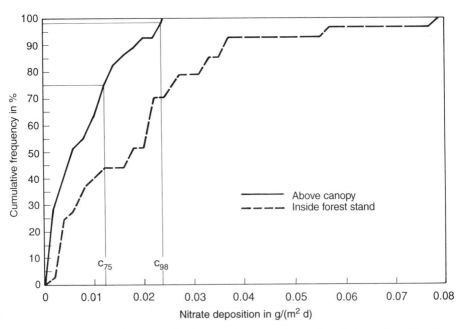

Fig. 6.12. Cumulative frequency distribution of a measuring series of nitrate deposition in a forest stand with 75 and 98 percentiles (c_{75} and c_{98}) [45]

is the concentration below which 98 % of all measuring values ar to be found, s. also Fig. 6.12. Thus, 2 % of all values are in excess of this concentration. Accordingly, the 98 percentile is, e.g., identical with the above-mentioned EU maximum concentration which is exceeded when 2 % of all values are in excess of it. Apart from the c_{98} value the c_{75} value has been marked in Fig. 6.12 which is the concentration which is not reached by 75 % of the measuring values or exceeded by 25 % of all values.

If the total number of the measuring values is known, it is possible to calculate how many measuring values exceed the threshhold concentrations (e.g., in excess of c_{75} or c_{98}). Percentile methods have become established in many institutions so that most of the time c_{98} values are shown in evaluations.

If the frequency distribution curve is known one use the threshold concentration on the abscissa as a basis; where the curve intersects with the ordinate one can see how many percent of the values are below this concentration. The percentage multiplied by the total number gives the number of the values which fall below the threshold concentration.

If one wants to determine excess frequencies then simply those values which lie above the desired concentrations must be counted in the evaluation. Table 6.3 gives an example of such an evaluation of SO_2 measurements in several measuring stations in a narrow valley where a factory is located [36]. The concentrations limits shown here are the threshold values of different organisations.

Other institutions, too, indicate excess frequencies in this tabular form, compare [48, 49].

If excess frequencies are to be represented graphically, then the *sum frequency curve* is not so clear. This latter shows how many values are below a certain concentration – thus, this curve primarily shows frequencies remaining below these concentrationsd. Of course, excess frequencies can be calculated upto 100 % by determining the difference. But it is also possible to enter the percentage excess over the concentration directly. Fig. 6.13 shows an example of excess frequencies of different pollut-

Table 6.3. Number of SO_2 half-hour values exceeding the listed maximum values in the measuring period from October 1985 to March 1986; 4 SO_2 monitoring stations in a Black Forest valley [36]

Measuring Site	Total number of measured values	Threshold Concentrations				
		0.075[a]	0.15[b]	0.25[c]	0.40[d]	1.0[e]
1 Stöckerkopf top	7327	349	162	50	16	3
2 Stöckerkopf middle	6142	462	206	89	44	8
3 Stöckerkopf bottom	4629	27	12	8	6	1
4 Surrbachkopf	6541	440	87	5	0	0

[a] Threshold value of IUFRO for fir.
[b] Threshold value of IUFRO for spruce.
[c] Threshold value acc. to VDI for highly sensitive plants.
[d] Threshold value acc. to VDI for sensitive plants.
[e] MIK value for the protection of human health.

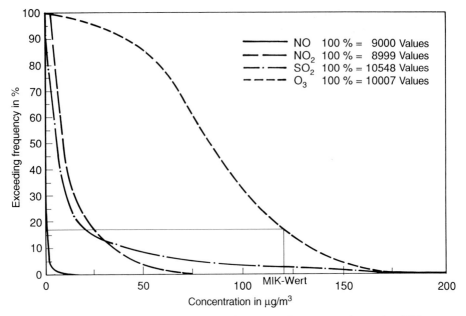

Fig. 6.13. Exceeding frequency of different pollutants at a forest measuring station [45]

ants at a forest measuring station. If one is trying to determine the number of measuring values having exceeded a certain threshold concentration in the measuring period, e.g., 120 µg/m^3, then one must, at the correct spot of the abscissa, go vertically upward until it intersects with the respective curve, and at this level of the ordinate one can read how many percent of the values exceeded the concentration. The total number of measuring values (= 100 %) is also indicated in the diagram.

As can be seen, a lot of low values occur for the component NO, whereas with ozone the concentrations frequently turn out considerably higher.

6.2.4 Spatial Distribution of Pollutants

An essential part of air quality control concepts is the determination and graphic representation of the spatial dispersion of pollutant concentrations in the investigated areas in air quality inventories (e.g., in polluted cities). In this representation with high pollution zones are to be identified so as to be able to take specifically directed air pollution control measures.

6.2.4.1 Method of Determination and Graphic Representation

For the determination of the spatial dispersion measuring trips are carried out. During these trips, the grid points of the 1 x 1 km Gauß-Krüger coordinate system are generally chosen as measuring points. According to German regulation [46] 26 or 13

6.2 Evaluation and Graphic Representation of Air Quality Measurements   347

half-hourly measurements per year are preferably to be carried out, with 13 of them being the minimum. The area mean value I1 and the 98 % value I2 (short-term value) are formed from the measuring results of the 4 corners of a 1 x 1 km area. Thus, each area value is based on 104 or 52 (with 13 measurements per measuring point) half-hour values. The drawbacks of this determination method with so few measuring points per year will be discussed in more detail in the next section.

The reliability of the random measuring values can be improved by comparison with the results of continuously measured values which have been carried out simultaneously at a few points. The measuring values can turn out different from year to year depending on whether there were smog situations in the year of measurement or not and whether these were registered at some of the measuring points and not at others. In addition to the uncertainties of the random sample measuring method caused by the small number of measuring values, there is also the natural fluctuation of pollutant concentrations from year to year. Therefore, to determine the initial degree of pollution of the area to be surveyed, pollutant concentrations are to be determined by averaging the measuring values of at least three successive measuring periods (3 years) [46].

When setting up air quality inventories for air quality control concepts measurements are confined to a one-year duration, usually with only 26 measurements per measuring point, mostly due to reasons of expense.

Air quality inventories where pollutant concentrations are color-coded for each square kilometer can be found in numerous air quality control concepts [19, 22, 28-

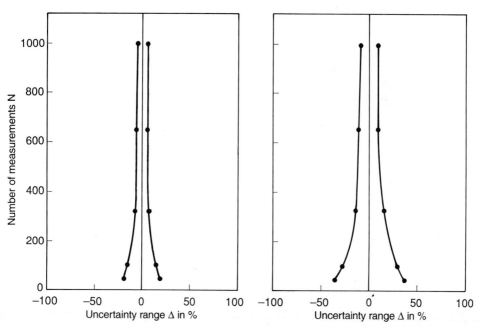

Fig. 6.14. Uncertainty ranges of the mean value (**a**) and the 95 percentile (**b**) as a function of number of measurements N [50]

6.2.5 Methods for the Investigation of Consistent Patterns in the Occurrence of Pollutants

6.2.5.1 Mean Diurnal Courses

The calculation and graphic representation of mean diurnal courses are an appropriate means of identifying certain regularities in the occurrence of daily pollutant concentrations. For this, mean values are calculated for each half hour of the day from the data of the measuring period to be investigated and they are entered above the time of day. In this way the diurnal course of one average day is obtained.

In streets with a lot of traffic pollution concentrations, e.g., depend on the traffic load on the one hand and on the atmospheric exchange conditions on the other. When representing mean diurnal courses the interrelationships become obvious. As the traffic load on Saturdays and Sundays is clearly different from the one on weekdays, it is advisable to represent the mean diurnal courses separately according to weekdays, Saturdays and Sundays. Fig. 6.15 shows that on weekdays the concentration maximum is distinctly higher in the morning (from 6:00 o'clock onwards) than in the afternoons, even though the traffic volume is greater in the afternoons. The influence of air exchange conditions which can be read off the mean diurnal course of the wind speed thus becomes obvious. In the morning, when the first rush hour is at its peak, the surface inversion is highly distinct and wind speed is at its minimum, which is why pollutant concentrations have their highest mean value then. With increasing wind speed in the further course of the day, pollutant concentrations decrease, even though the traffic load barely eases off. With the wind dropping in the afternoon, concentrations rise again. The widest maximum of concentrations is reached at a time when the traffic volume is already subsiding but the surface inversion is developing. At night the pollutants caused by traffic remain at a higher level than during the day because of the reduced change conditions over night. Two parameters, both of which influence the pollutant concentrations present, overlap here. If the cause of the pollution ceases to exist, e.g., the traffic maximum in the morning hours, then, despite inversion conditions, there is no concentration maximum of the polluting gases, as can be seen in the mean diurnal course of Sundays (Fig.6.15b). This course of events holds good for the majority of motor vehicle-related nitrogen oxides and carbon monoxide. With SO_2 the course of the heating activity, e.g., also forms part of the pollutant behavior.

The mean diurnal courses would be even more pronounced, if the days surveyed were grouped according to weather conditions, e.g., mean diurnal courses for weekdays with sunshine and nightly surface inversion on the one hand and days with overcast, windy weather on the other.

For the use of mean diurnal courses see also Figs. 3.19 and 3.23.

6.2 Evaluation and Graphic Representation of Air Quality Measurements

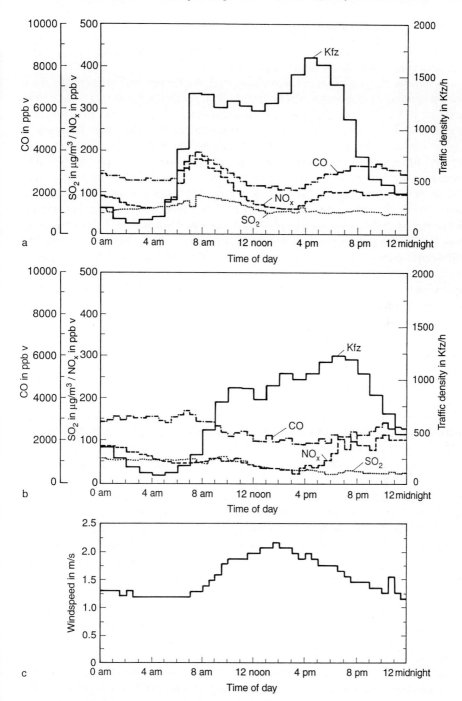

Fig. 6.15. Mean diurnal courses of the pollutants SO_2, NO_x, CO and of traffic density on working days (**a**) and Sundays (**b**) as well as wind speed on all days (**c**); measuring station at an arterial road [40]

6.2.5.2 Correlation Calculations

To establish the dependency of the occurring pollutant concentrations on other factors correlation calculations can be applied [57].

In forests, e.g., pollutant concentrations over the canopy and between the trees are determined by continuous measurements [45]. When the measuring values were assessed it was striking that concentration differences between the air above the canopy and the air within the trees were always particularly high when there was a high pollutant concentration up above. To examine the obvious dependencies more closely, all measuring values are entered in a diagram where the dependant variable parameter (in this case the concentration difference) is graphically represented over the independent variable parameter (in this case the pollutant concentration above the canopy). Fig. 6.16 shows two examples of this: the dependency of the concentration difference on those concentrations above the canopy of SO_2 and ozone. For the component SO_2 one can recognize a definite connection – all the values measured are positioned quite closely around the regression line and the calculated correlation coefficient is $r^2=0.998$. The component O_3, however, is an example of a less obvious dependency. Here, sometimes high concentration differences occur even when pollutant concentrations above the canopy are low, and vice versa there are also small differences when there is a large amouont of ozone ($r^2=0.287$). With ozone, chemical reactions which do not occur with SO_2 mask the dependency on the absolute pollutant concentration. Further information explaining pollutant behavior can be obtained from mean diurnal courses [45]. The dependencies shown here are not generally valid but only apply to the situation in the given measuring period at the measuring station mentioned.

In many cases it is advisable to investigate the dependencies of pollutant concentrations with the help of correlation calculations (further examples s. [45, 58]).

6.2.5.3 Pollutant Wind Roses: Pollutant-Wind Correlations

In air quality measurements wind direction and wind speed should always be measured to find out from which direction atmospheric pollutants are transported to the measuring site. How representative of a larger or smaller area the measured wind directions are depends on the local topography, and must be checked in each individual case. In the diagram the measured wind directions (the most frequent values per each half hour respectively) can, e.g., be divided into 12 or 16 sectors and the frequency (in %) of the measuring values occurring in the individual sectors can be plotted in the circular coordinates. In this way the wind rose of the measuring period is obtained which can be divided according to wind speed sections, [45, 60].

If as in [35] every measured pollutant value is matched according to the wind direction (in 12 or 16 sectors), if the concentration's mean value is formed for each sector and if then these mean values are entered in the circular coordinates over the wind direction, then the so-called pollutant wind rose is obtained, s. Fig. 6.17a, [45, 59]. If there are bulges in the concentration curve in certain sectors of the pollutant wind rose, then increased atmospheric pollutant concentrations are carried along with winds coming from these directions.

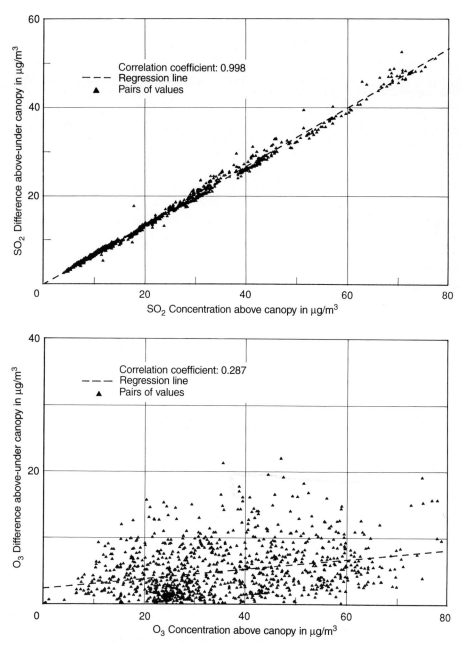

Fig. 6.16. Dependency of concentration difference between the canopy (above the tree tops) and the interior forest on the concentration level above the canopy. Graphic representation and verification by correlation calculation. **a**: SO_2; **b**: O_3 (IVD measuring station Waldenbuch-Betzenberg, June 1986 [45])

352   6 Evaluation of Air Pollution Measurements

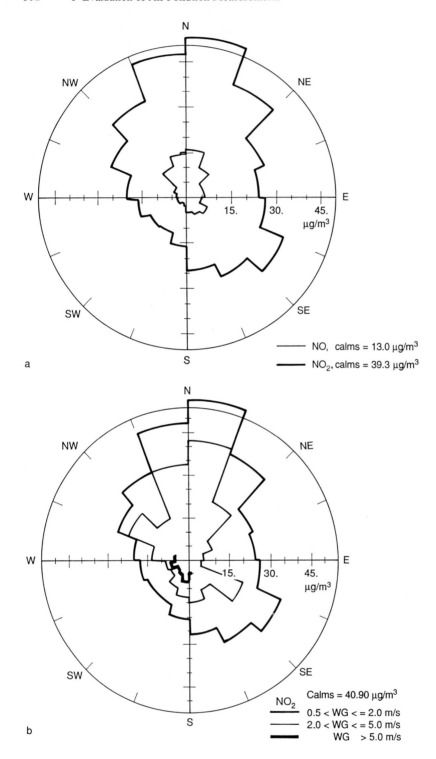

6.2 Evaluation and Graphic Representation of Air Quality Measurements 353

The following can be recognized from the pollutant wind rose shown in Fig. 6.17: the highest NO and NO_2 concentrations on the site of the measuring station (Schönbuch nature park) occur during winds from northerly directions. In this direction lies the greater Stuttgart area with its heavy traffic releasing nitrogen oxides which clearly pollute the Schönbuch nature park during northerly winds. In the mean, the lowest concentrations occur during southwesterly winds.

A slight increase in the nitrogen oxide wind rose occurs during southeasterly winds, implying sources in this direction.

Pollutant wind roses can also be drawn up for different wind speed classes. Fig. 6.17b shows a wind compass diagram divided into three wind speed classes [45]. It can be seen, that during week winds (0.5-2.0 m/s) higher nitrogen oxide concentrations occur than during stronger winds (2.0-5.0 m/s). During strong winds (> 5 m/s) which solely come from southwesterly directions in this area, only very low nitrogen oxide concentrations occur.

The pollutant wind roses shown here do not provide information on the frequency of the concentrations determined in the individual sectors. It could, e.g., well be possible that the wind directions during which high concentrations were measured occurred only very rarely.

If the concentrations of the pollutant wind roses are multiplied by the corresponding frequency of the wind direction (affecting duration) and then divided by the total concentration mean value from all sectors, a pollutant dose wind rose is obtained. The values thus calculated provide information on which percentage contribution is made by a wind direction sector to the mean pollutant load at the measuring site. Detailed information on the representation of wind direction-related pollutant loads including examples of such pollutant dose wind roses is given in [60] by Baumüller and Reuter.

One can also determine the total load by adding up all concentrations measured as half hourly mean values in the measuring period; also the concentrations of each wind direction sector can be added up and their contribution to the total load be determined. In this way, too, a pollutant dose wind rose is obtained representing the contribution of each wind direction to the total load. There is no difference between this pollutant dose wind rose and the one according to Baumüller and Reuter relating to the mean value. The pollutant dose wind rose of Fig. 6.18 is based on the same measurements as those represented in Fig. 6.17. In this pollutant wind rose low concentrations, too, if they occur often enough contribute to the bulging of the dosage curve (e.g., direction southwest) owing to the high frequency of this wind direction. By comparison, the elevated concentrations measured during southeasterly winds (s. Fig. 6.17) occur so rarely that they practically do not show up in the dose wind rose.

Fig. 6.17. Pollutant wind roses; the pollutant mean values of the entire measuring period are represented in the individual wind direction sections; IVD forest measuring station in the Schönbuch (south of the greater Stuttgart area, FRG) [45]; measuring period: 1/1/1986-31/12/1986. **a.** NO and NO_2 mean values of all wind speeds; **b.** NO_2 mean values of different wind speed classes (WS)

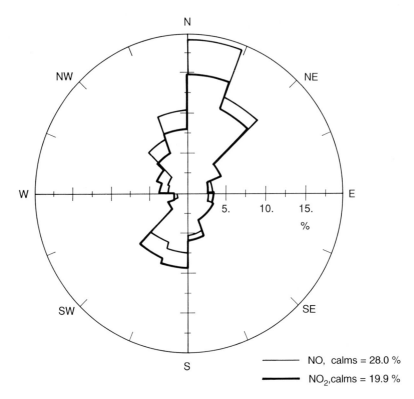

Fig. 6.18. Pollutant dose wind rose; NO and NO$_2$ data of Fig. 6.17 at the IVD forest measuring station in the Schönbuch [45]; measuring period from 1/1/1986 to 12/31/1986

Depending on the biological effectiveness of the pollutants the appropriate type of graphic representation must be chosen. If one is of the opinion that the frequent but very low concentrations do not have any effect, then the pollutant dose wind rose is certainly not a suitable means of representation as the low concentrations are overemphasized here by high wind direction frequencies. However, it would also be conceivable to draw a pollutant dose wind rose merely containing concentrations exceeding a certain threshold concentration. The criteria for this would have to be determined in cooperation with biologists and physicians. It must also be noted that pollutant concentration wind roses (Figs. 6.17a and b) do not lend themselves to short periods of measurement as individual sectors may only show with few measuring values which could be misleading. E.g., when evaluating just one month it is possible that, contrary to all theories, there are very high concentrations in some sectors. When checking this it can turn out that only one or two values caused this situation. A pollutant dose wind rose is better suited for short periods of measurement.

The German State Board for Air Quality Control in Essen applies another type of evaluation [61]: the flow density wind rose. It is formed by multiplying the pollutant concentrations in each sector with the wind speed prevailing at the time and by assessing them according to their frequency. In this way it is determined which pollutant mass flows through an imaginary vertical surface of 1 m^2 in one year at the measuring site. It must be emphasized that it is not a mass flow which is deposited on the ground or the vegetation. Apart from the frequency of the main wind direction the magnitude of the wind speed is included in this type of representation. As in Germany winds from westerly directions are the most frequent and also the strongest, these directions are strongly emphasized despite low concentrations.

It was shown that the representation of pollutant wind roses is an important means of determining the origin of atmospheric pollutants. However, even choosing a certain type of a pollutant wind rose can mean an interpretation of the measuring results in a certain way. One must therefore proceed with utmost care when applying and interpreting such wind direction-related representations.

6.2.5.4 Abatement Curves

It is often interesting to note how pollutant concentrations abate with increasing distance to the source, e.g., a street. Fig. 6.19 shows some abatement curves from measurements made on level terrain. When calculating these curves wind conditions may, of course, not be neglected. If, during the measurements, the wind blows against the street, then no traffic-related influence on measured concentrations can be detected even a short distance away (curve 4 in Fig. 6.19). On approximately level terrain the abatement curves usually have a profile like curves 2 and 3 [62]. The absolute level of the abating concentrations depends on the value of the initial concentrations, and these are again dependent on the source strength and on the weather-related exchange parameters. Curve 1 implies the abatement of the concentrations during low exchange weather conditions.

In street "canyons" (between rows of high buildings) atmospheric pollutants naturally behave differently [63]. Knowledge on abatement curves should have an effect on, e.g., development plans, suggesting residential areas near roads.

The dispersion of pollutants specific to motor vehicle traffic has been investigated in detail by Baumann [64] taking into account wind, time of day and street layout by using the example of a much frequented highway. Compared to a level roadway roads cutting below ground level are advantageous for the area in the immediate vicinity of the road at medium and high wind speeds; however, during weak wind conditions in general and low exchange weather conditions in particular, the load increases, as the air in the trough enriched with pollutants "overflows". This deep street bed layout of highways which can be found with increasing frequency represents a higher health hazard even for the road users themselves, as the motor vehicle emissions are channelled, cannot take part in the air exchange and thus accumulate.

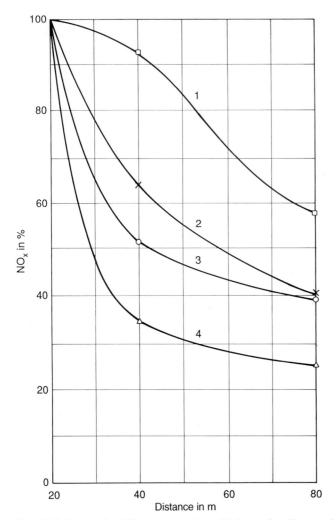

Fig. 6.19. Decreasing NO_x concentration with increasing distance from an arterial road during different meteorological exchange conditions [40]. *1* surface inversion, wind speed <0.1 m/s, wind contacting the highway at a small angle; *2* weak wind (<1 m/s) at almost right angles to the highway, snowfall; *3* turbulent fair weather, wind speed ~ 2 m/s, wind contacting the highway at a small angle; *4* turbulent fair weather, wind speed 2.5-3 m/s, wind slanting towards the highway

6.3 Bibliography

1 VDI: Richtlinie 2450, Bl. 1, Messen von Emission, Transmission und Immission luftverunreinigender Stoffe – Begriffe, Definitionen. Berlin: Beuth 9/1977
2 Stratmann, H.: Wirkungen von Luftverunreinigungen auf die Vegetation. LIS-Berichte Nr. 49, Landesanstalt für Immissionsschutz, Essen 1984
3 Umweltbundesamt Bericht 9/80: Das Abgas-Emissionsverhalten von Personenkraftwagen in der Bundesrepublik Deutschland im Bezugsjahr 1980. TÜV-Reinland im Auftrag des Umweltbundesamtes. Berlin: Erich Schmidt 1980

4 Umweltbundesamt Bericht 11/83: Das Abgas-Emissionsverhalten von Nutzfahrzeugen in der Bundesrepublik Deutschland im Bezugsjahr 1980. TÜV-Rheinland im Auftrag des Umweltbundesamtes. Berlin: Erich Schmidt 1983
5 Hauschulz, G.; Heich, H.J.; Leisen, P.; Raschke, J.; Waldeyer, H.; Winckler, J.: Emissions- und Immissionstechnik im Verkehrswesen. Köln: TÜV Rheinland 1983
6 Meier, E.; Plaßmann, E.; Wolff, C. et al.: Abgas-Großversuch. Abschlußbericht, Vereinigung der Technischen Überwachungsvereine e.V., Essen 1986
7 TÜV Rheinland/Umweltbundesamt: Emissionsfaktoren für Luftverunreinigungen. Materialien 2/80. Berlin: Erich Schmidt 1980
8 Fünfte Allgemeine Verwaltungsvorschrift zum Bundes-Immissionsschutzgesetz (Emissionskataster in Belastungsgebieten) − 5. BImSchVwV vom 30.1.1979, GMBl., S. 42 f
9 Umweltbundesamt: Daten zur Umwelt 1986/87. Berlin: Erich Schmidt 1986
10 Dritte Verordnung zur Durchführung des Bundes-Immissionsschutzgesetzes (Verordnung über Schwefelgehalt von leichtem Heizöl und Dieselkraftstoff − 3. BImSchV) vom 15.1.1975, BGBl. I, S. 264, geändert durch Fassung vom 18.2.1986, BGBl. I, S. 265 und durch Verordnung vom 14.12.1987, BGBl. I, S. 2 671f
11 Institut für wirtschaftliche Oelheizung e.V.: Die Ölfeuerung, Ein Beitrag zur Reinhaltung der Luft. Druckschrift, Hamburg 1986
12 Chae, J.O.: Aufstellung eines mathematischen Modells der NO_x-Bildung in eingeschlossenen turbulenten Erdgas-Diffusionsflammen. Dissertation Universität Stuttgart 1978
13 Schnell, U.: Die mathematische Modellierung der Stickstoffoxid-Emissionen von Kohlenstaubflammen. Diplomarbeit Nr. 2204 am Institut für Verfahrenstechnik und Dampfkesselwesen der Universität Stuttgart 1986
14 Breuninger, H.A.; Baumbach, G.: Untersuchungen zur Wirksamkeit von Additiven für schweres Heizöl. Fortschr. Ber. VDI-Z. Reihe 6 Nr. 113, 1983
15 Dröscher, F.; Rauskolb, J.: Luftverunreinigungen in einem Schwarzwaldtal bei Inversionswetterlagen, Teil 2: Untersuchung der Staub-Immissionen. Institut für Verfahrenstechnik und Dampfkesselwesen der Universität Stuttgart, Abteilung Reinhaltung der Luft, Bericht Nr. 3, 1986
16 Ministerium für Ernährung, Landwirtschaft, Umwelt und Forsten Baden-Württemberg: Emissionskataster Stuttgart, Quellengruppe Verkehr. Stuttgart 1986
17 Ministerium für Ernährung, Landwirtschaft, Umwelt und Forsten Baden-Württemberg: Emissionskataster Mannheim, Quellengruppe Verkehr. Stuttgart 1986
18 Ministerium für Ernährung, Landwirtschaft, Umwelt und Forsten Baden-Württemberg: Emissionskataster Karlsruhe, Quellengruppe Verkehr. Stuttgart 1986
19 Baumbach, G.; Baumüller, J.; Dröscher, F.; Reuter, U.: Lufthygienisches Gutachten für die Stadt Heilbronn. Amt für Straßenverkehr und Umwelt, Heilbronn 1986
20 Ministerium für Ernährung, Landwirtschaft, Umwelt und Forsten Baden-Württemberg: Emissionskataster Stuttgart, Quellengruppe Hausbrand. Stuttgart 1986
21 Ministerium für Ernährung, Landwirtschaft, Umwelt und Forsten Baden-Württemberg: Emissionskataster Mannheim, Quellengruppe Hausbrand. Stuttgart 1986
22 Dreyhaupt, F.J.; Dierschke, W.; Kropp, K.; Prinz, B.; Schade, H.: Handbuch zur Aufstellung von Luftreinhalteplänen. Köln: TÜV Rheinland 1979
23 Baumbach, G.; Cannon, T.; Bauer, L.; Bisinger, R.: Luftqualität in Waldenbuch. Institut für Verfahrenstechnik und Dampfkesselwesen der Universität Stuttgart, Abt. Reinhaltung der Luft, Bericht Nr. 1 − 1985
24 Vierte Verordnung zur Durchführung des Bundes-Immissionsschutzgesetzes (Verordnung über genehmigungsbedürftige Anlagen). 4. BImSchV vom 24.7.1985, BGBl. I, S. 1586 f
25 Elfte Verordnung zur Durchführung des Bundes-Immissionsschutzgesetzes (Emissionserklärungsverordnung). 11. BImSchV vom 20.12.1978, BGBl. I, S. 2027 f, geändert durch Verordnung vom 14.7.1985, BGBl. I, S. 1586 f
26 Ministerium für Ernährung, Landwirtschaft, Umwelt und Forsten Baden-Württemberg: Emissionskataster Stuttgart, Quellengruppe Industrie und Gewerbe. Stuttgart 1986
27 Ministerium für Ernährung, Landwirtschaft, Umwelt und Forsten Baden-Württemberg: Emissionskataster Karlsruhe, Quellengruppe Industrie und Gewerbe. Stuttgart 1986
28 Ministerium für Arbeit, Gesundheit und Soziales des Landes Nordrhein-Westfalen: Luftreinhalteplan Ruhrgebiet Ost − Dortmund − 1979−1983. Düsseldorf 1978

29 Ministerium für Arbeit, Gesundheit und Soziales des Landes Nordrhein-Westfalen: Luftreinhalteplan Rheinschiene-Süd — 1. Fortschreibung — 1982—1986. Düsseldorf 1984
30 Hessischer Minister für Landesentwicklung, Umwelt, Landwirtschaft und Forsten: Luftreinhalteplan Rhein-Main. Wiesbaden 1981
31 Ministerium für Soziales, Gesundheit und Umwelt des Landes Rheinland Pfalz: Luftreinhalteplan Ludwigshafen/Frankenthal 1979—1984. Mainz 1980
32 Kutzner, K.: Der Hausbrand als bodennahe Emissionsquelle — Flächendeckende Emissionen und ihre Bedeutung für die Lufthygiene. VDI Ber. Nr. 477 „Reinhaltung der Luft in großen Städten", Düsseldorf 1983
33 Friedrich, R.; Müller, T.; Scheirle, N.; Voß, A.: Feinmaschiges Kataster der SO_2- und NO_x-Emissionen in Baden-Württemberg und im Oberrheintal — Emissionsuntersuchungen im Rahmen des TULLA-Projektes, in Bericht über das 1. Statuskolloquium des PEF. S. 373—384, Kernforschungszentrum Karlsruhe 1985
34 Boysen, B.; Friedrich, R.; Müller, T.; Scheirle, N.; Voß, A.: Feinmaschiges Kataster der SO_2- und NO_x-Emissionen in Baden-Württemberg für die Zeit der TULLA-Meßkampagne, in Bericht über das 2. Statuskolloquium des PEF. Band 2, S. 481—492, Kernforschungszentrum Karlsruhe 1986
35 Vierte Allgemeine Verwaltungsvorschrift zum Bundes-Immissionsschutzgesetz (Ermittlung von Immissionen in Belastungsgebieten) vom 8.4.1975. GMBl. S. 358
36 Baumbach, G.; Konrad, G.; Minner, G.: Luftverunreinigungen in einem Schwarzwaldtal bei Inversionswetterlagen, Teil 1. Institut für Verfahrenstechnik und Dampfkesselwesen der Universität Stuttgart, Abteilung Reinhaltung der Luft, Bericht Nr. 3—1986
37 Umweltbundesamt: Monatsberichte aus dem Meßnetz. Sammelband 1983, Hrsg.: Umweltbundesamt, Fachgebiet II 6.3, Berlin 1984
38 Stern, A.C.; Boubel, R.W.; Furner, D.B.; Fox, D.L.: Fundamentals of Air Pollution. Orlando, Florida: Academic Press 1984
39 Verordnung der Landesregierung, des Ministeriums für Umwelt und des Innenministeriums zur Verhinderung schädlicher Umwelteinwirkungen bei austauscharmen Wetterlagen (Smog-Verordnung-Smog VO) vom 27.6.1988. Gesetzblatt für Baden-Württemberg 1988, S. 214f
40 Baumbach, G.: Derzeitige und zukünftige Schadgas-Immissionsbelastung an der B 10 in Stuttgart-Stammheim Süd. Gutachten für das Stadtplanungsamt Stuttgart 1981
41 Baumüller, J.; Hofmann, U.: Langjährige Schwefeldioxid-Messungen in Stuttgart 1965—1978. Chemisches Untersuchungsamt der Landeshauptstadt Stuttgart, Abt. Klimatologie, Mitteilung Nr. 2, Stuttgart 1980
42 Landesanstalt für Umweltschutz Baden-Württemberg: Zweiter Umweltbericht Baden-Württemberg 1983, Karlsruhe 1983
43 Baumüller, J.; Reuter, U.; Hoffmann, U.: Luftschadstoffe und Klima im Raum Stuttgart Vaihingen/Möhringen. Chemisches Untersuchungsamt der Landeshauptstadt Stuttgart, Abt. Klimatologie, Mitteilung Nr. 9, Stuttgart 1985
44 Umweltbundesamt: Monatsberichte aus dem Meßnetz 2/84. Berlin 1984
45 Baumbach, G.; Baumann, K.; Dröscher, F.: Luftverunreinigungen in Wäldern. Institut für Verfahrenstechnik und Dampfkesselwesen der Universität Stuttgart, Abt. Reinhaltung der Luft, Bericht Nr. 5—1987
46 Erste allg. Verwaltungsvorschrift zum Bundes-Immissionsschutzgesetz (Technische Anleitung zur Reinhaltung der Luft — TA Luft) vom 27.2.1986. GMBl. S. 95f
47 Statistische Berichte, Umwelt, Immissions-Konzentrationsmessungen. Hrsg. vom Statistischen Landesamt Baden-Württemberg, Stuttgart
48 Kenngrößen der Ozonkonzentration für die UBA-Meßstellen, Monatsberichte aus dem Meßnetz, Februar 1983. Umweltbundesamt Berlin 1984
49 Landesanstalt für Immissionsschutz des Landes NRW: Monatsbericht über die Luftqualität an Rhein und Ruhr 1/85. Essen 1985
50 Buck, M.; Doppelfeld, A.: Die Bedeutung des Stichprobenumfanges bei der Messung und Bewertung von Immissionen. Schriftenreihe der Landesanstalt für Immissionsschutz des Landes NRW, Verlag W. Giradet, Essen, H. 50 (1980) 31—40
51 Beier, R.: Zur Kennzeichnung von Immissionsbelastungen durch Quantile von Schadstoffverteilungen. Schriftenreihe der Landesanstalt für Immissionsschutz des Landes NRW, Verlag W. Giradet, Essen, H. 55 (1982) 7—14

52 Beier, R.; Doppelfeld, A.: Statistische Analyse von Spitzenwerten stichprobenartig untersuchter Schadstoffkonzentrationen in der Außenluft. Schriftenreihe der Landesanstalt für Immissionsschutz des Landes NRW, Verlag W. Giradet, Essen, H. 57 (1983) 7−14
53 Junker, A.; Kühner, D.: Meßhäufigkeit und Aussagesicherheit bei kontinuierlichen Immissionsmessungen. Staub-Reinhalt. Luft 39 (1979) Nr. 1, S. 22−25
54 Müller, H.G.: Statistische Methoden zur Beurteilung der Immissionsstruktur. Im Auftrag des Umweltbundesamtes Berlin, Abschlußbericht, Dornier System GmbH, Friedrichshafen 1977
55 VDI: VDI-Richtlinie 2310, Bl. 11, Maximale Immissions-Werte zum Schutze des Menschen (Schwefeldioxid). Berlin: Beuth 8/1984
56 Erste allg. Verwaltungsvorschrift zum Bundes-Immissionsschutzgesetz (Technische Anleitung zur Reinhaltung der Luft − TA Luft) vom 28.8.1974. GMBl A, 25 (1974) Nr. 24, S. 425−452
57 Kreyszig, E.: Statistische Methoden und ihre Anwendungen. Göttingen: Vandenhoeck & Ruprecht 1985
58 Baumbach, G.: Gleichzeitige Erfassung von Außenluft- und Innenraumkonzentrationen verschiedener Schadstoffe. VDI Ber. Nr. 608 (1987) 537−557
59 Baumbach, G.; Baumann, K.; Dröscher, F.: Luftqualität in Freudenstadt. Institut für Verfahrenstechnik und Dampfkesselwesen der Universität Stuttgart, Abt. Reinhaltung der Luft, Bericht Nr. 6 − 1987
60 Baumüller, J.; Reuter, U.: Hinweise zur Darstellung windrichtungsabhängiger Schadgasbelastungen. Staub-Reinhalt. Luft 44 (1984) Nr. 4, S. 183−186
61 Pfeffer, H.U.: Immissionserhebungen in quellfernen Gebieten Nordrhein-Westfalens. Staub-Reinhalt. Luft 45 (1985) Nr. 6, S. 287−293
62 Esser, J.: Schadstoffkonzentrationen im Nahbereich von Autobahnen in Abhängigkeit von Verkehr und Meteorologie. In: Abgasbelastungen durch den Kraftfahrzeugverkehr. Köln: TÜV Rheinland 1982, S. 165−183
63 Waldeyer, H.; Leisen, P.; Müller, W.R.: Die Abhängigkeit der Immissionsbelastung in Straßenschluchten von meteorologischen und verkehrsbedingten Einflußgrößen. In: Abgasbelastungen durch den Kfz-Verkehr. Köln: TÜV Rheinland 1982, S. 85−114
64 Baumann, K.: Schadstoffausbreitung im Nahbereich einer Autobahn. Institut für Verfahrenstechnik und Dampfkesselwesen der Universität Stuttgart, Abt. Reinhaltung der Luft, Bericht Nr. 8 − 1987

7 Emission Control Technologies

7.1. General Considerations

The tasks of air pollution control in the different emission source areas of traffic, industrial processes and industrial furnaces as well as domestic heating systems are manifold. Basically, emission reductions are possible by
- changing to lower-emission processes or low-emission fuels,
- improving the process, e.g., the combustion process: primary prevention measures,
- exhaust gas purification: secondary measures.

Which course of action should be taken in the individual case depends both on the degree of air pollution control that is to be achieved, and on the feasibility of the measure which usually shows directly in the cost.

The cost of pollution control is generally exponentially dependent on the degree of pollutant removal, s. Fig. 7.1.

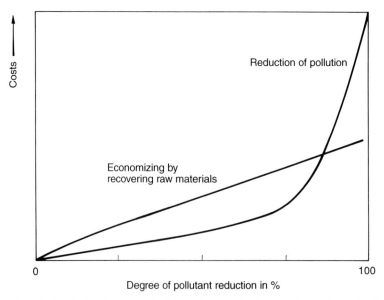

Fig. 7.1. Cost of pollution control dependent on the degree of pollution reduction [1]

If, by pollutant removal, raw materials can be recovered, expenses can be saved initially. With low degrees of removal savings can be greater than the cost of the emission reduction process, s. Fig. 7.1. With an increasing degree of pollutant removal, however, the cost of pollutant reduction will exceed the attainable savings.

It must be decided in each individual case which process of pollution control can best meet the necessary requirements while causing the least expense at the same time. However, not only short-term cost considerations should be weighed but long-term developments should also be taken into consideration. E.g., processes requiring less energy should always be given preference even if the currently low prices of certain fuels (e.g., fuel oil) seem to make low energy consumption unnecessary.

7.1.1 Process Modification

One *example* of modifying an industrial process to reduce emissions of air pollutants is *painting*. Most articles of daily use made of metal, from tin can to refrigerator and automobile undergo painting as surface treatment. Fig. 7.2 shows that in the traditional, mostly used spray painting process with compressed air at least 50 % paint-aerosol losses (solid pigments) occur solely due to process-related conditions [2] the majority of which pollute the surrounding air if no measures are taken. Depending on the workpiece painted losses can rise up to 90 %, e.g., when spray-painting a bird

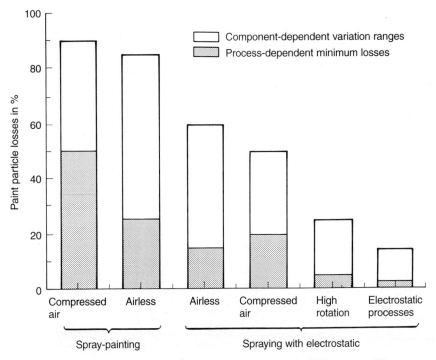

Fig. 7.2. Comparison of paint fog losses in different spray painting processes [2]

cage. By spray-painting without compressed air or by spraying with electrostatic processes paint-aerosol losses can be reduced considerably. It must be noted, however, that with electrostatic spray processes surfaces do not turn out as smooth as with spray-paint processes – something which is more a purely aesthetic difference rather than a loss of quality.

By using other painting systems air pollution caused by solvent vapors can be considerably reduced and work- and cost-intensive air purification measures could possibly be saved. Fig. 7.3 shows the solvent content of different varnishes. It is to be assumed that the greatest part of the solvents evaporates during painting and drying and is released into the air. Traditional fast-drying varnishes and synthetic resin varnishes have high solvent contents and are thus highly emission-intensive. One possibility of reducing the emission of organic solvents as much as possible while at the same time maintaining varnish quality almost at the same standard is the use of water-soluble varnishes. In part, these varnishes are already in use for painting automobiles. Problems of treating waste water produced due to the water solubility of the residual solvent may not be ignored. Coating powders which are applied electrostatically are completely free of solvents. With these powders, however, the varnish coat does not turn out as smooth as in the other systems.

There are numerous other cases where modifying the process can drastically reduce the emissions of air pollutants. A known example is the production of sulfuric acid where SO_2 emissions of originally 17 kg per ton of produced sulfuric acid could be reduced to 2-3 kg owing to the introduction of the double contact process. [3, 4].

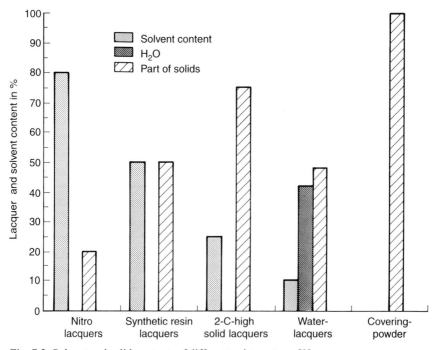

Fig. 7.3. Solvent and solid contents of different paint systems [2]

7.1.2 Emission Control for Furnaces

Industrial and domestic furnaces are a major source of air pollution. Before discussing special air quality control measures some general aspects on emission control will be pointed out.

Fig. 2.1 shows three groups of pollutants emitted by furnaces:
1. products of incomplete combustion: CO, soot, unburned hydrocarbons,
2. complete combustion by-product: NO_x,
3. products caused by fuel impurities: SO_2, SO_3, H_2S, NO_x, chlorides, fluorides, metal compounds.

The different pollutants are not emitted by all furnaces to the same extent. Composition and amount of pollutants vary depending on the size and type of furnace as well as the fuel used. Fig. 7.4 presents an overview of the main pollutant emissions from different furnaces.

While small furnaces have the problem of emitting pollutants caused by incomplete combustion and insufficient control, large-scale furnaces cause air pollution by burning low-priced but impure fuels which in turn cause emissions containing, e.g., particles, SO_2, NO_x and other inorganic components.

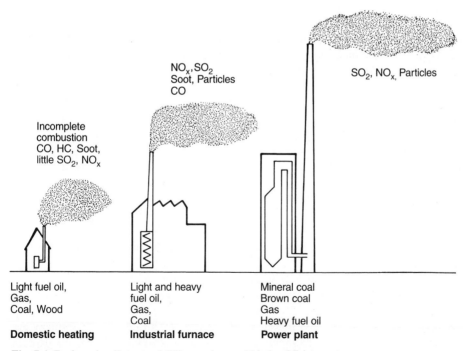

Fig. 7.4. Fuels and pollutants of different sizes and kinds of firing systems

7.1.2.1 Products of Incomplete Combustion

Due to uneven burning, insufficient possibilities for controlling and adjusting the load and the air-fuel ratio, small furnaces predominantly used for the heating of private homes or in small-scale industries, especially *solid fuel furnaces* (wood and coal), cause the major share of all CO, C_nH_m- and particulate emissions [5]. As a consequence, this often leads to an intensely annoying odour in residential areas. It is possible to reduce these emissions by changing the design of the furnaces and their operation. In fact, efforts are being made with the goal of keeping a steady flame burning despite varying charge volumes (different amounts of fuel). However, even well-designed furnaces cannot avoid being influenced by different types of operation. Using damp firewood, e.g., leads to low burning temperatures, and closed air flaps will inevitably lead to smouldering fires with higher emissions. The market also offers catalytic converters for the afterburning of CO and hydrocarbons. Regardless of the efficiency of such converters, however, the focus of emission removal in small furnaces will have to be directed to preventing the formation of pollutants during the combustion process.

In small oil furnaces with blowpipe burners emissions of products caused by incomplete combustion are normally negligible. Faults, however, such as the pollution of air pipes or the blocking of fuel nozzles can lead to incomplete combustion. Until the chimney-sweep finds these faults during his next annual check-up there may be emissions of unburned substances over a long time, possibly in combination with an annoying odor. This hazard is particularly acute with oil stoves for individual rooms operating with evaporator burners. Improvements can be achieved by optimizing the reliability of the burner systems used and by regular maintenance and control.

Pollution removal measures for small furnaces will not be discussed in further detail; instead the reader is referred to specialised literature on this subject which is compiled in [5].

7.1.2.2 Particulates

The highest emissions of particulate matter are caused by solid fuel furnaces due to the release of mineral fuel components. Except for few cases, even in large-scale industrial furnaces particle precipitators have not been installed either in oil furnaces operating with light fuel oil or in those operating with heavy fuel oil. At most particle emissions can pose a problem in heavy fuel oil furnaces. Here, e.g., combustion-promoting additives can improve the degree of combustion of soot particles to such an extent that the prescribed perticle emission threshold values can be adhered to [6, 7].

In coal furnaces the removal of suspended particles makes the use of particle precipitators necessary. In medium-sized furnaces cyclone collectors, among others, are used and in special cases filtration systems have also been installed. Practically all large coal-dust furnaces in power plants in the Federal Republic of Germany are equipped with electrostatic precipitators. The mode of action of the different particle precipitation processes will be treated in sect. 7.2.

7.1.2.3 Nitrogen Oxides

To reduce nitrogen oxide emissions caused by combustion processes there are basically two possibilities:

1. *Primary measures.* By modifying different parameters nitrogen oxide formation is reduced during combustion or already formed nitrogen oxide is re-converted while still in the combustion chamber.

2. *Secondary measures.* The nitrogen oxide formed during combustion is removed from the exhaust gases by purification processes.

The different possibilities of reducing nitrogen oxide emissions in combustion processes will be dealt with in more detail in sect. 7.3.

7.1.2.4 Sulfur Dioxide

Apart from a few special chemical processes stationary furnaces are the main source of sulfur dioxide (SO_2). There are basically three methods of reducing SO_2 emissions:
1. fuel desulfurization,
2. choice of fuel,
3. flue gas desulfurization.

In the FRG just under 50 % of all small furnaces which are used for domestic heating and small-scale industries run on light fuel oil. To reduce emissions of this source group the sulfur content of light fuel oil was progressively reduced by ordinance . Until 1988 0.3 % sulfur was permitted, from March 1988 on it was reduced to 0.2 % [8]. Diesel fuel is subject to the same ordinances as far as sulfur content is concerned. There were and are still attempts to make small flue gas desulfurization units for SO_2 reduction in domestic heaters commercially available. However, reducing the sulfur content of the fuel oil used is a means of efficient SO_2 reduction for *all light fuel oil furnaces.*

Fuel desulfurization

The refineries regulate the sulfur content of *light fuel oil* by mixing oils of different origins and varying sulfur contents. Besides this, light fuel oil is also desulfurized by adding hydrogen to the oil. Desulfurization is then carried out with specific catalytic converters – mostly of cobalt and molybdenum on an alumina carrier – at temperatures of 320-420 °C and pressures of 25-70 bar [9]. In this process the sulfur is split off by hydriding and escapes as hydrogen sulfide which is subsequently transformed into elementary sulfur in a Claus process. Fig. 7.5 shows the diagram of a plant for the catalytic desulfurization of oils. The oil charge is heated in an oven to the correct temperature, is mixed with pre-warmed hydrogen and then reaches a reactor containing the catalytic converter. The reaction product is cooled and freed from the excess hydrogen in a high pressure gas precipitator. The hydrogen is then fed back into the

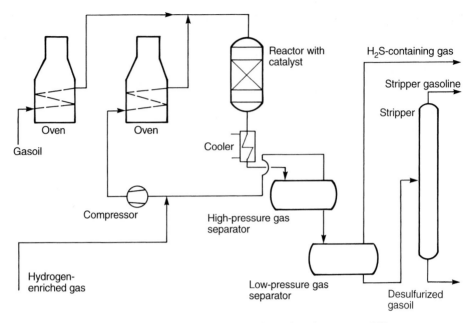

Fig. 7.5. Simplified diagram of a catalytic fuel desulfurization plant (acc. to [9])

process as cycle gas and is replenished with fresh hydrogen. Hydrogen sulfide is separated in a low pressure gas precipitator. The liquid product is freed from reaction products with a low boiling point in a stripper.

In principle, heavy fuel oil can also be desulfurized in this way; reaction conditions, however, must be considerably more extreme (higher temperatures and higher pressures) [10]. Heavy metal traces in heavy fuel oils pose problems as these can act as catalytic poisons [9]. So far, however, one has managed to do without this special fuel desulfurization when preparing heavy fuel oils with certain sulfur content (e.g., heavy fuel oil with 1 % sulfur or even lower values).

Another possibility of reducing SO_2 emissions from furnaces is *to choose a certain fuel or changing to a different fuel*. In the past few years, the use of natural gas, e.g., has increased considerably in small and in industrial furnaces. As natural gas contains practically no sulfur, SO_2 emissions have been further reduced in this way.

According to the German "Large Furnace Ordinance" of 1983 [11] flue gas desulfurization plants should have been installed in the last years in industrial furnaces of more than 100 MW thermal power operated with heavy fuel oil. For reasons of expense, however, most furnaces were converted to natural gas with light fuel oil as substitute fuel.

When *coal* is used as fuel, fuel desulfurization is hardly at all possible. In most large-scale furnaces, particularly in power plant furnaces, a conversion to lower sulfur and more expensive fuels is out of the question. For this reason flue gas desulfurization plants are used for SO_2 removal from these furnaces. Flue gas desulfurization will be dealt with in more detail in sect. 7.4.

7.1.3 Efficiency of Exhaust Gas Purification Systems

In many cases the problem of air pollution cannot be solved satisfactorily solely by changing the process or choosing a different fuel. Special processes for the purification of exhaust gases must be applied. The efficiency of exhaust gas purification processes is characterized by the degree of removal and the residual emission. The definition of this and further characteristic parameters of a precipitator are represented in a diagram in Fig. 7.6.

The degree of removal is the quantity of pollutant removed (M_{rem}) or the concentration S_{rem} in reference to the pollutant quantity (M_{raw}) or concentration S_{raw} entered:

$$\eta = \frac{S_{ab}}{S_{roh}} \cdot 100 \text{ in \%}. \tag{7.1}$$

If the entered pollutant quantity is unknown, then S_{raw} can also be calculated from the removed S_{rem} and the quantity (M_{pure}) or concentration S_{pure} still being emitted:

$$S_{roh} = S_{ab} + S_{rein}, \tag{7.2}$$

$$\eta = \frac{S_{ab}}{S_{ab} + S_{rein}} \cdot 100 \text{ in \%}. \tag{7.3}$$

The degree of removal can also be determined from the raw gas and pure gas balance:

$$\eta = \frac{S_{roh} - S_{rein}}{S_{roh}} \cdot 100 \text{ in \%}. \tag{7.4}$$

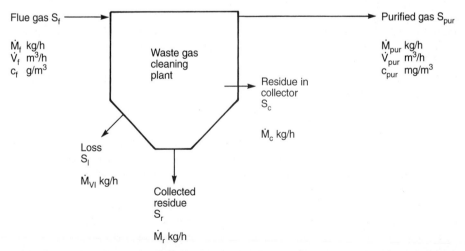

Fig. 7.6. Parameters for expressing properties and efficiency of exhaust gas purification plants. $S_{raw}, S_{pure}, S_{ab}, S_{Vl}, S_{Vb}...$ pollutant contents or amounts

Generally the pure gas concentration c_{pure} and the emission mass flow $\dot{M}_{pure}$ in the pure gas are the factors of interest where observing threshold values is concerned. Mostly, pure gas concentration can only be determined by measurements. To be able to compare the results, it is common practice to convert the concentration to standard conditions (0 °C, 1 013 mbar) with the general gas equation:

$$pV = mRT \tag{7.5}$$

or

$$\frac{pV}{mT} = \text{const}, \tag{7.6}$$

with

$$c = \frac{m}{V} \tag{7.7}$$

the following results are obtained:

$$\frac{p_n}{c_n T_n} = \frac{p_1}{c_1 T_1} \tag{7.8}$$

$$c_n = c_1 \frac{1\,013\,(t+273)}{273\,p_1}, \tag{7.9}$$

where p is pressure, V volume, m masses, R general gas constant, T temperature in K, t temperature in °C, c concentration, index n standard conditions (0 °C, 1 013 mbar).

Apart from the pure gas concentrations certain degrees of removal are required either due to regulations or for economic reasons. In balance considerations of this type losses (S_{lo}) and residues in the precipitator (S_{res}) may not be ignored (s. Fig. 7.6).

Merely determining the pure gas concentration c_{pure} is not sufficient for monitoring adherence to threshold values. In addition, either the emission mass flow $\dot{M}$ is calculated by multiplying the pure gas concentration c_{pure} with the exhaust gas volume $\dot{V}$

$$\dot{M} = c\dot{V}, \tag{7.10}$$

or the exhaust gas dilution is taken into account in the pure gas concentrations measured. The concentration could theoretically be lowered by merely mixing exhaust gas with air. In exhaust gases from combustion processes exhaust gas dilution is, e.g., taken into account by relating the measured concentration or emission to a fixed oxygen reference content [11, 12]:

$$E_B = \frac{21 - O_B}{21 - O_M} E_M, \tag{7.11}$$

where E_B is emission, related to the reference oxygen content, E_M measured emission, O_B referred oxygen content, O_M measured oxygen content.

7.2 Processes for the Removal of Particulate Matter from Exhaust Gases (H. Gross)

Dispersions of particles with a diameter of between approx. 10^{-3} and 10^3 µm of any form and density in gaseous media are called aerodispersions. Dispersing forces such as flow forces which are caused by turbulent movements of the gas delay or prevent the natural sedimentation process to which every particle is subject due to gravity.

Therefore, the basic principle of all particle removal is to transport the particles with suitable forces, e.g., centrifugal forces or electrical forces, into areas where dispersing forces are no longer of decisive influence. In this area the final separation is effected. This principle is illustrated with the example of some dust removal processes.

Inertial particle collectors. In the gravitational collector particles are transported by gravity into a flow stagnation zone. In the centrifugal collectors the gas flow laden with particles is diverted in a suitable manner. Particles which cannot follow this change of direction due to their inertia are flung against the outer wall of the gas flow in the collector. They accumulate there and under the influence of gravity sink into the particle bunker.

Scrubbers. The forces of inertia occurring while the gas is flowing around water droplets lead to the trapping of particles onto the surface of the liquid, e.g., onto the surface of droplets dispersed in the gas; subsequently these droplets are collected by gravitational or centrifugal forces.

Electrostatic precipitators. Particles are charged electrically and are transported to the precipitation electrode through an electric field. They are retained there by adhesive forces. The precipitation electrodes are either cleaned by periodic shaking or by a trickling film of a fluid.

Filters. Apart from the grid effect of a porous system (e.g., woven material) inertial and diffusing forces are mainly decisive for the transport of the particles. The particles are retained at the surface of the porous system by adhesive forces. The porous material is subsequently replaced or cleaned by shaking or rinsing with gas.

The performance of removal devices can be characterized by the following criteria:
- *Total degree of removal.* The total degree of removal is the ratio of removed quantity of particles per unit of time to the entering quantity of particles per unit of time.
- *Fractional degree of removal.* It represents the ratio of removed particle quantity of a certain class of size to the entered particle quantity of this particle size per unit of time.
- *Residual content.* It is characterised by the particle quantity per volume unit of the pure gas after leaving the particle removal device. This residual content can be related to the gas condition at the intake of the removal device.

A suitable particle removal process should be chosen according to the following criteria:
- Required total degree of removal or residual content: Inertial collectors can only achieve limited degrees of removal. In contrast to this, very low pure gas particle content can be achieved with electrostatic precipitators, scrubbers and fabric filters.
- Condition of the carrier gas: Mass or volume flow, temperature, humidity, dew point, chemical composition.
- Type of particulate matter: Amount of dust, density, grain size distribution, chemical behavior a.s.o..

The following sections will provide an overview of the different particle removal processes. Detailed descriptions of particle removal processes can be found in specialized literature on the subject [13-16].

7.2.1 Inertial Collectors

Even today, gravitational particle separators are still widely in use. Owing to their simple design they are robust and long-lasting, additionally they stand out for their relatively low cost of acquisition and low maintenance requirements. They are mainly used in areas where coarse particles are to be removed or where only a partial particle separation is necessary. The principle modes of operation and designs of gravitational partice separators have been dealt with in numerous publications and in the VDI guideline 3679 [19].

It is common practice to classify gravitational particle separators according to the forces of transport in effect:
- gravitational separators: gravitation is the active force.
- Under the influence of gravitation, particles sediment out of the gas flow with a sinking speed of v_s:

$$v_s = \frac{d^2(\varrho_2 - \varrho_1) \cdot g}{18 \cdot \eta_G} \qquad (7.12)$$

with: d being particle diameter, ρ_2 density of the particle, ρ_1 density of the gas, g earth gravitation, η_G viscosity of the gas.
- Inertial particle collectors: as centrifugal force is considered an inertial force, the centrifugal particle collector belongs to this group. Merely owing to its inertia the particle is flung against the wall of the collector in the diverted gas flow. Strictly speaking, only the observer actually rotating in the gas flow may speak of a centrifugal force. The smaller the particle diameter or the lower the density of the particle, the smaller the forces of inertia. Therefore, finer and lighter particles must be subjected to greater accelerations to achieve adequate separation.

7.2.1.1 Inertial Force Particle Collectors

Particulate matter can be separated from a gas by making use of their inherent forces of inertia. Obstacles are placed in the gas flow and divert it. Due to their mass inertia the particles carried along try to maintain their original direction. Thus, they either collide with the obstacle or reach an area outside of the gas flow and are thus separated from the gas. From there the separated particles must be removed to prevent reentrainment.

The Linderoth tube, Fig. 7.7, also called Aerodyne Tube, is composed of the stub end of a sheet metal cone with numerous shutter-like slits. The particle-laden gas enters at the wide end and is abruptly diverted at the slits. The particles keep on floating straight ahead, so that the gas flowing straight ahead is enriched with dust. This gas can then be treated in a cyclone and is subsequently fed back into the process.

7.2.1.2 Centrifugal Force Particle Collectors

If a particle-laden gas stream is forced into a closed circle, the particles are subject to a centrifugal acceleration. Cyclones, shown in Fig. 7.8, and rotational flow collectors, shown in Fig. 7.10, operate according to this principle.

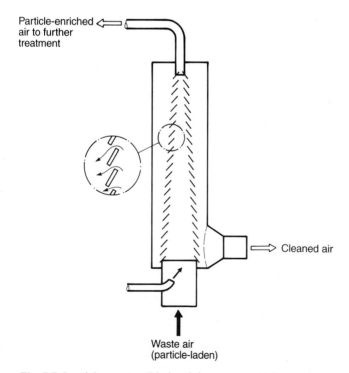

Fig. 7.7. Inertial separator (Linderoth louvre separator)

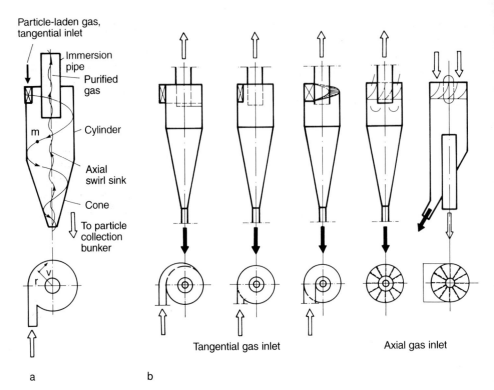

Fig. 7.8. Cyclone particle collector. **a.** Principle; **b.** different cyclone designs [13-15]

Cyclones

Muschelknautz, e.g., has investigated and described the cyclone in detail [20]. Fig. 7.8a shows the principle. A cyclone is composed of a cylindrical housing with a tangential inlet at the upper end and a gas outlet which is fixed concentrically in the lid as an immersion pipe.

The entering gas stream performs a spiralling trajectory downwards. The flow is a swirl sink. If a particle with the mass m is moving on a circular orbit around the axis of the cyclone with the radius r and the velocity v, then the particle is subject to centrifugal acceleration a:

$$b = v^2/r . \qquad (7.13)$$

At the usual circumferential speeds of 15-40 m/s and radiuses of curvature between 0.1 and 1 m a centrifugal acceleration is formed which is one hundred to one thousand times greater than the earth's acceleration. The radial centrifugal speed v_c formed under the influence of this acceleration is given as relative speed between solids and gas molecules:

$$v_f = \frac{d^2 \cdot b \cdot \varrho}{18 \eta_G} , \qquad (7.14)$$

where d is particle diameter, a centrifugal acceleration, ρ particle density, η_G gas viscosity.

As the centrifugal speed depends on the particle diameter d, a fractional particle separation is carried out in the cyclone, too. The particles flung against the wall slide down and accumulate in the lower part. As the cyclone is closed at the bottom the direction of the gas stream is reversed while the spiralling movement remains. The particle collection bunker must be positioned deep enough under the reversal area of the gas spiral so that the dust collected is not stirred up.

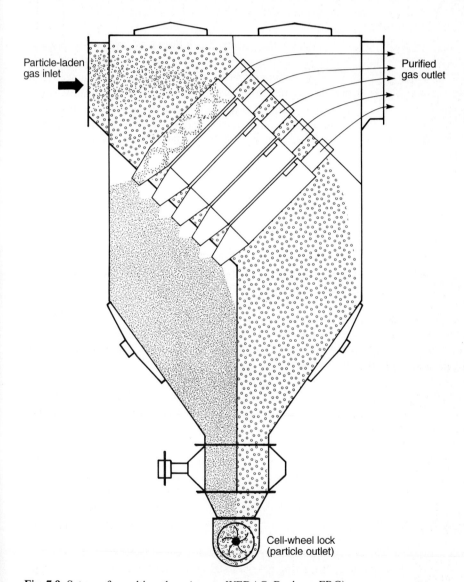

Fig. 7.9. Set-up of a multi-cyclone (acc. to WEDAG, Bochum, FRG)

The separation of the particles must be accomplished in such a way that no ambient air can enter the system. This can be achieved by a double-cone closure or a cell-wheel lock.

For normal requirements for which no particular degree of fine particle separation must be achieved, cyclones with diameters between 1 and 5 m are used. Their separation range covers particle sizes over 50 μm.

For the separation of smaller particle sizes centrifugal acceleration must be raised. According to equation (7.14) this is achieved by reducing either the track radius or the cyclone diameter. Raising gas speed is ruled out for economical reasons. As small cyclones can only handle small gas volumes, many small cyclones are combined to a multicyclone.

Fig. 7.9 shows a multicyclone collector which achieves high degrees of particle separation efficiency even with fine particles. The collector is composed of a number of very small cyclones with an outside diameter of 20 cm which are connected to form several blocks depending on the gas flow to be treated. Their resistance is between 9-11 mbar at approx. 3 m^3/min gas flow per cyclone. Depending on the particle size their efficiency is at 92-98 %. Currently, particles down to 12 μm are being separated in these cyclones. When designing the inlet channels of a multicyclone it is important to ensure that the surfaces of the individual cyclones will be loaded evenly.

Rotary flow particle collector

The physical basis and principle of the rotary flow particle collector have been described in detail by Schmidt [21].

In rotary flow particle collectors two concentric spin flows are at work which have the same sense of rotation but are axially opposed. The basic design is shown in Fig. 7.10. The inner spin flow is formed by the raw gas which is set in rotary motion by a swirl obstacle with fixed vanes. The outer spin flow is created by tangentially blowing in additional air. As a result of the centrifugal forces at work the particulate is flung out of the raw gas, reaches the area where the additional air is flowing downward and is transported along with it into the hopper. Part of the raw gas, ambient air or pure gas may be used as additional air.

Energy consumption and degree of removal depend on the raw gas/additional air ratio. Common additional air volumes range between 40 and 100 % of the raw gas qunatity. The total degree of removal improves with increasing additional air. It is not possible to determine the size of the smallest particulate to be removed by way of calculation. Test results give reason to assume that it is between 0.5 and 3 μm.

Klein [22] has pointed out the performance limits of the rotary flow particle collector

7.2.2 Wet Scrubbers

Wet scrubbing was developed from dry mechanical particle separation. Due to the small mass of the particles the acting gravitational or inertial forces are not sufficient to separate the particles from the gas stream. If the mass of the particle grains is en-

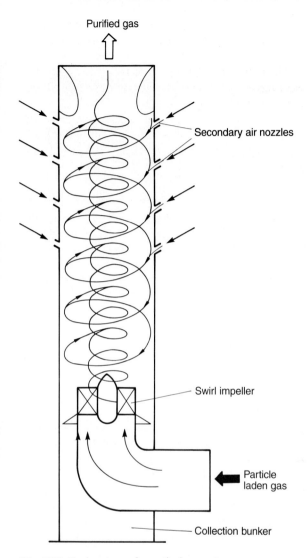

Fig. 7.10. Basic set-up of a vortical separator

larged by binding them to the drops of a scrubbing liquid, then a subsequent gravitational separation is more successful. For reasons of economy it is common practice to use water as scrubbing liquid.

As in wet particle collection the pollution problem is shifted to water there must be special reasons for employing this type of method. One important field of application is, e.g., the precipitation of flammable particles or the particle separation in exhaust gases containing sparks [23]. Wet particle precipitation is preferably used in the grain size range of 0.1-50 µm. The finer the particle and the lower its concentration, the more difficult it is to separate. The particle is considered separated when it has been brought into contact with the scrubbing liquid and is retained by it.

Usually the particle-laden fluid is subsequently collected in gravitational precipitators into sludge and purified fluid. The latter can be fed back into the cycle either totally or in part.

Basically, wet particle collection is divided into two processes:
- precipitation on fixed, flushed surfaces,
- precipitation on the surface of fluid droplets or gas bubbles.

Scrubbers with fixed flushed walls consist, e.g., of material with a large specific surface which is sprayed with the washing liquid and the particle flow passes through it in the same, opposite or lateral direction. It is also possible to use plates which are smooth or with a structured surface, which are positioned parallel to each other and wetted or sprayed. The particle-laden gas is passed through the passages between the plates.

7.2.2.1 Principles of Wet Scrubbing

When binding particles to the surface of drops or bubbles 5 zones can be distinguished.

1 Wetting zone

Bringing particles and washer liquid into contact can be carried out by the following methods, depending on the design and performance of the scrubber:

1.1 Spraying or nozzling

This method entails scrubber designs with large flow diameters and low flow speeds. The particle-laden gas is passed between horizontal or vertical walls and is sprayed with water from nozzles on its way.

1.2 Washing or whirling

If a gas-particle mixture is flowing through a layer of liquid and has its outlet below the liquid's surface, then the method is called washing. Removal efficiency is highly dependent on bubble formation speed, bubble size and the length of the bubble line which corresponds to the height of the liquid layer in this case. This method is unsuitable for stricter requirements pertaining to exhaust gas quality as the formation of bubbles cannot be influenced well and it is not reproducible.

1.3 Injection under pressure

In this method the gas stream is passed through a restriction of a tube, similar to a Venturi tube, where flow speed increases and pressure decreases. Water is injected at the place where speed is highest and pressure lowest. The high gas flow speed is necessary for a high removal efficiency as the relative speed between scrubbing liquid and gas considerably influences their contact. Fast acceleration of smaller droplets and with it decreasing relative speed have a negative effect on the removal efficiency.

2 Contact zone

The water injected into the wetting zone must be mixed thoroughly with the particulate matter so that a binding of the two components with each other can take place. This process is promoted by various effects.

2.1 Gravitation-related direct contacting of particles with the dispersed scrubbing liquid (impaction contact)

In most cases inertial forces together with technical flow-related effects influence a contact of particles and water droplets, which can be termed an impaction contact. Fig. 7.11 serves as an illustration of this process. A particle-laden gas stream with a speed v flows in streamlines around a spherical water droplet with the diameter d_{dr}. However, the particles, depending on their mass, follow the streamlines only in part. Under simplified conditions boundary streamlines can be calculated within which the particles do not follow the streamlines, but impact with the water droplet and, depending on its wetting ability, adhere to it or penetrate it [13, 14, 24, 25]. As little is known yet about the occurring droplet sizes, their distribution and the influence of diffusion and condensation, it is practically impossible to determine the efficiency of this type of scrubber by computation.

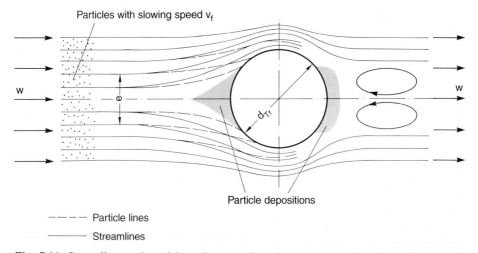

Fig. 7.11. Streamlines and particle paths around a sphere or sphere-shaped liquid droplet. d_{Tr} droplet diameter, w gas velocity, e distance between bordering lines

2.2 Condensation processes [14]

If cool water is injected into hot gas it vaporizes at first. Depending on the temperature gradient, the vapor thus formed mixes more or less quickly with the gas. As the gas cools down with increasing saturation, the vapor condenses whereby particles in the gas and also cold water droplets can act as condensation nuclei. During this, the vapor flow in the direction of the condensation nucleus can transport particles to the drop surface (diffusion effect).

2.3 Diffusion effects

Another cause for the binding of particles and scrubbing liquid can be found in the diffusion process. However, it is of more theoretical significance as there is not sufficient time for the formation of diffusion processes during the short retention times in the contact area.

Seen from the point of view of molecular kinetics, the molecules of a gas are in permanent, irregular motion which is perceived as heat. The molecules keep colliding and in doing so transfer a part of their energy onto particulate matter present in the gas. If the dust particles are large approx. the same number of molecules collide on all sides so that the resulting force equals zero. If the particles are small molecules hitting predominantly one side can cause a movement. If one imagines that the particles diffuse through a boundary layer close to the drop surface, then this process may be called a diffusion removal.

2.4 Sedimentation

Sedimentation is the precipitation of particles under the influence of gravitation. As the wet scrubbers described here are always based on flow effects with higher speeds, sedimentation has no significance due to its slow sedimentation speed.

2.5 Influence of wettability

Wettability is the adsorptive ability of liquid molecules on the surface of particles. As three phases (solid, liquid and gaseous) are in a turbulent state in these effects and an ordered adsorption seems practically impossible, wettability is defined as the affinity of a certain number of particles to a droplet of liquid [26].

On contact with the liquid the wetted particles immediately migrate into the inside of the drop. Particles, however, that do not wet well, settle on the surface of the droplets.

If the particle content is low the degree of removal is hardly influenced by the wettability due to the largely unoccupied drop surface. If the particle content is high and large part of the droplet surface is already occupied by particles, removal efficiency is expected to decrease in particles which do not wet well.

3 Separation Zone

In the separation zone the particle-laden scrubbing liquid is separated from the gas stream. Depending on the design and the geometric proportions of the scrubber, gravitational forces effect the separation of the scrubbing liquid. As, as a rule, the earth's gravitation is not sufficient, removal efficiency is increased by centrifugal forces similar to those in gravitational collectors. The turbid liquid collected in the separator is passed to the sludge zone for further treatment.

4 Conveyor zone

The conveyor zone takes care of the transportation of the gas-particle mixture. As a rule a fan of radial or axial design is installed which simultaneously produces the pressure difference which is needed to overcome the resistance of the scrubber and the pipes attached to it. The fan can be positioned before or behind the scrubber. Operating the scrubber on the suction side reduces hazards of erosion, fire and explosion as compared to the version with the pressure side.

5 Sludge zone

Depending on the gas streams to be separated, the wet scrubbers used today require considerable amounts of water. Fresh water operation, where the dirty water is not reused is therefore uneconomical in most cases; recycling water operation is common practice today. The sludge deposited in the separating zone is conducted into the sludge separators, e.g., settling ponds, where the particles sediment. The purified water is subsequently reused; only the amount of water which vaporized must be replenished.

7.2.2.2 Different Types of Wet Scrubbers

Different types of designs have been developed for wet scruubing. In their design they vary in their equipment for liquid distribution and for the homogenous mixing of the particle-gas stream with the scrubbing liquid. According to the German VDI guideline 3679 [25] wet separators are classified according to the following designs:
– wet scrubbers,
– whirl scrubbers,
– Venturi wet scrubbers,
– wet rotation scrubbers.

The different scrubber types are shown in a diagram in Fig. 7.12.

Wet Scrubbers

Wet scrubbers are devices usually containing components into which the scrubbing liquid is injected via nozzles in droplet form. These components (e.g., baffles or

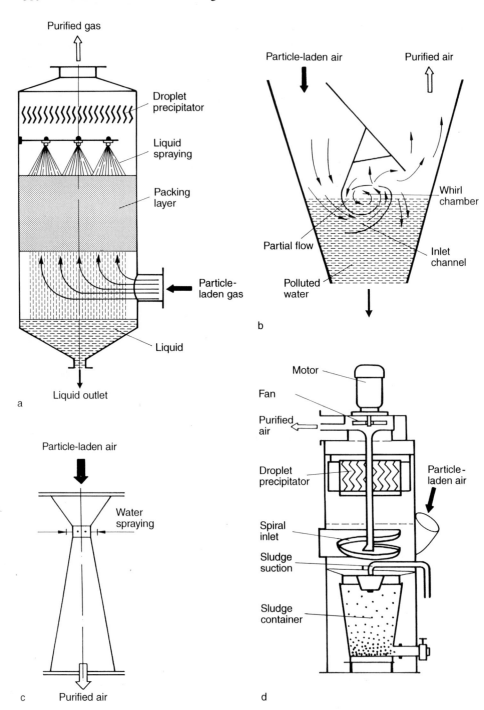

Fig. 7.12. Diagram of the most commonly used wet scrubber designs [23, 25].
a. wet scrubber; **b.** vortical scrubber; **c.** Venturi scrubber; **d.** wet rotational scrubber

layers of packing material) cause an even distribution of the scrubbing liquid. The particle-gas stream is usually conducted in a counterflow to the scrubbing liquid. Separation of the water droplets loaded with particles is usually carried out by gravitational or inertial forces, e.g., on collecting plates.

Whirl scrubbers

In whirl scrubbers the particle-gas mixture flows to the surface of a liquid and in doing so, is diverted. This already causes a particle removal. Subsequently, the gas is conducted through a channel during which the scrubbing liquid carried along is sprayed at the diversion edges of the whirl zone causing a homogenous mixing with the particle-gas stream. As in the normal scrubber droplet separation is carried out by gravitational forces , e.g., on plates, layers of packing material or centrifugal collectors.

Venturi wet scrubbers

The particle-gas mixture is conducted through a Venturi tube and at the smallest diameter of the tube, the throat, it attains its highest velocity and lowest pressure. In or before the throat the scrubbing liquid is added to the particle-gas flow. Owing to the high relative speed between gas and scrubbing liquid there is a good contact between gas stream and fluid causing atomization of the water. Gas velocity is reduced in the subsequent diffuser. During this, droplets containing particles are formed, as explained in the above paragraphs. Droplet separation is usually carried out by centrifugal forces, e.g., by a subsequently positioned cyclone. Venturi scrubbers are characterized by their high removal efficiency and the fact they require little floor space.

Wet rotational scrubbers

Wet rotational scrubbers have rotating components for the production of droplets and for the homogenous mixing of scrubbing liquid and gas stream.

Example

Depending on their application the most varying models have been constructed on the basis of the mentioned construction types. As an example, the reader will find a scrubber for the particle separation from flue gases of an asphalt mixing plant (max. 5 t/h of dust, 50,000 m^3/h exhaust gas at 0 °C, 1 013 mbar), s. Fig. 7.13. Wet scrubbing is carried out by spraying and nozzling.

To produce a rotational motion the flue gases are fed into the upper part of the interior tube tangentially. For a thorough whirling of the flue gases there are cover plates in the interior tube as well as full cone nozzles for intensive spraying. Flowing in the same direction as the scrubbing liquid the gases are wetted by numerous other

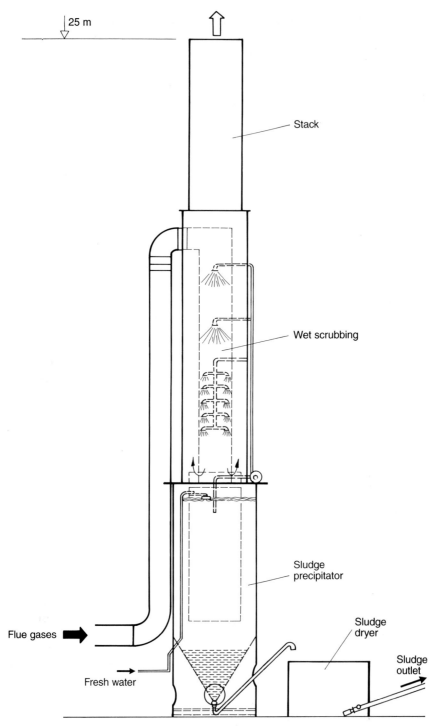

Fig. 7.13. Example of a wet scrubbing plant with spray nozzles

nozzles and at the foot of the interior tube they reach the separation tank. There, the droplets containing particulates are removed, while the purified flue gases rise in the exterior casing tube and leave the plant by way of the stack. Owing to its generous dimensions the plant operates at relatively low flue gas speeds; the plume dissolves very rapidly after having left the stack.

The dirty water drips from the lower casing edge into the outer ring gap of the sludge separator and sinks down. The particulates collect in the cone part of the tube and are pushed over a slide bar in the form of viscous sludge. The purified water is sucked in by a pump in the upper part of the interior tube and injected back into the flue gas flow via atomizer nozzles.

7.2.3 Electrostatic Precipitators

Electrostatic precipitators are some of the most efficient gas purification devices. They are characterized by high collection efficiencies and can be used for a multitude of applications. Their mode of operation is to a large extent dependent on physical and chemical factors. In general it can be said, that particle sizes down to fractions of a µm are covered and that high degrees of removal in excess of 99.9 % can be attained. The pressure loss of an electrostatic precipitator is generally in the range of 1-4 mbar. Individual electrostatic precipitators can handle gas volumes of more than 10^6 m^3/h (0 °C, 1 013 mbar).

Much has been published on the subject of electrostatic precipitators. The first detailed book was written by White [27]; in other books on particle removal electric particle precipitation is also given a lot of space [13-16]. In 1980 the German VDI issued a special guideline for electrostatic precipitators [28].

7.2.3.1 Operating Principle

The operating principle of an electrostatic precipitator is shown in a simplified form in Fig. 7.14. The particles distributed in the gas are electrically charged and collected on so-called collecting electrodes. The particles are charged by ions and electrons, which are generated by a corona discharge of the central wire electrodes carrying a dc voltage of 10,000-80 000 V. In the electrical field generated between the wire electrodes and the collecting electrodes the thus charged particles are attracted by the collecting electrodes [13-16, 27-29].

Fig. 7.15 shows the basic diagram of electric gas purification. There, the precipitator is indicated by a metallic tube in which a wire which is electrically insulated against this tube is axially installed. The inside of the grounded tube is the particle collection surface. The central wire electrode is negatively charged by the high-voltage plant. Apart from the tube-type electrostatic precipitator shown in the diagram there is the much larger group of the plate-type collectors with plate-like collecting electrodes.

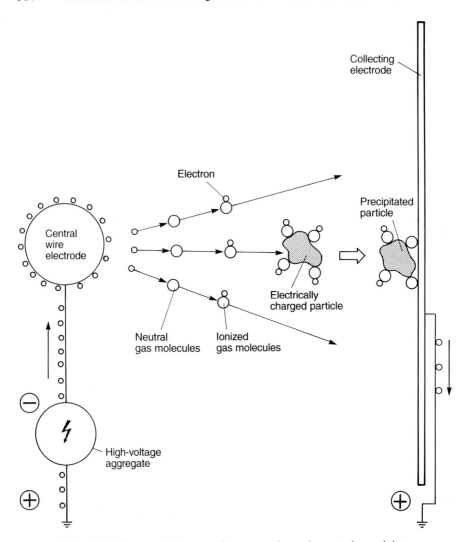

Fig. 7.14. Simplified diagram of the separating process in an electrostatic precipitator

7.2.3.2 Mode of Operation

The particle removal in an electrostatic precipitator can be divided into four basic sub-processes:
1. Charging of the particulates,
2. Migration of the particles to the collecting electrode,
3. Deposition of the particles on the collecting electrode,
4. Removing the collected particles into the storage hopper.

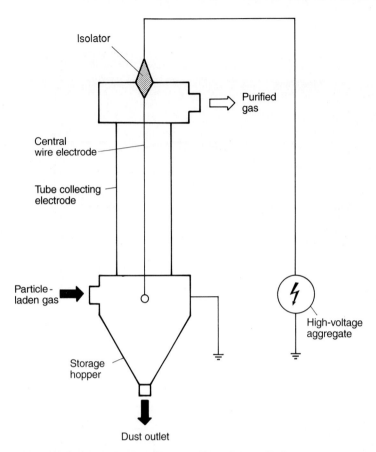

Fig. 7.15. Basic set-up of an electrostatic precipitator [30]

1 Charging the particulates

Of the different methods of charging particles artificially, charging them in an electrical field under the influence of corona discharge is the most efficient. Therefore, it is also used in the electrostatic precipitator. For this, the electrical discharge electrode is negatively charged. An electrical field is then formed between the electrical discharge electrode and the grounded collecting electrode. In the vicinity of the electrical discharge electrode, which either has a diameter of only a few millimeters or has sharp edges and points, the field intensity is so high that it exceeds a value characteristic of the gas – a gas discharge, called corona, is formed. The formation of the corona depends on the radius of curvature of the electrical discharge electrode, s. Fig. 7.16. In the corona the gas molecules in the vicinity of the electrical discharge electrode are ionized and are split into positive gas ions and electrons. The electrons settle on neutral molecules, sometimes forming negative gas ions at the same time. The positive gas ions remain close to the electrical discharge wire or migrate to it. Under the influence of the electrostatic force the electrons migrate with great speed

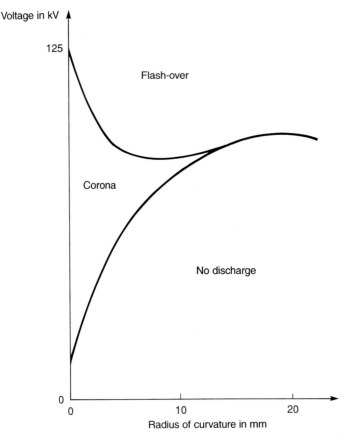

Fig. 7.16. Influence of surface curvature of the spray electrode and voltage applied on the type of discharge [31]

to the collecting electrode and ionize further neutral molecules, mainly by collision. Due to these processes a large number of additional electrons and ions is created avalanche-like in the vicinity of high field strength close to the electrical discharge wire.

On their way to the collecting electrode the negative gas ions and electrons collide with neutral gas molecules and give these a directed impulse. The flow created in this way is called "electric wind" [32].

2 Migration of the particles to the collecting electrode

Particles in the gas are charged by the taking up of electrons and gas ions. In the electric field they are then subject to force P_e directed to the collecting electrode.

$$P_e = qE, \tag{7.15}$$

where q is the total taken up load amount $A \cdot s$, E is the field strength in $V \cdot m^{-1}$.

7.2 Processes for the Removal of Particulate Matter from Exhaust Gases

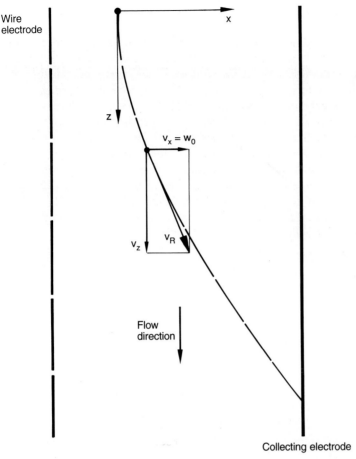

Fig. 7.17. Particle path between vertically arranged discharging and collecting electrodes

From this force and from the friction of the particles in the gas a migration velocity w_o of the particles on their way to the collecting electrode results:

$$w_0 = \frac{q \cdot E}{6\pi r \eta_G}, \qquad (7.16)$$

where r is particle radius, η_G is the viscosity of the gas in Nsm^{-2}.

Together with the flow velocity of the gas v_z, a resulting speed v_R of the particle in the precipitator can be calculated, s. Fig. 7.17.

This movement, however, is frequently disturbed by turbulences and electric wind, so that a calculation of the degree of collection based on the above formula (7.16) is not possible [33, 34] (s. sect. 7.2.3.3).

3 Deposition of the particles on the collecting electrode

Whether the particles arriving at the collecting electrodes adhere there depends on the extent of the adhesive forces acting on the dust. Electric adhesive forces are mainly influenced by the resistivity of the particles which determines whether the charge is drained slowly or quickly on to the collecting electrode. If the particles release their charge quickly, only a small electric field builds up in the layer of dust: the adhesive forces are small. Even a movement of charge can occur so that the particles are repelled again. If electric dust resistivity, however, is high, the electric adhesive forces can attain very high values due to the charge accumulation. The electric field building up in the dust layer can become so great, that this situation can lead to gas discharges in the pores of the particle layer and an electric disruptive discharge can occur which leads to the undesirable effect of back brush [15, 32]. If dust resistivity is too high, flue gases are sometimes conditioned with special gases, e.g., with SO_3, to improve collection efficiency [15, 35].

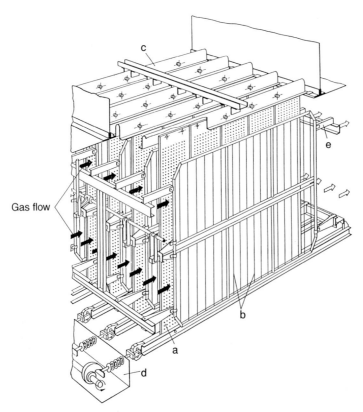

Fig. 7.18. Rapping of collecting electrodes by rappers (acc. to [36]). **a.** collecting electrodes; **b.** discharging electrodes; **c.** mounting of collecting electrodes; **d.** rapper; **e.** insulator

4 Removing the deposited particulate matter into the storage hopper

The dust adhering to the collecting plates must be removed from them and collected in the hopper. Dust removal is carried out in two steps:
- Removing or shearing the dust off the collecting plate,
- transport of the stripped dust into the storage hopper.

The dust layer on the collecting plate of a plate-type precipitator (s. sect. 7.2.3.4) is usually removed by vibrating the plate with one single rap while it is highly charged, s. Fig. 7.18. The time between two raps on the same collecting plate is called a rap interval. In practice, the rap interval is constant for all collecting plates of the same cleaning stage. However, for consecutive cleaning stages in the gas stream it varies.

The dust sheared off from the collecting plates falls by gravity into the hopper. While falling, part of the dust adhers anew to the collecting plates. Another part is carried by the gas stream out of the first cleaning stage into the following cleaning stage or to the pure gas side of the electrostatic precipitator. Collection efficiency is considerably influenced by the cleaning of the collecting plates [37]. In practice, rap intervals are varied until the quantity of the dust reaching the hopper has been optimized.

7.2.3.3 Precipitation Equation

There are a number of methods with whose help the precipitation equation for an electrostatic precipitator can be deduced. It is always assumed that there is a turbulent stream in the collecting chamber. This turbulent stream and the electric wind evenly distribute the particles over the area available, i.e., particle concentration is homogenous. Furthermore it is assumed, that velocity is the same everywhere in the filter and that there are no other disturbing effects.

Deutsch [38] and White [27] assume that a particle can only be precipitated if it enters the laminar boundary layer of the collecting electrode.

Calculating the degree of precipitation with the above assumptions yields the so-called Deutsch equation

$$\eta = (1 - e^{-Aw/\dot{V}}) \, 100\,\% \tag{7.17}$$

where η is degree of precipitation, A is the total collecting area of the precipitator in m^2, $\dot{V}$ is the volume flow rate in m^3/s, w is the index related to the technical process, also called migration velocity, in m/s.

As initially explained, the calculation of the collection efficiency of an electrostatic precipitator requires a series of assumptions usually not fulfilled by industrial units. The theoretical migration velocity calculated according to equation (7.16) is not identical with w in equation (7.17). The parameter w which is also simply called migration velocity is a characteristic index for any electrostatic precipitator; it is related to the technical process and only incidentally has the dimension of a velocity. In effect, it includes all measurable and unmeasurable influencing factors on the efficiency of an electrostatic precipitator. In the meantime, different authors have tried to modify the Deutsch equation [16].

The index w related to the technical process depends on a number of influencing parameters. These are mainly:
1. particle migration and dust adhesion,
2. dust composition (type, grain size, electrical resistivity),
3. gas velocity in the electrostatic precipitator,
4. gas composition, temperature and humidity,
5. geometric dimensions of the precipitator and design of electrical discharge electrodes and collecting electrodes (width of passageways),
6. reraising of the dust,
7. characteristics of current and voltage profiles.

The migration velocity w rises with increasing voltage between electrical discharge electrode and collecting electrode. The voltage is limited by the electric breakdown. This breakdown voltage, and with it the precipitation efficiency, depends on the gas composition. As the composition is not constant due to strand formations in the precipitator passageways, the breakdown voltage also varies [39]. In practice, voltage for each separate electric field is increased to such an extent that single short-term spark-overs are admitted, however without a steady arc being formed [28]. Nowadays, optimum current-voltage control is carried out by microcomputer [40].

In contrast to gravitational dust collectors electrostatic precipitators have a high collection efficiency for almost all particle size fractions. In certain particle size ranges collection effciency may be reduced [41]. Causes for this have not been fundamentally researched yet.

7.2.3.4 Constructions

There are tube-type and plate-type electrostatic precipitators. Both types can be used for dry, wet and misty gases.
- *Tube-type collectors* consist of a number of parallel vertical tubes with a circular or honeycomb-like cross-section, where insulated wire electrodes are installed. The wires are attached to the negative pole of a high-voltage unit, the tubes are connected with the other pole and are grounded. They form the collecting electrodes.
- In *plate-type collectors* (Fig. 7.19) many vertical, parallel plates serve as collecting electrodes, with insulated wire electrodes stretched between the plates. The collecting electrodes are smooth and profiled metal plates, in special cases box-shaped hollow electrodes. The filter housings, depending on application and gas properties, are available in sheetmetal, brick, concrete or plastic.

Mode of operation
- *Dry precipitators.* The dust collected on the electrodes is sheared off periodically by rapping, falls into the hopper and is removed through locks. Hammer mechanisms are used for rapping, rap intervals are adapted to the adhesive properties of the dust by control units.

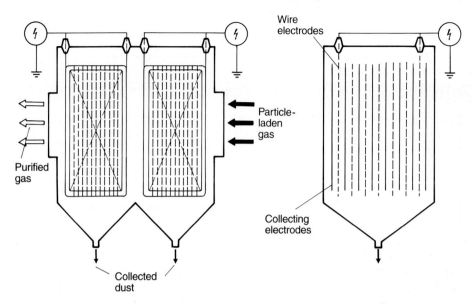

Fig. 7.19. Horizontal plate electrostatic precipitator (cross-section in vertical and horizontal direction to flow), [30]

- *Wet precipitators.* Dust from wet gases collects on the electrodes as sludge, from where it flows off or is rinsed off. Water, tar or acid vapors are also collected and flow off in the shape of a fluid film. Wet electrostatic precipitators are, e.g., used in the removal of explosive dusts or for particles, which are particularly difficult to clean off from the collecting plates, e.g., varnish particles.

Depending on the duct of the gas stream to be purified, one speaks of horizontal and vertical electrostatic precipitators. Dry electrostatic precipitators are usually built as plate-type collectors.

Shapes of electrical discharge electrodes and collecting electrodes

Fig. 7.20 shows the shapes of electrical discharge electrodes and collecting electrodes most commonly used today.

The width between the collecting electrodes usually amounts to 250 mm, and the width of a plate is between 400 and 700 mm. As the floor space available is often limited, large electrostatic precipitators are often very high. Plate height frequently peaks at 8-10 m. Modern collectors are divided into two or three consecutive fields, each of them possessing an independent power supply and plate rapper.

The length of a field is about 8-12 m. Gas velocity is between 1.5 and 2.0 m/s. The efficiency of collection is usually between 95 and 99.9 %.

A modern electrostatic precipitator for the dust collection of the flue gases of a 700 MW power plant block with 2.3 mio m^3/h is shown in Fig. 7.46. The precipitator has a height of approx. 35 m and a length of 31 m. In combination with the flue-gas desulfurization plant pure gas particle contents of approx. 10 mg/m^3 are achieved.

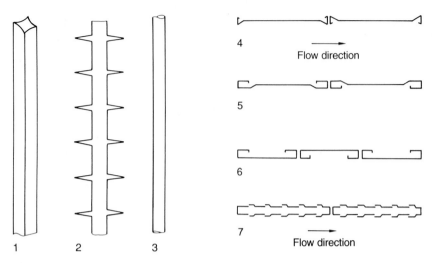

Fig. 7.20. Different types of discharging and collecting electrodes. **a.** discharging electrodes: *1* star wire electrode, *2* barbed wire electrode, *3* cylindrical wire electrode; **b.** *4-7* collecting plate electrodes, different types of tin plates, view from the top [36]

7.2.4 Filters

Filters represent one of the oldest methods of removing particulate matter from a medium. Particularly in the recent past, with higher levels of removal of especially fine particles being increasingly required, filters have gained more and more importance. Filter materials are different types of fabrics or fleeces.

The fundamentals of particle collection with fabric filters, filter media and filtering units have been treated in detail by Löffler, Dietrich and Flatt [42] and have also been represented in the German VDI guideline 3677 [43]. For an in-depth study of this subject we recommend this literature.

Both in former times and partly even today fabric filters have consisted of interwoven natural fibers. However, artificial fibers and fabrics made of fine metal threads have decidedly expanded the field of application of the fabric filter owing to their resistance against wearing and chemical influences.

The suitability of a fabric filter or fleece and its durability are determined by the operating temperature and the properties of the particles. Cotton can be used up to 80 °C, wool or wool felt up to 110 °C, polyacrylic nitride (orlon) or superpolyamide (nylon, perlon) up to 130 °C. In addition, the two latter fabrics are almost nonreactive to moisture and chemical influences. For temperatures up to 260 °C, glass fiber fabrics, e.g., are used. Table 7.1 lists the properties of different filter materials.

For trouble-free operation moisture content and temperature of the gas must be within a certain limited range. If, e.g., gas temperature falls below its dew point, clogging of the filter can ensue. Excessive gas temperatures have an unfavorable effect on the durability of the fibers. If there is not sufficient moisture, man-made fibers can become brittle and possibly tear, natural fibers also need a certain minimum moisture.

7.2 Processes for the Removal of Particulate Matter from Exhaust Gases

To obtain a rough estimate of the size of fabric filters, the filter ratio F_r must be determined, i.e., the ratio of the carrier gas volume flow $\dot{Q}$ to the gross surface of the filter elements F.

$$F_v = \frac{\dot{Q}}{F}, \tag{7.18}$$

where F_r is the filter ratio, $\dot{Q}$ is the carrier gas volume flow in m³/s and F is the gross surface of the filter elements in m².

Corresponding to its dimension F_r is also called surface velocity. In conventional filters, e.g., tube filters, F_r is between 1-2.5 m³/m²·s.

The filtering process is initiated by a layer of dust forming on the fabric threads. This first layer consists of the finest particles which are, theoretically, small enough to fit through the voids of the woven fabric. However, gravitational influences, processes involving molecular kinetics (diffusion, Brownian movement) and electrostatic forces effect the settling of these particles which, in turn, function as a layer filter. This process is also called depth filtration. The sieve-like filtering effect even removes particles as small as 1 μm, while depth filtration can separate particles smaller than 0.5 μm. With natural fibers such as wool or cotton this primary formation of layers is supported by the fine fibers projecting from the threads into the fabric. With the mostly smooth artificial and metal fibers this effect is lacking. Instead, electrostatic adhesive forces occur in synthetic fibers.

The performance of a fabric filter is limited by the consideration that the initially favorable deposited layer becomes too thick and must be removed because of the excessive pressure drop. The different types of fabric filters and their cleaning processes will be dealt with in more detail in the following sections.

7.2.4.1 Tube Filters

Design and mode of operation of a tube filter can be seen in Fig. 7.21. Depending on the performance required, a closed housing accomodates a number of filter tubes. The lower part of the housing functions as a hopper. During the filtering process dusty air enters it via the inlet. The coarse particles are immediately separated here and are removed with the already separated dust by a removing mechanism. Subsequently, the dirty air flows through filtering tubes, deposits the particles and leaves the collector, with the help of a fan, by the outlet. If the filtering layer becomes too thick and the pressure drop increases, the filter is shut down for cleaning. For this, the outlet is closed with a flap and pressurized air is blown in, washing out the tube in the opposite direction. Simultaneously, the tubes are vigorously rapped with the rapping mechanism, so that the filtrate layer falls off.

A filtering plant for continuous processes must therefore consist of several such filtering units as one filter is always shut down during the cleaning period.

So-called reverse air flow cleaning was developed to avoid interrupting the filtering process by periods of cleaning [42]. In this process a nozzle ring whose nozzles are blowing constant jets of air on the deposited dust glides up and down each single filter tube, thus continuously removing dust from the filter fabric. The dust blown off then falls into the hopper.

Table 7.1. Properties of different textilite yarns for filter fabrics [13]

Chemical structure	Natural fibers		Man-made fibers							
			Poly-vinyl-chloride	Poly-amide	Polyacrylonitrile			Copolymers	Poly ester	Poly-tetra-fluoro-ethylene	Glass
					Pure Polyacryl-nitril	Polyacryl-nitril (formerly Pan)	Orlon				
Brand name German	Wolle	Baum-wolle	PCU PeCe Rhovyl-Fibro	Nylon Perlon Phrilon	Redon	Dralon	Dralon	Dolan	Diolen Trevira	Hosta-flon Viton	
English American	Wool Cotton	Cotton Vinyon	Nylon	Nylon					Terylene Dacron	Teflon	
Density in g/cm³	1.32	1.47···1.50	1.39···1.44	1.13···1.15	1.17	1.14···1.16		1.14	1.38	2.3	2.54
Moisture absorption at 20 °C and 65% relative humidity in %	10...15	8...9	0	4.0···4.5 1.3		2	1	1	0.4	0	0
Resistance to acids	high to weak acids at low temp	poor	almost completely in any concentration	high to diluted and cold acids, little when warm			high		high to almost all acids	very high	is affected by some strong acids
to alkalies	poor	good	almost completely resistant	practically resistant		sufficiently resistant to weak alkalies			high at room temp.to weak	very high	is corroded by strong alkalies

7.2 Processes for the Removal of Particulate Matter from Exhaust Gases

	low when un-treated	low when un-treated	absolutely resistant			——excellent——			alkalies	will not be infested	
to insects and bacteria											
Temperature stability, continuous temperature in °C 250...	80...90	75...85	40...50	75...85	125...	125...	110...	110...		140...	200...
					135	135	130	130	160×	250	300
Max. temperature in °C (× = dry heat)	100	95	65	95	150	150	—	—	×	—	350

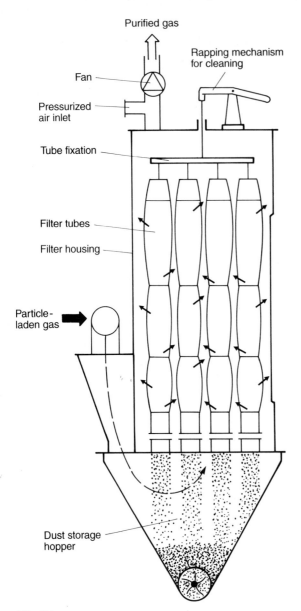

Fig. 7.21. Set-up of a tubular bag filter

7.2.4.2 Baghouse Filters

There is no difference between tube and frame filters in their basic mode of operation. The filtering elements consist of pillowcase-like bags pulled over tentering frames. The open slot on the one side is pulled over a flange and sealed. Filtering usually takes place from the outside to the inside. Cleaning is carried out by vibrating the frame or by alternating the air pressures to cause the filtering surfaces to flutter. During this, usually only the upper layers of the filter cake will fall off, so that the primary layer of dust with its favorable filtering properties is preserved. The timing for cleaning is regulated by automatic pressure drop controllers. In most frame filter systems operations must be shut down for cleaning. In this respect they are no different from tube filters, however, with the advantage of accommodating more frame filter surface in the same amount of space.

In processes yielding very fine and light dust, gravity settling chambers and wet collectors do not suffice. This is where dust filtering systems can best be used. Their collection efficiency is very high.

7.3 Nitrogen Oxide Reduction in Combustion Processes

In the following pages primary measures for the reduction of nitrogen oxide in combustion plants and secondary measures for large-scale furnaces and motor vehicles will be discussed.

7.3.1 Primary Measures in Furnaces

The formation of nitrogen oxides is classified as follows:
- thermal NO formation,
- NO formation from fuel nitrogen and
- prompt NO formation (s. also Chap. 2).

Thermal NO formation can be mainly influenced by primary measures, but partially also the transformation of fuel nitrogen. Prompt NO formation is merely of subordinate significance.

According to the formation conditons of nitrogen oxides in flames mentioned in Chap. 2, primary measures aim at the following:
- reducing the oxygen available in the reaction zone,
- lowering combustion temperatures,
- avoiding peak temperatures by even and quick mixing of the reagents in the flames,
- reducing the retention time at high temperatures,
- reducing already formed nitrogen oxides at the end of the flame.

These goals can be achieved with different technical measures.

7.3.1.1 Reducing Air Excess

Figs. 2.11 and 2.2 have shown the considerable dependency of the NO_x emissions on the air excess during combustion. Reduction measures aim at achieving complete combustion even with a lower air excess by fine-tuning the air-fuel ratio and improving the mixing in the reaction zone.

7.3.1.2 Stage Combustion, Multiple-Stage Mixing Burner and Top Air Nozzles

Corresponding to Fig. 2.11 NO emission drops to very low values when there is a lack of air on the one hand and high air excess in a flame on the other. High air excess as a solution is out of the question due to the energy loss involved – the excessive air would escape by way of the stack as heated flue gas and be useless.

The principle of stage combustion is to lower the air-fuel ratio to values below 1 in the main reaction zone of the flame where high temperatures occur, and to afterburn the residual products of incomplete combustion – CO, hydrocarbons, soot – at a low temperature (> 750 °C). The excess air in the center of the flame cannot be lowered at will because of the stability of the ignition. Special burner designs, however, achieve air staging as well as a stable flame [44].

Fig. 7.22 shows the principle of a multiple-stage mixing burner where besides air staging, fuel staging is additionally carried out. In a central burner a slightly understoichiometric primary flame is produced having a high ignition stability over the total range of performance owing to flow-induced inner recirculation and almost stoichiometric operation. At the periphery of the burner the residual fuel with carrier air is injected into the edge of the primary flame in such a way that a strongly sub-

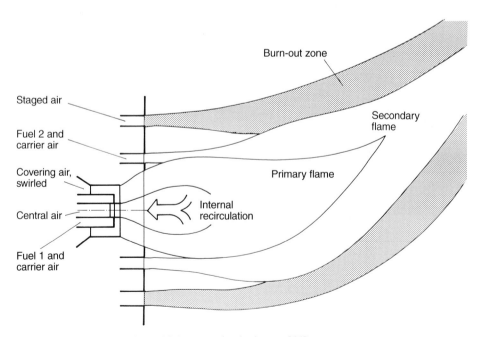

Fig. 7.22. Flame zones of a multiple-stage circular burner [44]

stoichiometric secondary flame results. In a third stage, the burn-out zone, the combustion products produced are remixed with a strong jet of more combustion air and are burned out.

Injecting the secondary fuel into the periphery of the primary flame produces an atmosphere where the already formed nitrogen oxide is reduced back to molecular nitrogen through gas components such as NH_3, HCN and CO. The formation of nitrogen oxide in the burn-out zone can be kept low by low combustion temperatures. If such burners are used as pulverized-coal burners, there is the danger of unburned coal particles and CO occurring in the flue gas. Very fine pulverizing of the coal can provide for an improved burn-out process here [44, 45].

A simplified form of staged combustion, particularly in already existing plants, consists of the installment of top air nozzles. For this, the burners are operated slightly substoichiometrically and complete burn-out is achieved by the upper air at low temperatures. This principle of nitrogen oxide reduction is, however, limited by the flame stabilities in substoichiometric operation and by insufficient mixing of the upper air with the flame gases.

7.3.1.3 Slight Air Preheating

Reduced preheating of combustion air results in a lowering of the combustion temperature and thus a reduction of nitrogen oxide. In gas and light fuel oil furnaces air preheating is not necessary to achieve a complete burn-out. Therefore, it has been done away with in most of the newer plants. The residual heat from the flue gas is, in this case, used for preheating water or for other purposes. Fig. 7.23 shows the considerable influence of combustion air temperature on NO_x emissions in gas furnaces.

7.3.1.4 Reduction of the Volume-Specific Combustion Chamber Load

If a lot of heat is released in a small space high combustion temperatures and thus high nitrogen oxide emissions result. Small, intensive flames lead to a lot of NO_x formation, whereas comparable larger flames produce less. The volume of the combustion chamber where the flames are confined also has an influence. This dependency becomes particularly apparent if the load dependency of the NO_x emissions are considered at certain given combustion chamber volumes. Fig. 7.24 shows the NO_x emissions of a steam boiler over the boiler load with the fuels heavy fuel oil and natural gas. As an additional parameter the air excess or the residual oxygen content in the flue gases is indicated. Due to the fuel NO, the combustion of heavy fuel oil causes higher nitrogen oxide emissions than the combustion of natural gas. Apart from the influence of the excess air already mentioned, one can see the strong dependency of the NO_x emissions on the load. If, from the outset, chamber and burners were designed larger and were not operated with such high loads, then NO_x emissions could be kept lower, particularly in the combustion of heavy fuel oil.

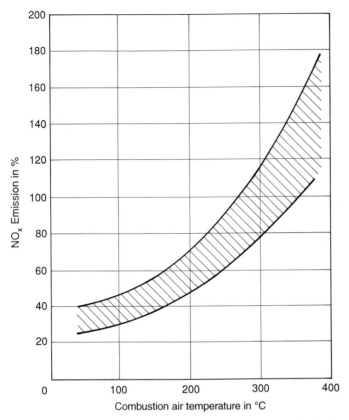

Fig. 7.23. Influence of combustion air temperature on NO_x emission in gas furnaces [46]

7.3.1.5 Flue-Gas Recirculation

By recirculating the flue gases into the combustion air of the burner, the inert gas part in the combustion zone is raised and the O_2 concentration reduced, leading to a fall in peak temperatures and to increased evenness of combustion. In this way, nitrogen oxide emissions can be reduced, especially in furnaces with high combustion results of tests with flue-gas recirculation in a heavy fuel oil furnace operated with approx. 500 kW thermal power. [47]. It can be seen that with an increasing degree of recirculation nitrogen oxide emissions drop, which is to be attributed to the sinking flame temperature. The heat radiation of the flame sinks owing to the recirculated flue gases and the convective part goes up. Due to this shifting of heat release, the use of flue-gas recirculation is limited in existing plants of steam boiler design. Recirculation streams in excess of 10-12 % would require considerable modifications to the steam generator.

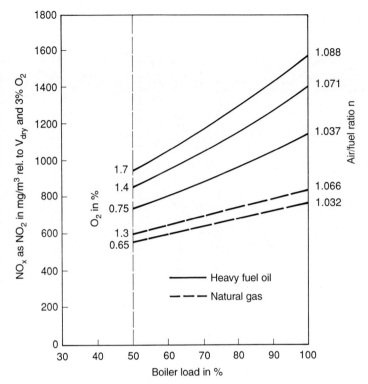

Fig. 7.24. Dependency of NO_x emissions on boiler load and excess air [46]

7.3.1.6 Low-NO_x Burners

Based on the knowledge of low-NO_x combustion several companies have developed burners which not only make a staged combustion possible, but also force a recirculation of flue gases in the flame. Such burners keep NO_x emissions low from the start [44, 48].

7.3.1.7 NO_x Reduction Potential of Primary Measures

Fig. 7.26 compiles the NO_x reduction potential of the individual primary measures. The effects of the measures sometimes overlap so that the NO_x reduction is altogether limited to approx. 40-70 % depending on the type of fuel.

As primary measures predominantly reduce the thermal NO, the greatest reduction potential is achieved in plants with little fuel nitrogen oxide, i.e., in natural gas furnaces. It is hoped that with gas and possibly also oil furnaces, the control measures listed will make it possible to stay below the prescribed threshold values even without additional, costly secondary measures.

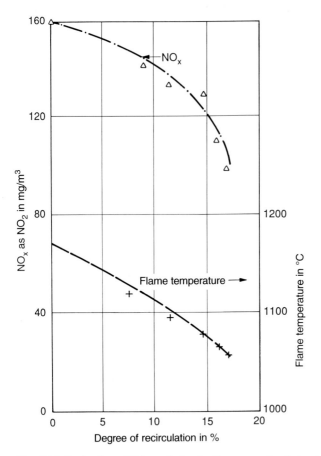

Fig. 7.25. Reduction of NO_x emission by flue gas recirculation in a light fuel oil furnace, 500 kW thermal power; dependency of flame temperature on degree of recirculation is also shown

7.3.2 Secondary Measures in Furnaces

If primary measures for the reduction of nitrogen oxides in furnaces are not sufficient, secondary measures must be applied. Their specific investment costs as well as their operating costs are higher than those of primary measures. Yet, for keeping emission threshhold values they are nevertheless unavoidable in large-scale coal and sometimes also oil furnaces.

The main problem with removing nitrogen oxides from flue gases is that nitrogen monoxide (NO) as the main component does not dissolve well in water. Simple scrubbing procedures are therefore not suitable for the separation of nitrogen mon oxide.

Basically, two process principles are employed in the removal of nitrogen oxides from flue gases:
– *Reduction processes.* NO is reduced to molecular nitrogen, with NH_3 generally

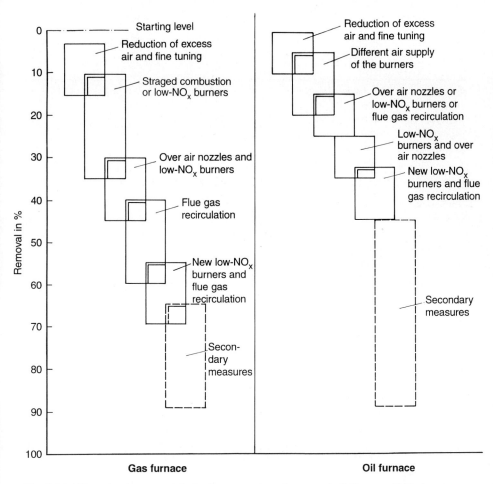

Fig. 7.26. NO_x reduction potential of primary measures in gas and oil furnaces [46]

being used as reductant for the absorption of oxygen. One distinguishes between non-catalytic and catalytic reduction.
- *Oxidation processes.* NO is oxidized, e.g., by radicals which are generated by electron beams or by ozone. The oxidation product NO_2 or nitric acid is generally transformed into ammonium salts by ammonia (NH_3).

In the following section the three most frequently used processes currently are described in brief.

7.3.2.1 Reduction Processes

In industrial furnaces reduction processes are most often used for the reduction of nitrogen oxides. NH_3 is generally employed as reductant. The reaction products

Table 7.2. Characteristics of NO_x reduction methods commonly used today

Method	Catalyzer	Temperature range in °C	NO_x reduction in %	Remarks	References
Selective non-catalytic reduction (SNCR)	No catalyzer, direct NH_3 or $(NH_2)_2CO$ (urea)- dosing into the furnace chamber	850...1000	30...80		[49]...[56]
Selective catalytic reduction (SCR)	Metal oxides on ceramic carriers	280...450 depending on catalyzer material	70...>90		[56]-[60]
	Molecular sieves (zeolites, solid ceramics)	380...480	70...90		[61]-[62]
	Activated charcoal	100...150	30...80	Simultaneous SO_2/NO_x removal	[63]...[65]

formed by this process – N_2 and H_2O – are unproblematic. Table 7.2 compares today's commonly used processes and their characteristic features.

Selective non-catalytic reduction – SNCR processes

A selective reduction of NO with NH_3 can proceed without a catalyst, too, if temperatures are high enough. Fig. 7.27 gives an overview of the different reaction mechanisms which play a role in the reduction with NH_3 [53-55].

In non-catalytic processes optimum nitrogen oxide reduction takes place within a relatively narrow temperature range. When temperatures are too high NH_3 reacts with oxygen to become H_2O and the unwanted NO (path B in Fig. 7.27). If temperatures are too low NH_3 slip (path C) occurs. The following conditions are important for the SNCR process to function well:
– thorough mixing of the flue gases with the NH_3,
– injection of the NH_3 at optimum temperature into all load ranges of the furnace,
– maintaining certain minimum retention time.

As these requirements cannot usually be realized to an optimum extent, NO reduction rates are either very low or a high NH_3/NO ratio is necessary to achieve an acceptable degree of reduction [56].

Newer developments point to employing aqueous urea solutions in place of ammonia. Handling these solutions is considerably less problematic than handling am-

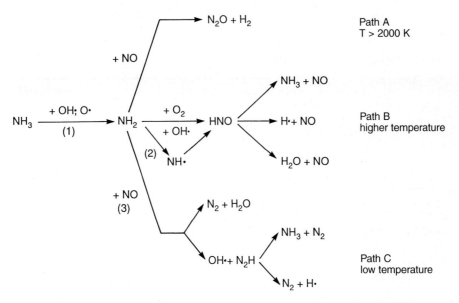

Fig. 7.27. Simplified reaction mechanism of NO reduction or NO formation with NH_3

monia. At present it is assumed that urea splits at such high temperatures and that equal parts of NH_3 and cyanic acid are formed [55, 56]:

$$\begin{array}{c} NH_2 \\ \quad \diagdown \\ \quad CO \\ \quad \diagup \\ NH_2 \end{array} \begin{array}{c} \diagup NH_3 \\ \diagdown HNCO \end{array}$$

The ammonia formed takes effect via the reaction paths shown in Fig. 7.27. It is not known by which reaction mechanisms HNCO also contributes to NO reduction.

When a reasonable quantity of reductant is used, the SNCR process does not have very high NO_x reduction rates. In furnaces with relatively low initial NO_x values, e.g., in industrial oil-fired units, it can actually be a sensible alternative to the SCR process for a further necessary lowering of emissions.

Selective catalytic reduction – SCR processes

For large-scale furnaces the SCR is the most frequently used process for the reduction of nitrogen oxide emissions. It is basically possible to reduce nitrogen oxides with gases such as CO, H_2 or CH_4. These gases, however, do not reduce nitrogen oxides selectively but are used up (oxidized) in large amounts by the residual oxygen present in the flue gases. As has been shown earlier, *ammonia* (NH_3), on the contrary, possesses a good selectivity for the reduction of nitrogen oxides. The reduction reaction achieves its greatest efficiency within the temperature range of between 900 and 1 000 °C. The decrease in temperature necessary for the reduction is accomplished by special catalysts.

The following main reactions take place at the catalyst during the reduction of nitrogen oxides with ammonia:

$$4NO + 4NH_3 + O_2 \rightarrow 4N_2 + 6H_2O$$

NH_3/NO molar ratio = 1:1 (7.20)

$$6NO + 4NH_3 \rightarrow 5N_2 + 6H_2O$$

NH_3/NO molar ratio = 2:3 (7.21)

Possibly present NO_2 (generally no more than 5 %) will also be reduced:

$$6NO_2 + 8NH_3 \rightarrow 7N_2 + 12H_2O$$

NH_3/NO molar ratio = 3:4 (7.22)

$$2NO_2 + 4NH_3 + O_2 \rightarrow 3N_2 + 6H_2O$$

NH_3/NO molar ratio = 2:1 (7.23)

Undesirable side-reactions at the catalyst are, e.g.,:

$$4NH_3 + 3O_2 \rightarrow 2N_2 + 6H_2O \quad \text{additional } NH_3 \text{ consumption} \quad (7.24)$$

$$2SO_2 + O_2 \rightarrow 2SO_3. \quad (7.25)$$

Excess ammonia reacts with sulfur trioxide (SO_3) to become ammonium-hydrogen sulfate and ammonium sulfate:

$$NH_3 + SO_3 + H_2O \rightarrow (NH_4)HSO_4 \quad (7.26)$$

$$2NH_3 + SO_3 + H_2O \rightarrow (NH_4)_2SO_4. \quad (7.27)$$

When temperatures fall below certain ranges these ammonium salts can lead to considerable deposits on the components after the catalytic converter, e.g., the air preheater.

The following *requirements* must be fulfilled by the catalytic converters:
- high activity over a wide temperature range,
- high selectivity,
- low SO_2/SO_3 conversion,
- sulfuric acid resistance,
- resistance to dust abrasion and catalyst poisons,
- long life.

The following *materials* can be used for SCR catalysts:
- titanium dioxide, aluminum or silicon oxides as porous basic materials (zeolites),
- adding vanadium pentoxide, molybdenum oxides and tungsten oxides as active substances or oxide and sulfate mixtures of iron, manganese and copper or other metal oxides [56, 57, 67].

To prevent abrasion of the active catalyst layer by dusty flue gases and to thus also prevent inactivation, the catalyst material is included in the carrier. Therefore, new active material takes effect with each abrasion.

The activeness of the catalytic converter and with it the degree of reduction is, among other factors, dependent on the temperature. Fig. 7.28 shows this dependency for a SCR converter of the kind commonly used today in power plant furnaces.

The NO_x reduction degree R is defined as follows:

$$R = \frac{NO_x \text{ prior to converter} - NO_x \text{ after converter}}{NO_x \text{ prior to converter}} \cdot 100 \text{ in \%}. \tag{7.28}$$

Other active materials have other temperature dependencies. In the presence of SO_2 the activity optimum shifts to higher temperatures due to partial sulfating of the catalyst material [67, 68].

A totally different type of catalytic converter is the molecular sieve catalyst containing no additions of active substances at all [61]. Its catalytic effect is based on its molecular structure. It consists of a crystal lattice traversed by a system of voids interconnected by pores (zeolites). At a defined pore size of 50-70 nm there is high selectivity, as only molecules of a certain size can diffuse into the system. According to the manufacturer's product information, SO_2, CO, CO_2 and halogenides cannot interfere with the actual reaction, as they cannot enter into the molecular sieve due to their size. High electrostatic forces in the pores and channels reduce the reaction potential to such an extent that the exothermic NH_3/NO_x reaction already takes place at 300-480 °C. The reaction products N_2 and H_2O are expelled from the molecular sieve by the heat released during the reaction. The basic material of molecular sieve catalytic converters is made from zeolites, which are in turn made up of silicon and aluminum oxides.

At given temperature conditions and given flue gas volumes three interdependant factors are important for the design of a catalytic converter [56]:

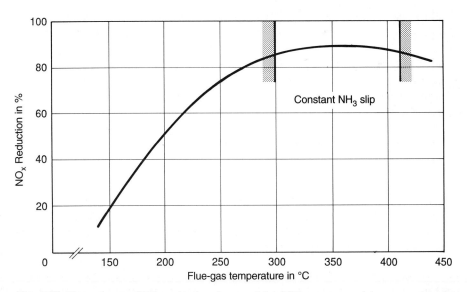

Fig. 7.28. Dependency of NO_x reduction degree of the SCR process used in power-plant furnaces on flue-gas temperature [56]

- the required degree of NO_x reduction,
- the admissible or justifiable NH_3 slip,
- the catalytic converter volume.

The basic assumption appears valid that under otherwise equal conditions a certain degree of reduction can either be attained by a large catalytic converter volume and a small ammonia slip or by a small catalytic converter volume and a higher ammonia slip.

The almost linear dependencies between reduction degree, NH_3 slip and catalytic converter volume shown in Fig. 7.29 are only valid up to a reduction degree of 80-85 %. In excess of that, the required catalytic converter volume and/or the NH_3 Schlupf rise disproportionately due to various factors such as irregularities in the flue gas stream (velocity, temperature, local NO_x and NH_3 concentration) [56].

Apart from the temperature, the molar ratio of NH_3/NO and the so-called space velocity are *influential factors* pertaining to NO_x reduction.

Depending on whether reaction (7.20) or reaction (7.21) prevails at the catalytic converter, maximum reduction ensues at a NH_3/NO molar ratio of 1 or of 0.66. Which reaction preferentially takes place depends, among other factors, on the material of the catalytic converter and on the oxygen content of the flue gases [57, 68].

The space velocity (SV) indicates the amount of flue gas put through per catalytic converter volume:

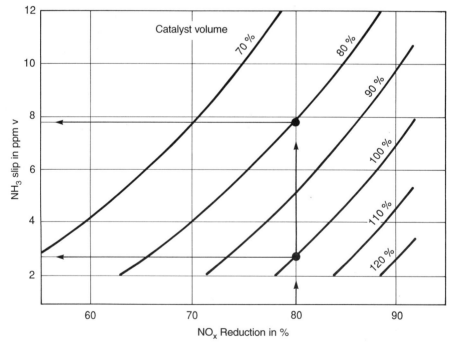

Fig. 7.29. Correlation between NH_3 slippage, catalyst volume and NO_x reduction degree (acc. to [56])

$$\text{SV}[1/\text{h}] = \frac{\text{flue gas volume } V_A}{\text{catalytic converter volume } V_K} \left[\frac{m^3/h \ (0\,°C, \ 1\,013\,mbar)}{m^3} \right]. \qquad (7.29)$$

Thus, space velocity is the inverse value of retention time. The catalytic converter volume does not indicate anything about the size of the active surface. Catalytic

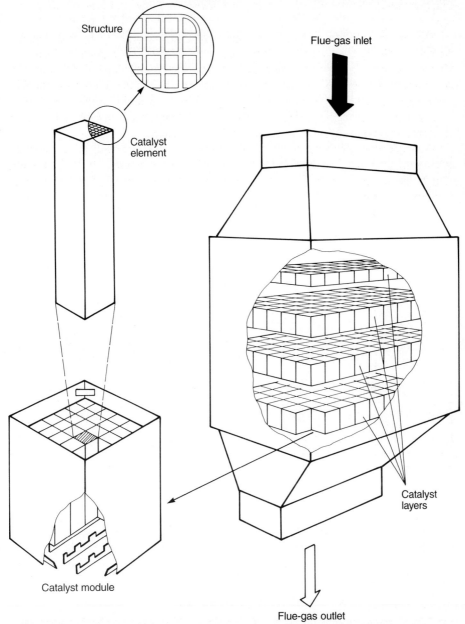

Fig. 7.30. Set-up of an SCR catalyst unit for NO_3 reduction in large furnaces [56]

converters with a large interior surface can be operated with a higher space velocity than converters with a smaller interior surface.

An SCR converter generally consists of individual elements which are combined to form modules. The modules are built into the catalytic converter housing at several levels, s. Fig. 7.30. A multi-level set-up is required as a certain flow velocity as well as a certain space velocity must be maintained.

Fig. 7.31 shows the configuration of a SCR converter unit for the reduction of nitrogen oxide emissions in a power plant furnace. In the flue gas path the SCR reactor must come directly after the steam boiler so that a temperature range between 350 and 400 °C can be exploited; air preheater, electrostatic precipitator and flue gas desulfurization unit follow. At present, electrostatic precipitators are not operated at high temperatures, they must therefore be preceded by the air preheater. This means that the SCR converter is exposed to the entire dust load of the flue gases, a so-called "high-dust set-up".

There are also set-ups where the catalytic converter is positioned at the "cold" end, which means it is positioned subsequent to the flue gas desulfurization unit; one then speaks of a "low dust set-up". These versions are used mainly in existing plants where it is technically impossible to install the catalytic converter between the steam boiler and the air preheater. In this case the flue gas must be heated again to achieve the required reaction temperature which means the use of additional fuel [56].

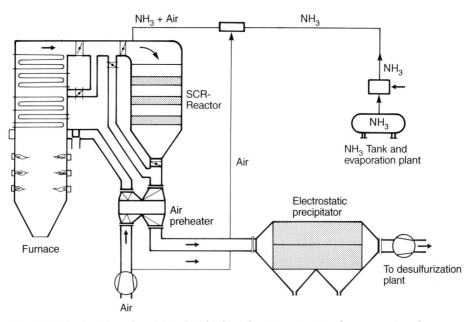

Fig. 7.31. Construction of an SCR plant in the exhaust gas system of a power plant furnace [56]

Activated coke process

In this process the flue gas is conducted through a moved bed reactor operated with char (activated coke) at 100-150 °C during which SO_2 and NO_x are adsorbed to the charcoke. The SO_2 oxidizes to sulfuric acid and by adding NH_3, NO_x is reduced catalytically to elementary nitrogen and water on charcoke.

The bound SO_2 or the H_2SO_4 is expelled again by thermic desorption and further processed to become SO_2 liquid gas, sulfuric acid or elementary sulfur.

During the charcoke process the following reactions take place:
- SO_2 adsorption

$$SO_2 \xrightarrow{O_2,\ H_2O} H_2SO_4 \rightarrow H_2SO_{4\ ads} \tag{7.30}$$

$$H_2SO_{4\ ads} \xrightarrow{Act.\ coke,\ NH_3} NH_3(NH_4)HSO_{4\ ads},\ (NH_4)_2SO_{4\ ads} \tag{7.31}$$

(unwanted reaction: unnecessary NH_3 consumption)
- NO_x reduction

$$6NO + 4NH_3 \xrightarrow{Act.\ coke} 5N_2 + 6H_2O \tag{7.32}$$

- *regeneration* at 350-600 °C

$$H_2SO_{4\ ads} \xrightarrow{Act.\ coke} SO_2,\ H_2O \tag{7.33}$$

$$(NH_4)_2SO_{4\ ads} \xrightarrow{Act.\ coke} SO_2,\ H_2O,\ N_2 \tag{7.34}$$

Fig. 7.32 depicts the circuit diagram of a two-stage adsorber for the simultaneous removal of SO_2/NO_x from furnace flue gases.

There are also several other processes for NO_x reduction which are not yet very widespread [65].

7.3.2.2 Oxidation Processes

There are different processes during which nitrogen monoxide (NO) is oxidized to nitrogen dioxide (NO_2). In most cases the NO_2, together with SO_2, is absorbed by substances such as ammonia or calcium hydroxide which react on an alkaline basis with nitrates and sulfates being formed, or else the NO_2 is separated in aqueous absorption solutions.

In all these processes the problem is how to achieve inexpensive oxidation of the NO to become NO_2. NO reacts sufficiently fast with oxygen to become NO_2 only in very high concentrations. For NO concentrations such as those occurring in flue gases, other oxidants are required – e.g., ozone (O_3) or chlorine dioxide are used [65]. Another possibility of making the NO more reactive so that it can, e.g., react with the residual oxygen of the flue gases, is by excitation with high energy radiation. Electron beams are used for this purpose.

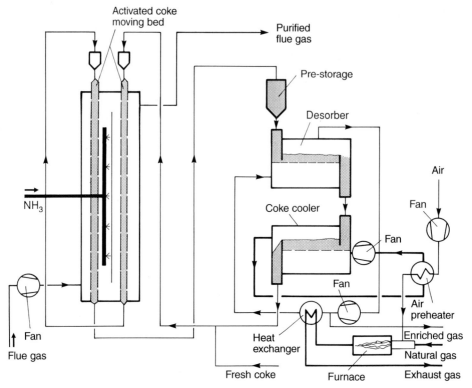

Fig. 7.32. Principle of the Bergbauforschung/Uhde process for simultaneous removal of SO_2/NO_x from furnace exhaust gases [64]; flue gas is generally used as desorption medium, part of which is discharged as gas laden with SO_2

The so-called *electron-beam process* is a dry simultaneous process for the removal of SO_2 and NO_x from flue gases [69] whereby the flue gas stream is continuously exposed to intense electron beam radiation. During this, flue gas components such as H_2O, O_2 and others are split into the radicals OH·, O·, HO_2· and N·. In this way the compounds SO_2 and NO are first oxidized to SO_3 and NO_2, and finally with the water vapor of the flue gases to sulfuric acid and nitric acid. Subsequently the acids are neutralized by adding NH_3, whereby the final crystalline products ammonium sulfate and ammonium nitrate are formed. The end products are collected on filters and can be used as fertilizer. Flue gas temperature for optimum NO_3 reduction and desulfurization is between 70 and 120 °C in this process. In test and pilot plants NO_x and SO_2 collection efficiencies of over 70 % were measured. The present trial period in a German power plant will show whether this technical process will be used on a large-scale basis in the future [69].

7.3.3 Catalytic Converter Technology for the Reduction of Nitrogen Oxides and Other Components in Automotive Exhaust Gases

Apart from the products of incomplete combustion – CO, hydrocarbons and soot – nitrogen oxides, constitute a main problem in automotive exhaust gases. Basically, diesel-powered engines come off more favorably in this respect than gasoline-powered engines as they are operated with a high excess air and thus lead to fewer NO_x emissions [70-72]. Technical modifications to the engine have been able to lower NO_x emissions in diesel passenger cars to such an extent that so far no further exhaust gas purification measures have been necessary to keep the threshold values pertaining to NO_x. The soot problem of the diesel engine will not be dealt with in detail here. To solve this problem not only the engine is being worked on but also the development of exhaust gas purification systems (e.g., soot filters), particularly in truck engines [73].

In the *gasoline-powered engine* technical modifications to the engine alone are generally not enough. Therefore, the three-way catalytic converter is nowadays being used as a "secondary measure". The term "three-way" implies that the three pollutants CO, HC (hydrocarbons) and NO_x are transformed.

Construction of A Three-Way Catalytic Converter

Catalytic converters consist of honeycomb-shaped ceramic or metal monoliths coated with the active substance, primarily with the noble metals platinum and rhodium [74-77]. An intermediate layer, the so-called wash-coat, is the carrier of these metals and it consists of aluminum oxide (Al_2O_3) applied to the walls of the ceramic body or metal carrier. The wash-coat has a high specific surface and thus increases the effective surface many times over.

Oxidation reactions preferably take place on the platinum, whereas rhodium assists reduction of oxides of nitrogen. The optimal platinum-rhodium ratio is approx. 5:1 [75, 76]. As the raw material rhodium is considerably more expensive than platinum, efforts are being made to raise the Pt/Rh ratio or to find replacement substances for rhodium.

The catalyzer bodies are embedded in a high-grade steel housings which are installed in the exhaust path of the motor vehicle before the mufflers. Fig. 7.33 shows a diagram of the design of a catalytic converter for motor vehicles. When designing such a converter the following problems must be taken into consideration, [75-78]:
- the ceramic catalyzer must not be exposed to too much vibration to prevent it from breaking,
- the hazard of overheating on the one hand and fast starting during cold starts on the other must be taken into consideration – the optimal temperature range is between 300 and 850 °C [78],
- resistance to fast changes in temperature;
- resistance to corrosion,
- lead, but also other fuel and oil additives can have a disactivating effect due to chemical reactions and deposits.

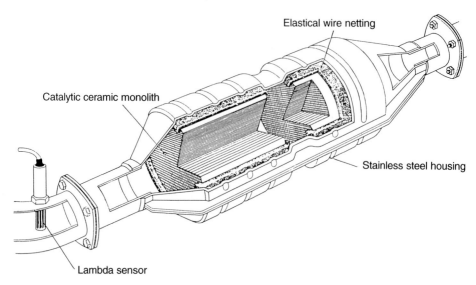

Fig. 7.33. Set-up of a motor vehicle catalytic converter [78]

Reactions in the catalyzer

In contrast to catalyzer technology in large-scale furnaces there is no one particular reductant used for the transformation of nitrogen oxides in automotive catalytic converters. This task is carried out by the residual CO and residual hydrocarbons present in the exhaust gases. The following gross reactions occur in the three-way catalyzer [78]:

Hydrocarbon transformation:

$$C_nH_m + (n+m/4)O_2 \rightarrow nCO_2 + m/2\ H_2O \qquad (7.35)$$
$$CH_m + 2H_2O \rightarrow CO_2 + (2+m/2)H_2 \qquad (7.36)$$

Carbon monoxide transformation:

$$CO + 1/2\ O_2 \rightarrow CO_2 \qquad (7.37)$$
$$CO + H_2O \rightarrow CO_2 + H_2 \qquad (7.38)$$

Reduction of nitrogen oxides:

$$NO + CO \rightarrow 1/2\ N_2 + CO_2 \qquad (7.39)$$
$$2(n+m/4)NO + C_nH_m \rightarrow (n+m/4)N_2 + m/2\ H_2O + nCO_2 \qquad (7.40)$$
$$NO + H_2 \rightarrow 1/2N_2 + H_2O \qquad (7.41)$$

Undesirable side-reactions:

$$SO_2 + 1/2\ O_2 \rightarrow SO_3 \quad \text{with excess air} \tag{7.42}$$

$$SO_2 + 3H_2 \rightarrow H_2S + H_2O \quad \text{with lack of air} \tag{7.43}$$

$$NO + 5/2\ H_2 \rightarrow NH_3 + H_2O \tag{7.44}$$

$$2NH_3 + 5/2\ O_2 \rightarrow 2NO + 3H_2O \tag{7.45}$$

$$NH_3 + CH_4 \rightarrow HCN + 3H_2 \tag{7.46}$$

$$H_2 + 1/2\ O_2 \rightarrow H_2O. \tag{7.47}$$

Fig. 7.34 depicts the hydrocarbon, CO and NO_x concentrations before and after a catalytic converter depending on the air ratio.

With air ratios $\lambda < 1$ (*rich mixture*) the residual oxygen in the exhaust gases is not sufficient to completely oxidize the CO and hydrocarbons. The reason for the hydrocarbons decreasing more in this range than CO is that the hydrocarbons are more easily oxidized than CO, and that a partial oxidation of these substances with the residual oxygen takes place before the CO oxidation does [74].

However, as a result of the partial oxidation in the catalyzer in this rich range new substances can form which are conspicuous for their odour intensity [e.g., H_2S, s. equation (7.43)] or their irritating effect (e.g., aldehydes). On the other hand, the NO_x present is reduced almost completely in this rich range, as CO and hydrocarbons are abundantly available.

During *air excess operation* (*lean mixture* $\lambda > 1$) emissions of CO and hydrocarbons are low, as these substances are post-oxidized by the residual oxygen of the exhaust gases in the catalyzer. Emissions of nitrogen oxides, however, are not lowered in this oxidizing atmosphere as there is no reductant (CO or hydrocarbons) available.

If the *mixture is stoichiometrical* ($\lambda = 1$) all emission components are at a minimum. To which extent these substances can be decreased depend, on the one hand, on the properties and the condition of the catalytic converter used but, on the other hand, mainly on how precisely the stoichiometric mixture can be adjusted in all of the vehicle's modes of operation. To adjust the exact stoichiometric mixture injections with lambda (λ) control are used [79]. In some cases controlled carburetor systems also exist. In both cases continuous measurement of the air-fuel ratio is carried out by a lambda sensor.

In motor vehicles having catalytic converters with lambda control, emissions of nitrogen oxides are appreciably reduced, as also emissions of the different, in part not limited exhaust gas components such as aldehydes, phenols, alkanes, alkenes, aromatic compounds such as benzene and toluene, polycyclic aromatic hydrocarbons, cyanides a.s.o. [80].

It is obvious that pollutant reduction cannot show such satisfactory results in catalyzer vehicles without lambda control as conditions with the above emission ratios frequently occur in the fuel-rich lack-of-air range as well as in the excess-of-air range (Fig. 7.34). The efficiency of emission reduction in motor vehicles without lambda control depends on the individual case, namely on how well a stoichiometric

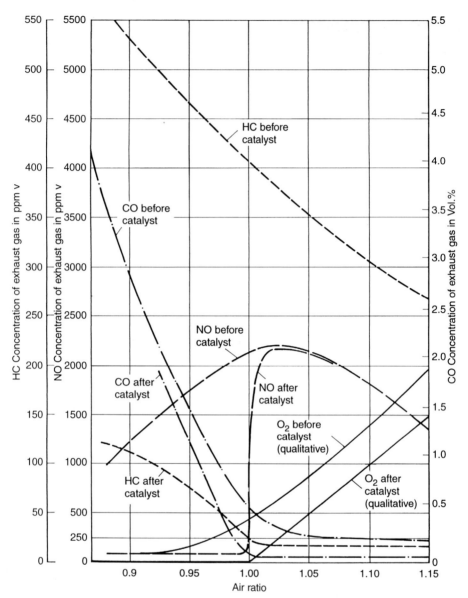

Fig. 7.34. Hydrocarbon (HC), CO and NO_x concentrations in relation to the air ratio before and after passing through the catalyst unit [74]

mixture can be made available by the fuel treatment system (generally the carburetor).

7.4 Flue Gas Desulfurization

Processes for the desulfurization of flue gases of furnaces can be categorized as follows:

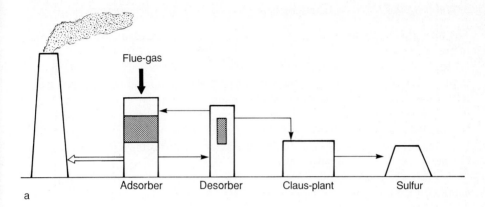

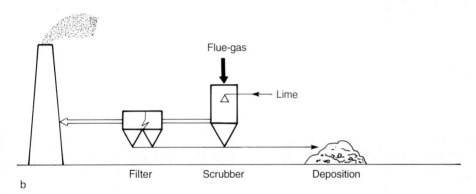

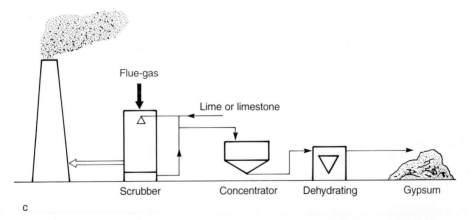

Fig. 7.35. Flue gas desulfurization systems – diagrams of the most frequently used methods. **a.** dry adsorption process; **b.** dry additive or spray absorption process; **c.** lime scrubbing

1. Dry processes: dry absorption processes, absorption-desorption processes.
2. Semi-dry processes: spraying scrubbers, other scrubbers.
3. Wet processes: lime scrubbers.

The most frequently used flue gas desulfurization systems are depicted in a diagram in Fig. 7.35.

In *dry flue gas desulfurization* SO_2 and SO_3 sorption takes place purely physically (adsorption) on fixed sorption agents or through absorption and reaction processes (chemisorption). In dry *ad*sorption processes one frequently works with continuous moving-bed operation whereby the adsorption agent is conducted to meet the gas to be desulfurized. Subsequently the agent is reactivated in the regenerator and is fed back into the adsorption process. In dry *ab*sorption processes the absorption agent is blown into the flue gas in pulverized form.

In *semi-dry processes* an alkaline or alkaline-earth suspension is brought into contact with the flue gas for the removal of SO_2. A dry reaction product is formed in the hot flue gases which is subsequently collected in particle precipitators.

Wet flue gas desulfurization is mainly carried out in scrubbers. The principle is that alkaline or alkaline-earth absorption agents (solutions or suspensions) are dispersed in the gas to be desulfurized or are brought into homogenous contact with it, so that the oxides of sulfur are washed out by absorption. Gas scrubbing with simple water would basically also be possible, but the tendency of the gases to be removed to solve in water is poor.

Much research has been done in the field of flue gas desulfurization. There are numerous processes applying these different principles. They are frequently based on the materials available for use and differ only minimally from each other. An overview of the most important, currently used flue gas desulfurization processes is given in Table 7.3. Some processes are designed only for large-scale firing plants (>100 MW Feuerungswärmeleistung thermal capacity), others for small to medium-sized plants (<50-100 MW).

7.4.1 Dry Flue Gas Desulfurization

One possibility of dry desulfurization of flue gases is with solid sorbents by physical adsorption. The charcoke process for the combined removal of SO_2 and NO_x works on this principle. Another possibility, mainly for small to medium-sized plants (<50-100 MW_{th}), is the chemical absorption of the SO_2 by dry, reactive additives. Adding dry, pulverized additives on a calcium or magnesium basis for the emission reduction of acidic gas components (SO_2, but also, e.g., HCl and HF) has been known for a long time [81, 82] and is in the meantime common practice in numerous furnaces. Additives can be added at different points of the firing or in different points in the flue gas path. Fig. 7.36 gives an overview of this. Processes where the additive is dosed in between boiler and filter are the most common, as in this way the fewest changes to the existing plant components such as combustion chamber and boiler construction are necessary [83]. There are also units which are positioned at the rear end, i.e., between the particle collector and the stack [8]. Additional particle filters are required for this.

In brown coal and fluidized-bed furnaces good results have been obtained with additive dosing in the fuel [85, 86]. For mineral coal Chughtai and Michelfelder suggest dosing in the additives with the combustion air of the burners [87].

Binding SO_2 and also other acidically reacting gases such as HCl and HF depends on several factors. One very important parameter is the temperature. Fig. 7.37 shows the degree of SO_2 absorption of different dry additives in relation to temperature. As can be seen, the effectiveness of the additives declines strongly both at low and at high temperatures, with magnesium compounds exhibiting smaller degrees of absorption and narrower temperature ranges than additives on a calcium basis.

When temperatures are too low there is not sufficient activation energy to initiate the absorption reactions, if temperatures are too high the reaction equilibrium is on the gas side [87], and surfaces of the additive particles sinter at high temperatures, so that there is no effective surface left available for SO_2 absorption. It is evident from the curves that MgO is hardly effective by itself. As the compound $MgSO_3$ is not formed the presence of a catalyzer to oxidize the SO_2 to SO_3 is necessary. Adding 2 % of Fe_2O_3 as catalyst distinctly improves the effect of the MgO.

With raw limestone ($CaCO_3$), whose temperature behavior is not indicated in Fig. 7.37, the reaction with SO_2 starts only at temperatures in excess of approx. 600 °C, with raw dolomite ($CaCO_3$) in excess of approx. 500 °C [89].

Apart from the temperature, the effectiveness of SO_2 absorption with dry additives depends on several other factors, e.g., on the residence time, on the mixing of the additives with the flue gas, on the particle size and on the pore structure. Pore structure which plays an important role, is, among other factors, influenced by the combustion process of the limestone [83]. Weisweiler et al. [89] have carried out fundamental research on SO_2 absorption by optimizing the pore structure. Modified sorbent types can be produced by impregnation with reagents, calcination, hydration as well as by pelletizing. They can differ considerably in their SO_2 sorption ability depending on the reaction temperature and reaction time. The reaction sequences occurring during the calcination, hydration and sulfating (SO_2 absorption) of, e.g., dolomite ($CaCO_3 \cdot MgCO_3$) are shown in Fig. 7.38. Hydration generally increases the specific surface which improves the SO_2 absorption pertaining to burnt limestone or dolomite. The CaO part is hydratized more than the MgO and thus contributes more to the SO_2 absorption. The type of hydration strongly influences the sorption ability [89].

The different dependencies of SO_2 absorption explain why, depending on the required application or dose inlet location (according to Fig. 7.36), an optimized additive must be used.

By additionally contaminating the hydrated lime particles with moisture, a wet particle surface favorable to SO_2 sorption can be created. In moist surroundings SO_2 solubility increases exponentially with sinking temperatures. This being the case, it is sensible to cool down the flue gases as much as possible – something which can also be achieved by adding water. When employing this process the additive cannot simply be blown into the existing flue gas channels. Additive preparation, conditioning

Table 7.3. Survey of the flue gas desulfurization processes used today

Principle of process	Process designation/name	Examples of suppliers/producers	Maximum achievable rate of desulfurization (%)	Final product
Dry processes				
Absorption on calcium basis	Dry additive added to the fuel	Babcock, EVT, Steinmüller	Lignite: 75	Flue ash with $CaSO_4$, CaO
	added to the flame area			
	added above the flame area		Mineral coal: 85 (including natural binding)	
	Fluidized-bed firing	Ahlström, Babcock, EVT Götawerken, Lurgi, Steinmüller, Thyssen, VKW, Waagner-Biro	90 / Up to 95 / 95	Flue ash with $CaSO_3$, $CaSO_4$, CaO
Absorption on	Dry additive added to exhaust gas calcium basis	BMD, Fläkt, H. Lühr, Standard Filterbau, Micropul Ducon	60	$CaSO_3/CaSO_4$
Adsorption on activated coke	Bergbau-Forschung	Bergbau-Forschung, Babcock, Krantz, Hugo Petersen Perflutiv Consult, Uhde	95	SO_2-rich gas, S elementary
Chemisorption/catalyzer	SFGT	Salzgitter Lummus	90	SO_2 rich gas, S elementary
Semi-dry processes				
Absorption on calcium or sodium basis	Spray absorption	Fläkt, Niro Atomizer, EVT	95	$CaSO_3/CaSO_4$ (Na_2SO_3/Na_2SO_4)
Wet processes				
Absorption on calcium basis	Lime scrubbing processes	Steinmüller, Babcock, Bischoff, GEA, KRC-EVT, Kroll-Zeppelin, Lentjes-Liesegang, Thyssen, Insuma, Saarberg-Hölter-Lurgi	95 to 97	$CaSO_4 \cdot 2H_2O$

Absorption on sodium basis	Citrate methods Boliden process Wellman-Lord	Fläkt, Mannesmann Huge Petersen Davy McKee	95	SO$_2$-rich gas, S elementary
Absorption with MgO		Babcock & Wilcox Chemico, Steinmüller, Waagner-Biro	95	SO$_2$-rich gas
Absorption with NH$_3$	Walther	Walther	95	(NH$_4$)$_2$SO$_4$
Oxidation/absorption with H$_2$O$_2$		Degussa/Plinka, Techno Trade Schira, Toschl	95	H$_2$SO$_4$
Oxidation/absorption with catalytic converter	WAS	Haldor Topsoe	95	H$_2$SO$_4$
Condensation/absorption	Fumex	Air Fröhlich	90	CaSO$_3$, CaSO$_4$
Physical scrubbing	Solinox	Linde	95	SO$_2$-rich gas residue
Double alkali processes		Sulzer	90	CaSO$_3$/CaSO$_4$

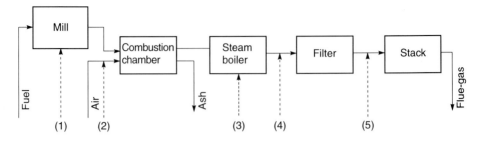

Fig. 7.36. Overview of sorbent dosing possibilities in furnaces [83]

Sorbent addition	Process	Sorbents
(1)	Direct desulfurization by addition to fuel	$Ca(OH)_2 \cdot (CaCO_3)$
(2)	Direct desulfurization by addition to combustion air	$Ca(OH)_2 \cdot (CaCO_3)$
(3)	High temperature process, non regenerative	$Ca(OH)_2 \cdot (CaCO_3)$
(4)	Low temperature process, non regenerative	$Ca(OH)_2$
(5)	Low temperature process, regenerative	Activated charcoal

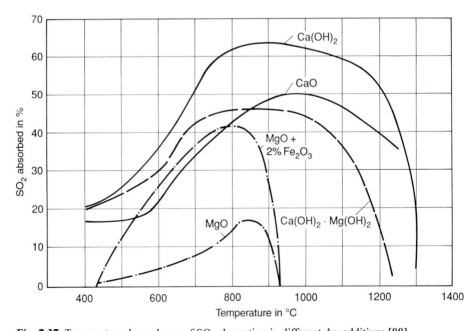

Fig. 7.37. Temperature dependence of SO_2 absorption in different dry additives [88]

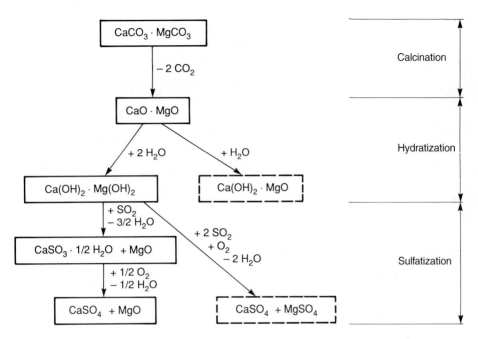

Fig. 7.38. Reaction sequences of calcination, hydratization and sulfatization of dolomite [89]

the flue gas, reaction and disposal of the used additive require a separate equipment [84].

To achieve sufficiently high degrees of desulfurization, additives must be added in hyperstoichiometric quantities in all dry additive processes [83-86].

Residues

Dry additive processes produce no waste water. Usability of the dry residual products depends on their chemical make-up and on the pre-separation of the airborne dust [84, 90]. The possibilities of recovery or disposal of the residues of dry additive desulfurization is shown in Fig. 7.39.

7.4.2 Semi-Dry Processes – Spray Absorption Technique

The semi-dry processes were developed based on the knowledge that SO_2 absorption on limestone particles can be improved by making them moist. In these processes the absorption agent – generally a lime or sodium carbonate suspension, or caustic lye of soda – is sprayed into the hot flue gas in a superfine dispersion. Depending on the manufacturer of the process, the spray mist is either produced by spraying nozzles [91-93] or with the help of a spinning-disk atomizer [94]. When the spray mist contacts the flue gas the water of the absorbent vaporizes, and the SO_2 reacts with the

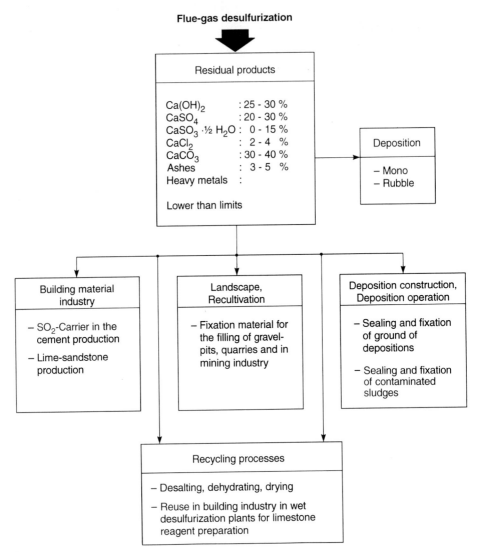

Fig. 7.39. Utilization or disposal of residual products of dry flue gas desulfurization [84]

absorption agent. The reactions taking place during this are shown in a diagram in Fig. 7.40 for the limestone process. The vaporization process is sustained until the ensuing reaction products take on the form of dry powder. This can then be removed from the flue gas with a standard dust collector (electrostatic precipitator or, when requirements are exacting, a fabric filter).

Fig. 7.41 shows the diagrammatic representation of flue gas desulfurization according to the spray absorption process.

The main advantage of the *spray absorption process* is that the water required vaporizes and that there is no waste water. Generally, the flue gases also do not need

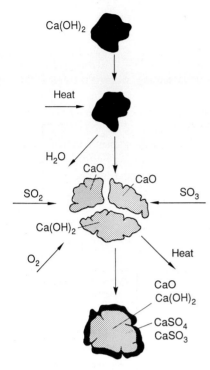

Fig. 7.40. Possible mechanism of SO$_2$ absorption in the spray absorption process

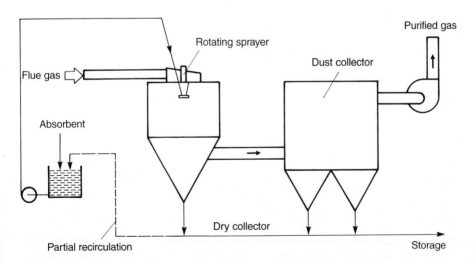

Fig. 7.41. Diagram of a spray absorption process for flue gas desulfurization [92]

to be reheated again. Compared to the dry additive process somewhat higher degrees of SO_2 absorption with a smaller excess of sorbent are achieved; expenditure for equipment, however, is higher.

Dosing of the sorbent must be precise and is generally adapted to the load of the furnace. When dosing is too high there is the danger of scaling, when it is too low the degree of SO_2 removal drops.

The *final product* not only consists of gypsum ($CaSO_4 \cdot 2H_2O$) but also has a relatively high calcium sulfite ($CaSO_4 \cdot 1/2\ H_2O$) and calcium oxide (CaO) content. Therefore thermic aftertreatment is sometimes carried out to raise the gypsum content [92]. Products with a higher gypsum content can be used better in the building material and concrete industry. Otherwise the residues have the same possibilities for use as in dry additive processes, s. Fig. 7.39.

7.4.3 Wet Desulfurization Processes

Wet absorption processes can be divided into processes with:
- alkaline sorbents,
- alkaline-earth sorbents,
- other sorbents.

The water solubility of alkaline sorbents is considerably greater than that of alkaline-earth ones; the same is also true of the reaction products after chemisorption of SO_2. Correspondingly, the sorbate of a process with an alkaline sorbent is, as a rule, a homogenous solution, whereas the sorbate of a process with an alkaline-earth sorbent is a suspension.

In the group of processes involving other absorption agents the scrubbing liquid is essentially an acid in an aqueous solution. The common factor for all these processes is the use of gas-liquid contact equipment. These devices have the ability to create a large boundary surface and to promote the mass transfer of the SO_2 to the liquid phase.

7.4.3.1 Lime Scrubbing Processes

The development of lime scrubbing processes to remove sulfur dioxide from power plant flue gases started, in particular in the USA, with trying to carry out desulfurization and wet particle removal in one procedural step. In this way, the Venturi scrubbers in the wet systems were primarily used for flue gas desulfurization processes. This combination of particle removal and desulfurization resulted in many cases in reactions between the dust from the coal firing and the lime or limestone, and thus led to heavy scaling and plugging in scrubbers, pipes and pumps [95]. Thus, common practice was to connect the traditional electrostatic precipitator before the flue gas desulfurization. This process separation "dedusting-desulfurization", is now general practice. Only with this process can a sufficiently pure final product (mostly gypsum) be achieved.

Spray tower absorber

In addition, Venturi scrubbers were not generally accepted because of their high pressure loss (high energy consumption). In the meantime, different scrubber types have been developed, e.g., scrubbing towers with an integrated Venturi spray module or other components [96, 97]. Spray towers with uncomplicated components to minimize scaling and obstructions have proved to be the best solution [98, 99], s. Fig. 7.42.

Such spray towers of newer construction do not contain any other packs apart from a few spacious spraying levels and the horizontally installed droplet precipitators. This has made it possible to minimize the pressure loss. The spray tower's

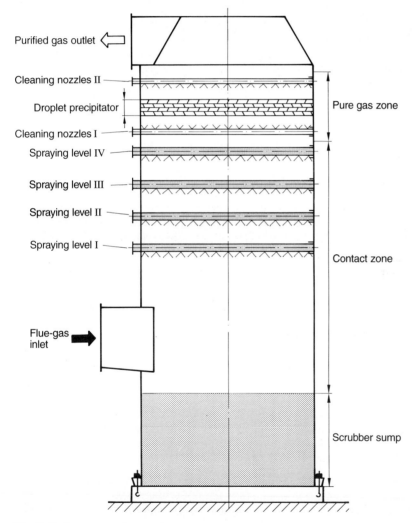

Fig. 7.42. Spray tower for wet flue gas desulfurization [98]

smooth walls prevent some areas from being poorly sprayed or not at all. This keeps the formation of scaling at a minimum. Spray tower lay-out can be divided into three sections:
- the scrubber sump,
- the gas/liquid contact zone and
- the pure gas zone.

The *scrubber sump* is where the scrubbing suspension is collected, stirred, vented and replenished with fresh sorbent. The volume of the scrubber sump is determined mainly by the dissolving velocity of the sorbent as well as by the amount of SO_2 to be removed.

In the middle zone of the scrubber, the *gas/liquid contact zone,* the flue gases are brought into contact with the scrubbing suspension in a counterflow and are thus washed. The scrubbing liquid is evenly dispersed over several spraying levels. Thus a homogenously mixed gas/liquid space is created where the mass transfer from flue gas to scrubbing liquid takes place.

In the upper area of the scrubber, the *pure gas zone,* the flue gases are conducted through a droplet separator and are freed from the fine droplets of fluid which have been carried along. For cleaning, the droplet separator is rinsed sector by sector with water from top to bottom in a certain cycle [98].

Chemical reactions during desulfurization with lime or limestone [95-99]

The first flue gas desulfurization equipment in the FRG was based on the use of burnt (caustic) (CaO) or hydrated $(Ca(OH)_2)$ lime as highly alkaline sorbent. In this way, it was hoped to accelerate the purely physical solution of the SO_2 in water by the chemical reaction. Calcium sulfite is primarily formed in the alkaline to slightly acidic medium, which is difficult to dehydrate. The waste sludge was disposed of in landfills in the USA, where large dumping areas are available. In the FRG only the first flue gas desulfurization equipment in a power plant produced this type of sulfite sludge. All other plants produced gypsum [95]. For this, the calcium sulfite must be postoxidized to calcium sulfate. This postoxidation takes place in separate containers by blowing in air. The oxidation reaction proceeds via hydrogen sulfite which achieves its maximum at pH values of around 4, s. Fig. 3.22.

In the oxidation containers the pH value of the sulfite suspension is set at values of between 4-4.5 by adding sulfuric acid.

The main reactions of the lime scrubbing process are as follows (gross):

$$SO_2 + Ca(OH)_2 + H_2O \xrightarrow{pH\ 6-7} CaSO_3 \cdot 1/2H_2O + 3/2H_2O \qquad (7.48)$$

$$CaSO_3 \cdot 1/2H_2O + 1/2O_2 + 3/2H_2O \xrightarrow{pH\ 4-4,5} CaSO_4 \cdot 2H_2O. \qquad (7.49)$$

In the meantime it has been established that effective SO_2 sorption is also possible with pH values between 5 and 6, i.e., in a slightly acidic medium. At these values ground limestone $(CaCO_3)$ can be used instead of the burnt lime (CaO). With CO_2 separation occurring in the aqueous, acidic phase and not, as is the case with lime burning, by applying energy at high temperatures. Accordingly, natural limestone ($CaCO_3$) is far more inexpensive than burnt lime (CaO).

Limestone solubility in a slightly acidic medium is, however, poorer than that of lime and, in addition, further deteriorates with increasing chloride content. This poorer solubility of the limestone is compensated for by very fine pulverizing and longer residence time which is achieved by enlarging the scrubber sump.

Fig. 7.43 shows the influence of the pH value on the rate of oxidation. As can be seen, operating with a slightly acidic medium has the additional advantage that the

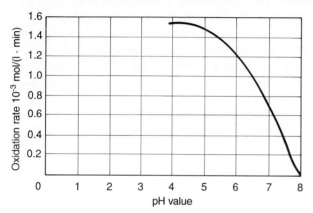

Fig. 7.43. Influence of the pH value on the oxidation rate [96]

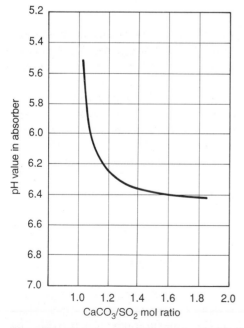

Fig. 7.44. pH value in relation to the $CaCO_3/SO_2$ molar ratio in limestone desulfurization processes [98]

oxidation can take place in the same medium and in the same container (scrubber sump) without additional acidification with sulfuric acid.

In limestone scrubbers the pH value of the SO_2 scrubbing cycle is between 5.5 and 6.0 at stoichiometric operation, s. Fig. 7.44, which means it is just above the value of <4.5 which is the optimum for oxidation. As, on the other hand, the scrubber sump of the limestone scrubber is much larger than those in lime scrubbers, the slightly poorer rate of oxidation is overcompensated by the longer residence time.

There are also two-stage scrubbers where, in the lower section, the pH value is kept low by an SO_2 excess which promotes oxidation. The upper zone (pure gas zone), on the other hand, is operated with limestone excess so that the total molar ratio limestone to SO_2 almost equals 1. Scrubbers of this type are of advantage when operating conditions change and SO_2 concentrations fluctuate [97].

The main reactions occurring in a spray scrubber (Fig. 7.42) operating with a limestone suspension are as follows [98]:

Contact zone

In the contact zone the SO_2 of the flue gas passes into the scrubbing liquid by physical solution and, in a pH operating range of 5.5-6.0, occurs mainly as hydrogen sulfite (sodium hydrogensulfite) (s. Fig. 3.22):

$$SO_2 \text{ (gas)} + 2H_2O \rightarrow HSO_3^- + H_3O^+. \tag{7.50}$$

A part of the hydrogen sulfite ions reacts with hydrocarbonate ions to become sulfite, and CO_2 is formed:

$$HCO_3^- + HSO_3^- \rightarrow SO_3^{2-} + H_2O + CO_2 \tag{7.51}$$

$$SO_3^{2-} + Ca^{2+} \rightarrow CaSO_3 \text{ (dissolved)}. \tag{7.52}$$

The calcium sulfite ions thus formed remain almost exclusively solved or, if crystals are formed, go back to a solved state in the scrubber sump by venting.

Scrubber sump

In the scrubber sump calcium carbonate (limestone) is solved in the slightly acidic medium and forms hydrocarbonate and calcium ions; precondition for this is very fine grinding:

$$CaCO_3 \text{ (solid)} + H_2O \rightarrow CaCO_3 \text{ (dissolved)} + H_2O \tag{7.53}$$

$$H_3O^+ + CaCO_3 \text{ (dissolved)} \rightarrow HCO_3^- + Ca^{2+} + H_2O. \tag{7.54}$$

The hydrogen carbonate ions which were not consumed in the contact zone by/with?? hydrogen sulfite ions break down in the scrubber sump to OH ions and CO_2, which remains temporarily solved:

$$HCO_3^- \rightarrow OH^- + CO_2 \text{ (dissolved)}. \tag{7.55}$$

Oxygen of the gaseous phase is solved in the scrubber sump by venting:

$$O_2 \text{ (gas)} + H_2O \rightarrow 2O \text{ (dissolved)} + H_2O. \tag{7.56}$$

The hydrogen sulfite ions of the contact zone react predominantly in the scrubber sump with solved oxygen to become sulfate:

$$HSO_3^- + O \text{ (dissolved)} + H_2O \rightarrow SO_4^{2-} + H_3O^+ \tag{7.57}$$

$$H_3O^+ + OH^- \rightarrow 2H_2O \tag{7.58}$$

$$SO_4^{2-} + Ca^{2+} \rightarrow CaSO_4 \text{ (dissolved)} \tag{7.59}$$

$$CaSO_4 + 2H_2O \rightarrow CaSO_4 \cdot 2H_2O. \tag{7.60}$$

Apart from supplying the oxygen, sump venting contributes to CO_2 removal from the solution and thus raises the solving velocity of the limestone. This is important, as chloride ions separated from the flue gas counteract limestone solubility. The existing, easily soluble chloride ions are finally washed out of the gypsum and must be removed from the system with the waste water.

The sulfite ions which were formed in the contact zone are also transformed into hydrogen sulfite and oxidize to become sulfate:

$$SO_3^{2-} + H_2O \rightarrow HSO_3^- + OH^-. \tag{7.61}$$

In a steady-state balance just as much sulfite is oxidized per unit of time in the scrubber sump as in desulfurization. Thus, despite constant deposits of new sulfite ions, the composition of the suspension in the scrubber sump remains the same due to the simultaneous and equally constant degree of oxidation in the sump.

The occurring reactions can be represented by the following gross equation:

$$SO_2 \text{ (gas)} + CaCO_3 \text{ (solid)} + \tfrac{1}{2} O_2 \text{ (from air)} + 2H_2O$$
$$\rightarrow CaSO_4 \cdot 2H_2O + CO_2. \tag{7.62}$$

Apart from the mentioned co-reactants the suspension also contains other substances such as Cl^-, Mg^{2+}, clayish components, iron and traces of other heavy metals which enter the system either by way of the flue gas or the sorbent.

Process diagram

Fig. 7.45 shows the diagram of a modern limestone flue gas desulfurization prlant with integrated oxidation. Plants working on this principle are in operation in new German power plants [98, 99].

The raw gas cooled down by the gas/gas regenerative heat exchanger flows through the spray tower absorber in a counterflow. To avoid moisture oversaturation in the stack the pure gas is reheated in the heat exchanger to approx. 90 °C. The degrees of efficiency of the scrubber amount to 92-98 %, so that the 400 mg SO_2/m^3 prescribed in the Ordinance for Large-Scale Furnaces and other more restrictive requirements pertaining to the desulfurization of the total flue gas stream can generally be observed without difficulty.

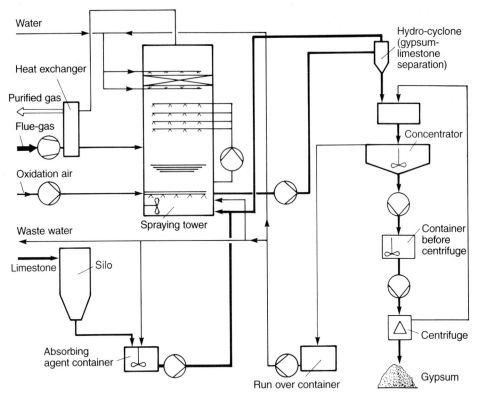

Fig. 7.45. Diagram of a limestone flue gas desulfurization process with integrated oxidation [98, 99]

In the absorber the scrubbing liquid is conducted in a cycle. Finely ground limestone is taken from the silo and prepared with water (waste water) as a suspension which is fed continuously into the scrubber sump. On the other hand, gypsum suspension is removed from the scrubber. The separation of relatively large gypsum crystals for further thickening and processing is, e.g., carried out in a hydrocyclone [99]. Fine grains of gypsum which cannot be dehydrated so well are fed back into the absorber cycle as "seed crystals".

The excess of *waste water* which is removed from the cycle must be specially treated due to its ion and metal content.

Table 7.4 shows the concentrations of components of untreated waste water from flue gas desulfurization plants based on lime scrubbing in comparison to achievable residual concentrations after being treated in waste water purification plants.

Minimum requirements pertaining to waste water quality in flue gas desulfurization plants are being worked out within the framework of the Water Resources Act [101]. The treatment of flue gas desulfurization (FGD) waste water can, e.g., be carried out by way of chemical conditioning, separation, recirculation, ester interchange, evaporation crystallization and recovery of useful material [199]. In this way waste water volume can be considerably reduced, and in some processes even com-

Table 7.4. Ranges of concentrations of effluent compounds in flue gas desulfurization plants (FDP) [100]

	FDP effluent untreated	FDP effluent treated
	Concentrations in mg/l	
COR[a]	0.05 ... 1[b]	<80
Chloride	5 ...200[b]	5...200[b]
Sulfate	0.8 ... 5[b]	0.8...2[b]
Fluoride	10 ...150	<30
Iron	10 ... 50	< 1
Cadmium	nd ... 0.2	< 0.05
Mercury	nd ... 1.0	< 0.05
Lead	0.1 ... 3.0	< 0.1
Chrome	nd ... 2	< 0.5
Copper	0.2 ... 2	< 0.5
Nickel	0.2 ... 4	< 0.5
Zinc	2.0 ... 4	< 1.0

[a] chemical oxygen requirements
[b] $\cdot 10^3$
nd = non detectable

pletely prevented. E.g., the useable residual materials of vaporization can be removed and reused or be disposed of as minimized residues [102].

Applications of the final product gypsum

Gypsum has manifold applications in the building materials industry. The volume of the German gypsum market amounts to 3-5 million tons per annum. Being a substitute for natural gypsum, FGD gypsum is in competition with it, a competition made

Table 7.5. Gypsum specifications of the National Association of the Gypsum and Gypsum Wallboard Industry and values of a flue gas desulfurization gypsum (Acc. to [103], [104])

Compound	Requirement	Values obtained
	Mass Proportion in %	
Residual Moisture	<10	7...10
$CaSO_4 \cdot 2H_2O$	>95	98...99
$CaSO_3 \cdot 0.5H_2O$	< 0.5	0... 0.5
MgO	< 0.1	<0.1
Na_2O	< 0.06	0.01
Cl^-	< 0.01	<0.01
pH	5...9	–
Color, degree of whiteness	>80	–
Odor	neutral	–

keener by the depletion of existing deposits and the difficulties of developing new ones. For the gypsum used in the building material industry, there are certain standards of quality required which can generally be easily fulfilled by FGD gypsum, as is to be seen in Table 7.5.

Inert impurities have the same effect on FGD gypsum as on natural gypsum; reducing the degree of purity up to approx. 80 % by inert impurities can be advantageous.

The differences in chemical composition and in trace element content between natural gypsum and FGD gypsum are irrelevant from a health point of view. Analytical results permit the assessment that both the FGD gypsums examined as well as natural gypsums can be used in the production of building materials without fearing a health hazard [105].

7.4.3.2 Other Wet Flue Gas Desulfurization Processes

For the desulfurization of large-scale furnaces lime scrubbing processes with the final product gypsum are offered by different producers and in various constructions [98, 99, 106-113].

There are also other wet flue gas desulfurization processes differing in the materials used and in their final products.

E.g., in the Walther Process SO_2 is bound to ammonium sulfite $(NH_4)_2SO_4$ with ammonia water. The ammonium sulfite is oxidized with air to become ammonium sulfate, $(NH_4)_2SO_4$, which can be used as fertilizer [114].

Regenerative flue gas desulfurization is, e.g., accomplished with the *Wellman-Lord Process* [115, 116]. The process uses a concentrated, circulating sodium sulfite solution as desulfurization medium which absorbs the SO_2 in a scrubbing tower from the flue gases while forming sodium acid sulfite. During the thermal regeneration of the circulatory solution the SO_2 is desorbed again and the sodium sulfite is reconverted at the same time:

$$Na_2SO_3 + SO_2 + H_2O \underset{\text{Regeneration}}{\overset{\text{Absorption}}{\rightleftharpoons}} 2Na\,HSO_3. \tag{7.63}$$

Finally, the SO_2 is on hand in a highly concentrated form and can, e.g., be converted in the Claus Process to elementary sulfur, sulfuric acid or to liquid SO_2. Side products are sodium sulfate and sodium thiosulfate pentahydrate [116].

The regenerative Wellman-Lord Process is particularly advantageous in the following cases:
- high SO_2 content in the flue gas and high degrees of desulfurization,
- when use of gypsum as final product is not possible,
- in power plants of chemical works where elementary sulfur, sulfuric acid or liquid SO_2 can be directly reused for other purposes. E.g., the process is integrated in BASF power plants in this manner [117].

The *Citrate Process* [118] is also desulfurization of the regenerative type. Sodium citrate which is fed to a scrubbber in dissolved form serves as absorbing agent. This saline solution absorbs the SO_2 of the flue gas, forming citric acid and sodium hydrogensulfite in the process. In thermal regeneration elementary sulfur and sodium citrate are formed when H_2S is added.

Various other flue gas desulfurization processes are available, and used particularly in small and medium-sized furnaces. Table 7.3 and special literature offer further information on this subject; overviews are to be found in [106, 117, 119].

7.5 Diagram of the Flue Gas Purification Units of a Power Plant

Fig. 7.46 shows the diagram of the lay-out of a modern coal power plant (750 MW_{el}) with the flue gas purification units with SCR catalytic converter for the removal of nitrogen oxides, electrostatic particle precipitator and flue gas desulfurization equipment [58, 120, 121].

After leaving the steam boiler the flue gases first reach the SCR catalytic converter. Subsequently, cooling takes place in the air preheater, then particle removal in the electrostatic precipitator. Further cooling of the flue gases is required for wet flue gas desulfurization (FGD). The purified flue gases are reheated by the heat released before being conducted to the stack. Wet flue gas desulfurization has an additional dedusting effect, so that the particle content of the flue gases is very low.

It is obvious that plants for the treatment of 2.3 million m^3 of flue gases per hour take on enormous dimensions. The volume streams and flue gas composition of this power plant are shown in Fig. 7.47.

7.6 Present Situation of Flue Gas Purification in West German Power Plants

Owing to the strict regulations governing air quality control in the FRG, particularly the Ordinance for Large-Scale Furnaces (s. also chap. 8) in effect since 1983, West German power plants have consistently been equipped with units for the removal of sulfur and nitrogen oxides from flue gases. Power plant furnaces had already been equipped with particle collectors even earlier.

The development of SO_2 and NO_x emissions resulting from these air quality control measures and the corresponding emission reduction in public electric power industry are shown in Fig. 7.48 for the period from 1982 to 1990. In 1990 159 flue gas desulfurization plants at 72 locations were in operation, removing approx. 90 % of the sulfur dioxide contained in flue gases. Despite the growing electricity generation from coal-fired power plants this program was able to reduce SO_2 emissions from a total of 1.55 million tons in 1982 to 0.18 million tons in 1989 [122].

The program for the reduction of nitrogen oxides in West German power plants was concluded at the end of 1991; 131 catalytic converter units have been taken into

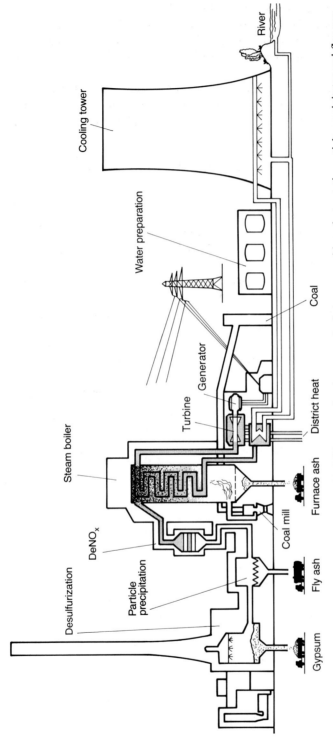

Fig. 7.46. Coal-fired 700 MW power plant block with SCR system for the removal of nitrogen oxides, electrostatic particle precipitator and flue gas desulfurization plant ($2.3 \cdot 10^6$ m^3/h flue gas) [120]

7.7 Removal of Organic Substances from Flue Gases

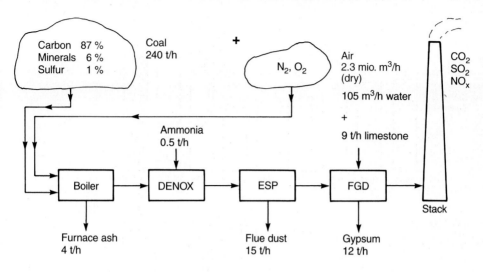

Fig. 7.47. Mass flows and flue gas composition of the 700 MW power plant block shown in Fig. 7.46 [120]

operation to remove nitrogen oxides at 60 locations. In some of the power plants NO_x emissions could be lowered by primary measures. Thus, NO_x emissions of the public electric utilities were reduced from 0.74 million tons per annum in 1982 to 0.24 million tons per annum in 1990. This is analogous to a reduction of 67.6 % (s. Fig. 7.48).

7.7 Removal of Organic Substances from Flue Gases

Air quality measures aim at preventing emissions of organic compounds in the first place. Possibilities supporting this were touched upon in sect. 7.1. There are many cases, however, where measures at the source alone do not suffice and removal of organic substances from the flue gases becomes necessary. To this aim, the following sections will describe some fundamental processes [123].

7.7.1 Overview of the Processes

An overview of the processes for the removal of organic substances from flue gases appears in Table 7.6 [124]. Which processes are best for the removal of certain com-

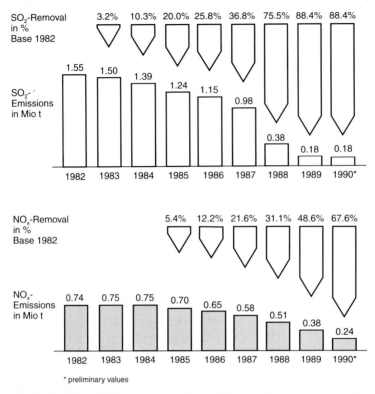

Fig. 7.48. SO_2 and NO_x reduction of the public electric power plants of the Federal Republic of Germany from 1982 to 1990 [122]

ponents is described in detail in the comprehensive work of Baum [125] and in the VDI guideline 2280 [126].

7.7.2 Condensation

In the condensation process vapors are cooled down below the dew point and are thus made to pass into the liquid state. The lower the temperature and the higher the vapor pressure, the lower the vapor concentration remains after its condensation. Because of this one tries, if possible, to condense under pressure to increase the purifying effect. Condensation can be classified into direct or indirect cooling. If, during cooling, temperatures reach dew point or fall below it, the condensate, e.g., in indirect cooling condenses on the cooling surfaces or falls in small droplets in counterflow. The condensate is separated from the flue gas stream and collected. The flue gas returns into the cycle of the process or is removed partly or wholly and, if necessary, enters one of the purification processes described below.

The condensation process is generally used only for pre-separating and recovering organic substances [125].

Table 7.6. Process techniques for the precipitation, recovery or transformation of organic pollutants from exhaust gases [124]

Process	Limitations, Resulting Problems
Condensation	Limited concentrations
Absorption – scrubbing	
physical	
chemical	Regeneration, waste water
biological	Construction size, waste water
Adsorption	
physical	Regeneration
chemical (chemisorption)	
Incineration/oxidation	
thermal	Energy consumption, exhaust gases: HCl, SO_x, NO_x
catalytic	exhaust gases, contamination of catalytic converters
Diaphragm Processes	High energy consumption, limited concentrations

Examples of applications:
- tri- or perchloroethylene in degreasing plants,
- methylene chloride in foil industry,
- carbon disulfide in the viscose industry.

7.7.3 Absorption

Absorption is the complete or selective molecular diffusion (sorption) of gases and vapors and their accumulation in scrubbing liquids, whereby gas solubility can be effected by physical (absorption) or chemical forces (chemisorption) which become active between gas and absorbing medium molecules [127, 128]. Furthermore, there is biological gas scrubbing; in which the absorbed substances are transformed by microorganisms.

Physical absorption is generally reversible, i.e., organic substances can be retrieved (by subsequent desorption). In chemisorption and biological scrubbing, however, mostly irreversible substance transformations take place.

The degree of efficiency of absorption plants basically depends on the type and concentration of the compounds to be removed and the individual process conditions such as temperature and pressure, residence time, the ratio of gas-liquid throughput, vapor pressure and solubility of the component to be removed. Different scrubber types are used as absorbers, whereby one aims at creating a boundary surface layer which is as large as possible (for scrubber types and producers of scrubbers s. [125, 129]; a clear overview of processes and equipment appear in [130]).

7.7.3.1 Physical Absorption

For absorption scrubbing liquids are used which chemically react only negligeably or not at all with the organic compounds, but which, however, attract the latter selectively and solve them physically. Maximum absorption capacity (loading) of the scrubbing liquid is expressed by Henry's law at low concentrations:

$$c_L = L p_i, \tag{7.64}$$

where
- c_L is the concentration or the molar part of dissolved components in the scrubbing liquid in mol l^{-1}
- L is the specific solubility constant which drops with higher temperatures in mol l^{-1} bar^{-1} and
- p_i is the partial pressure of the component i in the gaseous phase in bar.

In German literature the Henry constant $H = \frac{1}{L}$ is commonly used.

Thus, the gas volume absorbed by a volume unit of the fluid at a given temperature is proportional to the partial pressure of the gas component remaining in an undissolved state above the fluid.

The substance flow m moving from gas to fluid depends on the difference between the gas concentration c_G and the concentration at the interface surface c^* between gas and fluid, the overall mass transfer coefficient β and the interface surface A:

$$m = \beta A (c_G - c^*), \tag{7.65}$$

where
- m is the mass flow rate of gas into the liquid in g/s,
- β is the overall mass transfer coefficient relative to the gaseous phase in m/s,
- A is the boundary surface layer in m^2,
- c_G is the concentration in the gaseous phase, e.g., in g/m^3 and
- c^* is the equilibrium concentration between gas and fluid, e.g., in g/m^3.

The two phases – gas to be purified and scrubbing liquid – are brought into contact with each other while in motion, generally counterflow motion. In this way, a concentration gradient can be maintained as a driving force owing to the constant removal of loaded scrubbing liquid.

Fig. 7.49 shows an example of a scrubbing plant in the frequently used shape of a scrubbing tower filled with packings.

The waste air loaded with organic solvents is, e.g., conducted through the absorber from the bottom to the top. At the top, the scrubbing liquid is sprayed and trickles towards the waste air while passing over the packing. The loaded scrubbing liquid collected at the bottom of the absorber is pumped off from there, heated and transported to the regenerator. There, the substances washed out are separated from the scrubbing liquid by distillation. The separation is carried out at a higher temperature than during scrubbing. The dissolved solvents escape from the heated absorption

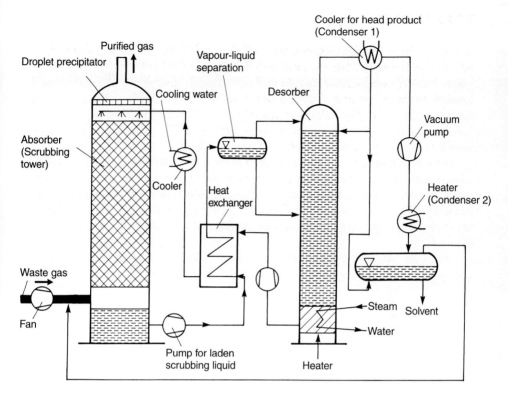

Fig. 7.49. Installation of physical absorption system with solvent recovery (acc. to [126 and 131])

liquid in a vacuum, are condensed on the outside and are available again in pure form. The desorbed absorption liquid is cooled in the heat exchanger and fed back into the absorber [131].

Possibilities of application [125, 126, 129-131]

Physical absorption can be used wherever high quality requirements are to be met by the recovered solvent, where complex and water-soluble solvent mixtures are used and whenever solvent loads are high. The process can be applied to almost all commercially available solvents such as aliphatic compounds, aromatic compounds, halogenated hydrocarbons, alcohols, ketones, esters, glycols a.s.o. [131].

According to [131] plants operating on the principle of physical absorption are characterized by the following advantages:
– high operational safety,
– small floor space requirements,
– waste water-free operation in most cases,
– minimized energy consumption.

7.7.3.2 Chemical Absorption

In this process a chemical reaction outweighs physical absorption. The substances to be removed from the waste gas react with the scrubbing liquid to form nonvolatile or barely volatile compounds [128]. In this way, there is always a high concentration gradient between gas and scrubbing liquid and the mass transfer takes place very fast. In addition, the physical absorption is promoted by chemical forces. Thus, very low concentrations can be attained in the exhaust gas. The scrubbing liquid is consumed and in many cases cannot be regenerated further.

Organic compounds reacting acidically are washed out with alkaline solutions, compounds with alkaline reactions are washed out with acidic solutions. However, mainly oxidational scrubbing processes (oxidizing gas scrubbing) are used for low-reactivity organic substances. The following oxidizing agents are used in scrubbers: ozone, hydrogen peroxide, potassium permanganate, chlorine, chlorine dioxide and hypochlorite.

There are many different types of scrubbers [125, 129] which have the task of creating a large boundary surface layer just as in physical absorption.

Possibilities of application

Removal of odor-intensive waste air:
- oxidation of amines, mercaptans, hydrogen sulfide, aldehydes, phenols a.o. with ozone or hypochlorite solution,
- washing out of organic acids, phenols, mercaptans with caustic lye of soda or caustic potash lye,
- washing out of pyridine and similar nitrogenous and alkaline organic substances with diluted sulfuric acid.

The process is, e.g., used for the purification of waste gases from foundries, industrial coffee roasters, coating (in carpeting), urea production, formaldehyde production, driers (sewage sludge), animal carcass utilization a.s.o. [129].

The scrubbing water is mostly used in cycles; however, in most cases it is necessary to treat the scrubbing water removed from the process.

7.7.3.3 Biological Waste Air Purification – Bioscrubbing and Biofiltration

Biological waste air purification has been described in detail by Fischer et al. [132]. In *bioscrubbing* the raw gas is brought into contact with scrubbing water where the components of the waste air are absorbed. Regeneration of the scrubbing liquid is carried out by biological degradation with microorganisms as in the waste water technology of bio-aeration plants. The microorganisms used for the regeneration of the scrubbing liquid either attach themselves permanently on the scrubber packings as a bacterial film (biological filtering process) or they are suspended in the shape of activated sludge in the absorbent [132]. The principles governing gas absorption as explained in sect. 7.7.3 also apply to the bioscrubber.

In the *biofilter* the pollutant-laden waste air flows through biologically active material, is absorbed there and subsequently transformed by the microorganisms settled there. The biological material – e.g., compost – must be moist. To this extent this is one type of pollutant absorption in liquid films with subsequent biological transformation. Adsorption processes on solid substances can also play a role (this is a question of definition, s. sect. 7.7.4). The inorganic nutrients necessary for microorganic life, such as nitrogen, phosphorus, potassium and trace elements are normally to be found in the filter material in adequate amounts.

According to [132, 133] the following preconditions must be fulfilled to be able to use biological waste air purification:
- waste-air components must be water-soluble,
- waste-air components must be biodegradable,
- waste-air temperature should be between 5 and 60 °C,
- the waste air must be free of substances which are in any way toxic or otherwise harmful to the microorganisms,
- the waste air must be moist (biofilter).

If possible, oxidation should take place as completely as possible so that only CO_2 and H_2O remain. Organic compounds containing, e.g., chlorine, bromine, nitrogen or sulfur can also be degraded. However, non-volatile substances are formed in the process and gradually accumulate [133].

The temperature range must remain limited as most microorganisms are not viable at high temperatures. If temperatures are too low, degradation becomes too slow. The optimum temperature range is between 10 and 40 °C.

Fig. 7.50 shows the layout of a bioscrubber with an activated-sludge tank where most of the degradation of the absorbed waste-air components takes place. Apart from the biological filter process and activated-sludge process there are also other biological methods for the regeneration of scrubbing liquids, e.g., a biocatalytic process [131].

The basic layout of a biofilter is shown in Fig. 7.51. Biofilters have a number of process variations [132, 133]:
- surface filters with filter heights of between 0.5 and 1.5 m,
- tower filters with filter heights of up to 6 m and mechanical circulation of the filter material,
- portable container filters,
- multi-layered biofilters to save floor space.

Examples of application for biological waste air purification [132]

- Foundry waste air from core-making departments with odor-intensive substances: phenols, formaldehyde, amines, ammonia etc.,
- waste air containing formaldehyde from the hardboard industry,
- waste air from paint shops, plastics processing, adhesive production a.s.o.,
- waste air from various food processing plants,

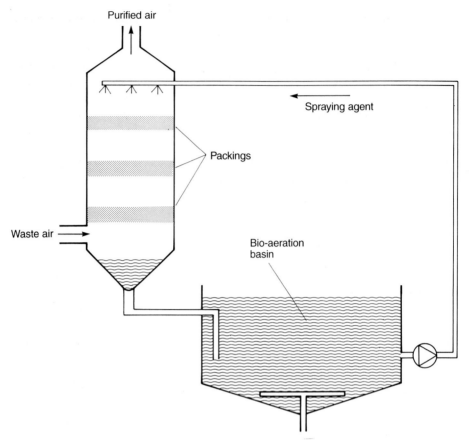

Fig. 7.50. Principle of a bioscrubber acc. to the activated-sludge process [132, 133]

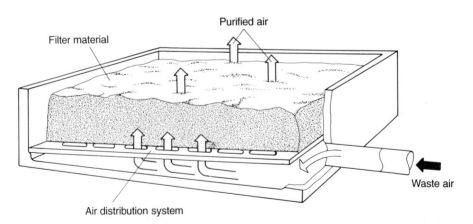

Fig. 7.51. Set-up of a biofilter [133]

- odor-intensive waste air from animal carcass disposal plants, bone processing, dung drying, waste recovery plants (compost works), sewage treatment plants a.s.o..

Biological waste-air purification offers the following advantages [132, 133]:
- in the most favorable cases the organic waste-air components are degraded to CO_2 and H_2O,
- it has no other adverse effects on the environment, as additional chemicals are unnecessary,
- capital costs and especially operating costs are low,
- biofilters are especially suited to filtering small amounts of organic substances out of large volumes of waste air. However, this requires relatively large building volumes as filter loads cannot be very high.

7.7.4 Adsorption

Adsorption is one of the most important processes for the removal of gaseous pollutants from exhaust gases, waste air or even ambient air, if pure air quality must meet special standards, e.g., in clean-room technology. The principles and technical processes of adsorption from the gaseous phase have been described in detail by Kast [134] and in the VDI guideline 3674 [135]. This literature is recommended for more intensive study. An overview of the practical application of adsorption processes including information on plant manufacturers is given by Baum in [125].

Essentially, adsorption is the accumulation of gases on unsaturated-valence solid boundary layers. The larger the surface of a solid adsorption agent, the more efficient it is. The inner surface of the best known adsorption agent, charcoal, is 500 – 1500 m^2/g. In addition to the adsorption of condensable gases and vapors on porous substances there is, from a certain surface pressure on, capillary condensation as the adhesive forces in the capillaries of the adsorbing agent effect the condensation. During capillary condensation heat is released, which means that it is an exothermic process. Finally, accumulation can be reinforced by chemisorption for which impregnated charcoal is used.

Some terms will be clarified [135] before the adsorption process is treated in more detail:

adsorbate	substance to be adsorbed,
adsorbent	solid substance on whose surface adsorption takes place,
(also adsorbing agent)	
desorption	reverse adsorption process, endothermic/endothermal??,
desorbate	substance removed during desorption.

According to [136] the process of adsorption takes place in the following steps:
1. substance transport through the boundary layer around the adsorbent grain in the flow,
2. substance transport into the pores of the particle by gas diffusion,

3. adsorption (and condensation) with simultaneous heat release in the particle,
4. adsorption heat transport in the particle,
5. heat transport through the boundary layer to the surrounding gas.

During desorption the steps are followed in reversed order.

Apart from charcoal other adsorbing agents such as aluminum oxide, silica gel, and zeolitic and other molecular sieves are used [134, 135, 137].

For optimum adsorption the choice of a suitable adsorbing agent is decisive, but so are further parameters such as temperature, molecular weight, concentration and boiling point of the adsorbate, relative humidity and the presence of other attending substances capable of adsorption and leading to mixed and displacement adsorption [135, 137]. Dust and aerosols can cover up the active surfaces of the adsorbing agents and must therefore be separated beforehand.

The most important aid for the characterization of the sorption equilibrium between different gases and adsorbing agents are adsorption isotherms. The curves show the dependency of the loading of the adsorbing agent (adsorbent) over the partial pressure saturation vapor pressure ratio (proportional to the concentration of the gaseous pollutant to be removed) for a certain, constant temperature. Adsorption isotherms are generally determined by experiments, but they can also be calculated thermodynamically [134]. With the help of adsorption isotherms the best adsorbing agent can be chosen for the waste air problem at hand, s. Fig. 7.52.

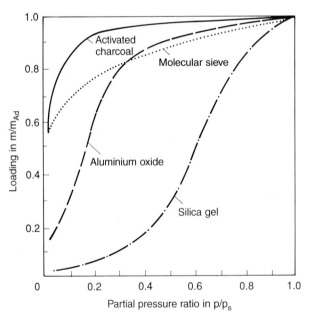

Fig. 7.52. Adsorption isotherms of benzene on different adsorbents [125, 134]. m/m_{Ad} is charge of the adsorber: adsorbed mass related to the mass of the adsorber; p/p_s relative saturation of the adsorbate: relation of the partial pressure of the substance to be removed to its saturation vapor pressure in the carrier gas (proportional to the concentration)

Curve types 1 and 3 are optimum for attaining low pure gas concentrations, as the load of the adsorbent initially increases very fast when concentrations are raised slightly. These processes are based on higher interaction forces between adsorbing agent and adsorbate than in curve 4. It is only at high concentrations that the adsorbing agent behaving according to curve 4 adsorbs appreciable amounts of benzene. Therefore low pure gas concentrations can hardly be achieved with this adsorbing agent with reasonable adsorber sizes. While charcoal has the best adsorption properties for organic substances, as for benzene here, curve profiles are reversed for other gases under certain circumstances. In low concentrations water vapor, e.g., is adsorbed much better by silica gel than by activated charcoal [125, 135]. In both cases, for benzene as well as for water, molecular sieves have excellent adsorptive properties.

Adsorption isotherms are set up for equilibrium conditions, i.e., the initial concentration before the adsorbing agent corresponds to the exit concentration after the adsorbing agent. In practice, there will be no constant concentrations over an adsorbent bed in the flow, but the initial portion of the bed will be loaded with a high concentration and will therefore adsorb a lot (higher load). In the further course of the bed the gas concentration decreases and along with it the loading. During the constant intake of the adsorbate there comes a point when the adsorbing agent is saturated, first of all at the initial portion of the bed. With continued supply this saturation continues on through the bed until finally the adsorbate can no longer be adsorbed, and a breakthrough has been reached. The determination of "breakthrough curves" is a helpful means for the practical interpretation of adsorbers [135]. The following example includes such breaktrough curves.

The diagram in Fig. 7.53 shows the adsorptive process with recovery of solvent mixtures.

First of all, the waste gas is sent through the adsorber (A) filled with activated charcoal. If there is a breakthrough after a certain period of time (s. breakthrough curves in Fig. 7.53) and a set threshold concentration is exceeded, which is, e.g., monitored by a flame ionization detector (FID), the waste gas is conducted through a second adsorber (B) and the first one is regenerated with water vapor as flushing medium (temperature 100-110 °C). The desorbate vapors which consists of water vapor and solvent vapors and come from the activated charcoal are partially condensed. Uncondensed water vapor expels the solvents from the hot condensate into the evaporation tank. The solvent vapors expelled are condensed together with the non-separated water vapor, upon which a low-water solvent phase and a low-solvent aqueous phase are formed. The latter flows back to the evaporation tank. Depending on its intended use, the solvent phase must, if necessary, be dehydrated further by distillation. The recovered volume of such a process is between 95 and 99 %.

Fields of application

The activated charcoal adsorption process is, e.g., used for the following applications in air quality control:
- solvent recovery in the solvent-processing industry, e.g., after paint driers [137-142],

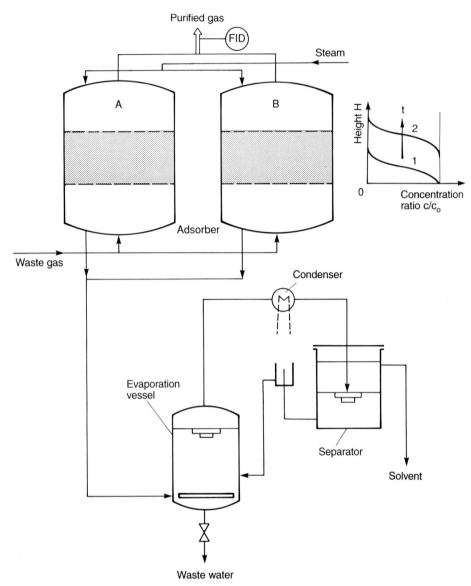

Fig. 7.53. Diagrammatic set-up for removal and recovery of solvents by adsorption (acc. to [129, 130] with breakthrough curves through the adsorber. *1* is concentration distribution in the absorber acc. to time t_1; *2* concentration distribution in the absorber acc. to time t_2

- removal and recovery of gasoline depot vapors [129, 138],
- removal and recovery of halogenated hydrocarbons in degreasing plants and dry cleaners [125, 143],
- purification of waste air containing odorous substances [125, 144].

Problems in solvent recovery can occur when waste air concentrations are very low. This is the case, e.g., in spraying booths for automotive painting.

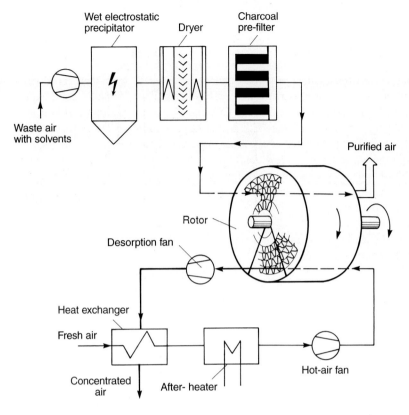

Fig. 7.54. Activated charcoal-rotor system for the concentration of waste air containing solvents [145, 146]

To make solvent recovery nevertheless possible, a concentration boost unit precedes the activated charcoal plant. To this end rotating adsorbers are used of late. For this, fiber-like activated charcoal is applied to fibrous papers which are combined to form element blocks. Fig. 7.54 shows a rotor plant for the reconcentration of solvents from spraying booth air. Adsorption takes place in one section of the rotor, which turns slowly, while desorption is carried out, e.g., with hot air, in another section of the rotor. The desorptive air stream is much smaller than the waste air stream, so that the solvents are present in concentrated form in the desorption air. For their recovery they are conducted to an activated charcoal plant as shown in Fig. 7.53.

Owing to the moving rotor, adsorption/desorption takes place in a continuous process. To protect the activated charcoal of the rotor the waste air is pre-cleaned with a dust collector, and an activated charcoal filter is included to even out the concentration.

7.7.5 Combustion

7.7.5.1 Thermal Afterburning

If, for whatever reasons, other processes cannot be applied, then there is always combustion as a waste gas purification process. As most organic compounds (containing C, H and O) can be oxidized to CO_2 and H_2O at temperatures of between 750 and 1000 °C. The problem generally is that large amounts of air must be freed from small amounts of pollutants, which means that the concentrations of the organic substances in the waste air are so low that they are far below the auto-ignition levels, meaning that they do not burn by themselves but only when energy is added. The degree of conversion and combustion are dependent on:
- mixing of the reactants,
- combustion temperature,
- retention time at this temperature,
- type of the compound to be oxidized.

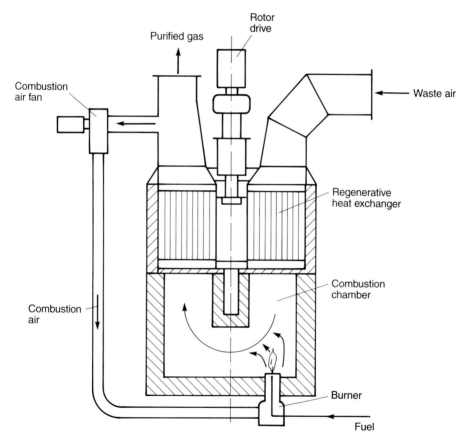

Fig. 7.55. Example of a thermal afterburning plant with integrated heat exchanger [129, 148]

The temperatures and retention times required are chosen according to the type of compound to be burned (activation energy). Combustion-inhibiting substances such as halogenated hydrocarbons can require higher combustion temperatures. As the compounds are frequently complex the necessary temperatures and retention times cannot usually be calculated in advance but must be determined experimentally.

Organic compounds containing, apart from C, H and O, other elements such as N, S, P, halogens or metals, emit undesirable reaction products when burned, e.g., halogen hydrogen, nitrogen oxides and sulfur oxides. Substances such as these can be formed additionally depending on the supplemental fuel used.

As an example Fig. 7.55 shows function and constructional layout of a thermal afterburning plant with heat recovery. The waste gas is conducted through a regenerative heat exchanger and then pre-heated indirectly by the hot pure gas. Then, oxidation of the organic components takes place in the combustion chamber at temperatures around 800 °C. To avoid a further increase of the gas volumes the burner is operated with as little excess air as possible. The need for additional fuel (fuel oil, gas, waste solvents or waste oil) is reduced due to the combustible substances in the waste gas.

If large amounts of additional fuel must be added one cannot do without further heat recovery for reasons of economy. For this, there are the following possibilities:
- pre-heating of the combustion air,
- heating up of the waste gases to be purified,
- heating up of heat carriers.

The principle of thermal afterburning and the requirements to be met have been described in detail by Baum [125] and in the VDI guideline 2242 [147]. Afterburners are available from several producers [125, 129].

7.7.5.2 Catalytic Afterburning

Energy requirements for afterburning can be reduced when catalysts are used. Noble metals (Pt, Pd) on metallic carriers or metal oxides on oxidic ceramic carriers are used as catalysts. The principles and applications of catalytic afterburning have been described in more detail by Schmidt [159] and Baum [125]. Kick-off temperature of the catalysts is at 300-500 °C. The layout of catalytic afterburning plants basically does not differ much from thermal afterburners [149]. Before entering the catalytic converter the waste gas is pre-heated by mixing with hot flue gases or by heat exchange.

As an example Fig. 7.56 shows the layout of a plant for the catalytic afterburning of waste gases.

In this case, the waste gas is pre-heated in a gas or oil furnace to the reaction temperature, then reaches the vertical uptake which houses the catalyst elements. There it is burned catalytically, flamelessly and completely. In this example, the purified waste air releases a part of its energy content in a subsequent heat exchanger to the still unpurified waste gas. The economy of this process depends mainly on the type of heat utilization (reentrainment into the process; other heat users). Under cer-

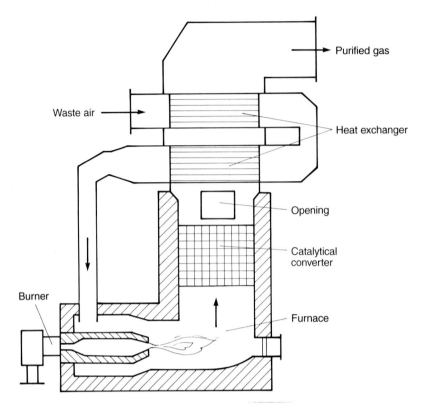

Fig. 7.56. Example of a catalytic afterburning plant (Lurgi, Frankurt, FRG)

tain circumstances (at sufficiently high concentrations of organic compounds) and if implemented optimally the process can even be operated autothermally, i.e., without additional energy [124, 150, 151].

The advantage of catalytic as compared to thermal afterburning lies without a doubt in the lower operating cost, due to the lower temperature. The process cannot be run at all or only at higher costs for reactivation measures if catalyzer poisons such as halogen or metal compounds are present in the waste gases to be purified.

7.7.6 Membrane Processes

A relatively new process for waste air purification has been introduced by the use of membranes. Their effect is based on the ability of certain membrane materials to be selectively permeable to different gases or vapors. This causes a separating effect which, in the ideal case, strictly selects the components.

The principles and possibilities of applications of the membrane process have been described in more detail in [152-155].

In gas separation by membrane (diagram s. Fig. 7.57) the waste air stream is conducted past the membrane where a partial stream branches off and flows through the

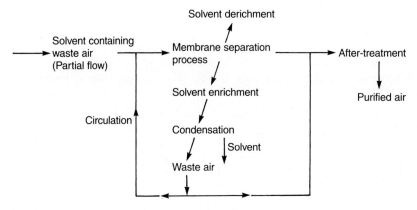

Fig. 7.57. Simplified procedure for solvent recovery by membrane separation

membrane. Those substances which can permeate the membrane faster accumulate beyond the membrane . For this membranes are used through which vaporous solvents can permeate considerably faster than air components. This partial waste air stream loaded with solvent is conducted to a condensation unit and stripped there. The solvent is collected in liquid form. Depending on the degree of seperation the partial waste air stream can then be mixed with the so far untreated partial waste air stream or can be refed into the membrane process (closed circuit).

If the total waste air stream were conducted through the membrane, nothing would be gained, as no stripped waste air stream would remain; the condensation units would have to be designed for the entire waste air stream as before. So it is essential to find an optimum ratio between total waste air stream and the partial waste air stream conducted through the membrane.

The remaining partial stream stripped of the permeated substance will have to undergo a secondary treatment (if the stripping was insufficient). This can be one of the processes described before.

Fields of application

Gas separation with membranes has already proved its efficiency in several fields [153]:
- bio-gas/landfill-gas recovery (separation of methane CH_4 and carbon dioxide CO_2 for the recovering CH_4),
- natural gas processing (removal of carbon dioxide CO_2, hydrogen sulfide H_2S and water vapor H_2O),
- economical use of special oil recovery processes,
- hydrogen recovery and accumulation,
- nitrogen and oxygen production by splitting air,
- recovery of hazardous materials from gasoline depots,
- recovery of solvents from foundry waste air.

Except for recovering gasoline vapors from gasoline depots plants in this field have not gotten beyond the stage of pilot plants. Basically, however, nothing stands in the way of a broader application in this field [153]. The membrane process is also being tested in biological waste air purification. Here the "membrane reactor" makes solvents with poor water-solubility or highly volatile solvents accessible to biological degradation [132, 133].

Additional investments and operating costs are the disadvantages of membrane processes. Depending on the quality of the membrane separating process of gas permeation the stripped waste air stream must go through conventional purification processes. Whether the load placed on subsequent processes is lightened by the use of membrane technology remains to be tested in each case under real conditions.

7.8 Bibliography

1. Stern, A.L.: Fundamentals of Air Pollution. London: Academic Press 1984
2. Ondratschek, D.: Luftreinhaltung beim Spritzlackieren. Vortrag am 24.5.1989, Institut für industrielle Fertigung und Fabrikbetrieb der Universität Stuttgart
3. Wiesner, J.: Umweltfreundliche Technik, Verfahrensbeispiele Chemie. Broschüre aus dem DECHEMA-Lehrprogramm „Chemie und Umwelt", Frankfurt 1978
4. Sander, U.; Rothe, U.; Kola, R.: Schwefelsäure. Ullmann Encyklopädie der technischen Chemie, Bd. 21, S. 117–165, Weinheim: VCH-Verlagsgesellschaft 1982
5. Struschka, M.; Straub, D.; Baumbach, G.: Schadstoff-Emissionen von Kleinfeuerungen. Fortschr. Ber. VDI Z. Reihe 15, Nr. 60, Düsseldorf: VDI 1988
6. Baumbach, G.; Breuninger, H.A.: Untersuchungen zur Wirksamkeit von Additiven für schweres Heizöl. VDI Ber. Nr. 498, Düsseldorf (1983) 209–214
7. Hoenig, V.; Baumbach, G.: Schadstoffminderung bei Schwerölfeuerungen durch Additive: Ruß, SO_3, NO_x. VGB-Tagung Kraftwerk und Umwelt 1989, Tagungsband, VGB-Kraftwerkstechnik GmbH, Essen 1989
8. Dritte Verordnung zur Durchführung des Bundes-Immissionsschutzgesetzes (Verordnung über Schwefelgehalt von leichtem Heizöl und Dieselkraftstoff) vom 15.1.1975 (BGBl. I S. 264 f), geändert durch Gesetz vom 18.2.1986 (BGBl. I S. 265 f) und durch Verordnung vom 14.12.1987 (BGBl. I S. 2 671 f)
9. Deutsche BP (Hrsg.): Das Buch vom Erdöl. Hamburg: Reuter und Klöckner 1989
10. Forck, B.; Lange, G.: Systemanalyse Entschwefelungsverfahren, Teil B, Bd. 1 „Brennstoff-Entschwefelung". VGB Technische Vereinigung der Großkraftwerksbetreiber, Essen 1974
11. 13. Verordnung zur Durchführung des Bundes-Immissionsschutzgesetzes (Verordnung über Großfeuerungsanlagen) vom 22.6.1983, BGBl. I S. 719 f
12. Erste Allgemeine Verwaltungsvorschrift zum Bundes-Immissionsschutzgesetz (Technische Anleitung zur Reinhaltung der Luft – TA Luft) vom 27.2.1986, GMBl. S. 95–202
13. Batel, W.: Entstaubungstechnik – Grundlagen, Verfahren, Meßwesen. Berlin: Springer 1972
14. Weber, E.; Brocke, W.: Apparate und Verfahren der industriellen Gasreinigung, Bd. 1: Feststoffabscheidung. München: R. Oldenbourg 1973
15. Maschinenfabrik Beth: Beth-Handbuch der Staubtechnik. 2. Aufl. Lübeck 1964
16. Löffler, F.: Staubabscheiden. Lehrbuchreihe Chemieingenieurwesen und Verfahrenstechnik. Stuttgart: Thieme 1988
17. Meldau, R.: Handbuch der Staubtechnik Bd. 1. Düsseldorf: VDI 1956
18. Knop, W.; Heller, A.; Lahmann, E.: Technik der Luftreinhaltung. Mainz: Krausskopf 1972
19. VDI: Richtlinie VDI 3676, Massenkraftabscheider. Berlin: Beuth 1980
20. Muschelknautz, E.: Auslegung von Zyklonabscheidern in der technischen Praxis. Staub-Reinhalt. Luft 30 (1970) 187–195

7.8 Bibliography

21 Schmidt, K.R.: Physikalische Grundlagen und Prinzip des Drehströmungsentstaubers. Staub-Reinhalt. Luft 23 (1963) 491–501
22 Klein, H.: Entwicklung und Leistungsgrenzen des Drehströmungsentstaubers. Staub-Reinhalt. Luft 23 (1963) 501–509
23 Schmidt, K.G.: Naßwaschgeräte aus der Sicht des Betriebsmannes. Staub-Reinhalt. Luft 24 (1964) 485–491
24 Weber, E.: Stand und Ziel der Grundlagenforschung bei der Naßentstaubung. Staub-Reinhalt. Luft 29 (1969) 272–277
25 VDI: Richtlinie VDI 3679, Naßarbeitende Abscheider. Berlin: Beuth 1980
26 Glowiak, B.; Kabsch, P.: Verfahren zur Bestimmung der Benetzbarkeit von Stäuben. Staub-Reinhalt. Luft 32 (1972) 9–11
27 White, H.J.: Entstaubung industrieller Gase mit Elektrofiltern (deutsche Fassung). Leipzig: VEB Deutscher Verlag für Grundstoffindustrie 1969
28 VDI: Richtlinie VDI 3678, Elektrische Abscheider. Berlin: Beuth 1980
29 Hesselbrock, H.: Physikalische Vorgänge im Elektrofilter. Mitteilungen der VGB (1960) H. 64, S. 13–26
30 Lurgi GmbH: Entstaubungstechnik. Firmendruckschrift. Frankfurt 1969
31 Quack, R.: Probleme der elektrischen Abgasentstaubung. DECHEMA-Monographien, Bd. 64, 1970
32 Kercher, H.: Elektrischer Wind, Rücksprühen und Staubwiderstand im Elektrofilter. Fortschr. Ber. VDI Z. Reihe 6, Nr. 27, Düsseldorf: VDI 1970
33 Plato, H.: Untersuchungen zur Staubabscheidung im Elektrofilter. Dissertation Universität Stuttgart 1969
34 Gross, H.: Zur Auswirkung der Turbulenz in elektrischen Abscheidern. In: Technik der Wärmekraftwerke. DFG Deutsche Forschungsgemeinshaft, Hrsg. von R. Quack und J. Wachter, Weinheim: VCH Verlagsgesellschaft 1987
35 Gomoll, H.J.: Verbesserung der Abscheideleistung durch SO_3-Konditionierung. VDI-Ber. Nr. 495 (1984) 219–222
36 Brandt, H.: Entstaubungseinrichtungen für Rauchgase industrieller Anlagen. Energie 15 (1963) H. 11, S. 177–216
37 König, W.: Zum Verhalten des Staubes im Elektrofilter. Dissertation Universität Stuttgart 1985
38 Deutsch, W.: Bewegung und Ladung der Elektrizitätsträger im Zylinderkondensator. Ann. Phys. 68 (1922) 335–344
39 Gross, H.: Messung der Durchbruchspannung in einem Elektroabscheider. Staub-Reinhalt. Luft 41 (1981) 458–460
40 Braun, W.: Stand und Entwicklungstendenzen der elektrischen Rauchgasentstaubung. VDI-Ber. Nr. 495 (1984) 223–228
41 Gross, H.: Messung des Fraktionsabscheidegrades von Elektroabscheidern mit Impaktoren. Staub-Reinhalt. Luft 41 (1981) 461–465
42 Löffler, F.; Dietrich, H.; Flatt, W.: Staubabscheidung mit Schlauchfiltern und Taschenfiltern. Braunschweig: Vieweg & Sohn 1984
43 VDI: Richtlinie VDI 3677, Filternde Abscheider. Berlin: Beuth 1980
44 Rennert, K.D.: Möglichkeiten der Stickstoffoxidreduzierung in Feuerräumen. Fachreport Rauchgasreinigung, Sonderteil der VDI-Zeitschriften BWK und Umwelt. Düsseldorf: VDI 1986
45 Käß, M.; Brodbek, H.; Spiegelhalder, R.; Pfau, B.: Einflüsse von Ausmahlung und Lufteintrittsbedingungen auf die Strömungsverhältnisse, die NO_x-Bildung und den Ausbrand in Kohlestaubflammen. 4. TECFLAM-Seminar, DLR-Stuttgart 1988
46 Diehl, H.: Entwicklung von Primärmaßnahmen zur NO_x-Minderung aus Kesselfeuerungen. Daimler-Benz AG, Werk Sindelfingen, Abt. HKW, persönliche Mitteilung 1985
47 Hoenig, V.: Untersuchungen zur Stickstoffoxidminderung durch Rauchgasrezirkulation bei der Verbrennung von Heizöl EL und Heizöl S. Diplomarbeit am Institut für Verfahrenstechnik und Dampfkesselwesen der Universität Stuttgart 1984
48 Deutsche Babcock Werke AG: Der ASR-Brenner, Versuchs- und Betriebsergebnisse. Firmendruckschrift, Oberhausen 1985

49 Hurst, B.E.: Exxon Thermal DeNO$_x$ for Stationary Combustion Sources. In: Air Pollution by Nitrogen Oxides, ed. Schneider, T.; Grant, L. . Amsterdam: Elsevier 1982
50 Warnatz, J.: Elementarreaktionen in Verbrennungsprozessen. BWK 37 (1985) Nr. 1—2, S. 11—19
51 Mittelbach, G.; Voje, H.: Anwendungen des SNCR-Verfahrens hinter einer Zyklonfeuerung. In: VGB-Handbuch NO$_x$-Minderung bei Dampferzeugern für fossile Brennstoffe, VGB-B-301, VGB-Kraftwerkstechnik GmbH, Essen 1986
52 Just, Th.; Kelm, S.: Mechanismus der NO$_x$-Entstehung und Minderung bei technischer Verbrennung. if-Die Industriefeuerung, Folge 38 (1986) 96—102
53 Lyon, R.K.: Thermal DeNO$_x$. Environmental Science and Technology, Vol. 21 (1987) No. 3, S. 231—236
54 Arand, J.K.; Muzio, L.J.: Urea Reduction of NO$_x$ in Combustion Effluents. US Patent #4, 208, 386, 1980
55 Fuel Tech GmbH: NO$_x$ Out — Das neue Verfahren. Firmenprospekt und Verfahrensbeschreibung, Eschborn 1988
56 L. & C. Steinmüller GmbH: Minderung der NO$_x$-Emission, Sekundärmaßnahmen. Firmenprospekt P 8711-10-04/2. L, Gummersbach 1987
57 Weber, E.; Hübner, K.: Ergebnisse reaktionskinetischer Untersuchungen zur katalytischen NO$_x$-Reduktion mit Ammoniak. In: VGB-Handbuch NO$_x$-Bildung und NO$_x$-Minderung bei Dampferzeugern für fossile Brennstoffe, VGB-B 301, VGB-Kraftwerkstechnik GmbH, Essen 1986
58 Schönbucher, B.; Fritz, P.: Auslegung, Anordnung und Funktion der DENOX-Anlage für Block 7 im Kraftwerk Heilbronn. VGB-Kraftwerkstechnik Nr. 3, 1987
59 Marnet, Chr.; Kassebohm, B.: SCR-Versuchsanlagen hinter einer Schmelzkammerfeuerung mit Flugstaubrückführung. In: VGB-Handbuch NO$_x$-Bildung und NO$_x$-Minderung bei Dampferzeugern für fossile Brennstoffe, VGB-B 301, VGB-Kraftwerkstechnik GmbH, Essen 1986
60 Koenig, J.; Derichs, W.; Hein, K.: Versuche zur SCR-Technik für Braunkohlenfeuerungen. In: VGB-Handbuch NO$_x$-Bildung und NO$_x$-Minderung bei Dampferzeugern für fossile Brennstoffe, VGB-B 301, VGB-Kraftwerkstechnik GmbH, Essen 1986
61 Das Mannesmann/Steuler Verfahren zur Minderung von NO$_x$-Emissionen aus Kraftwerken. Firmenprospekt Nr. 3376, Mannesmann Anlagenbau AG, Düsseldorf, und Steuler, Höhr-Grenzhausen 1987
62 Poller, J.: Versuchsanlage zur katalytischen Minderung von Stickstoffoxid-Emissionen nach dem Mannesmann-Steuler-Verfahren hinter einem steinkohlebefeuerten Schmelzkessel im VEW-Kraftwerk Westfalen. In: VGB-Handbuch NO$_x$-Minderung bei Dampferzeugern für fossile Brennstoffe, VGB-B 301, VGB-Kraftwerkstechnik GmbH, Essen 1986
63 Jüntgen, H.: Abgasseitige Minderungsmaßnahmen für NO$_x$ — Stand der Entwicklung, VDI-Ber. Nr. 495 (1984) 127—132
64 Uhde: Anlagen zur Rauchgasreinigung, Verfahren Bergbauforschung/Uhde. Firmenprospekt, Uhde GmbH, Dortmund 1983
65 VGB Technische Vereinigung der Großkraftwerksbetreiber e.V.: VGB-Handbuch NO$_x$-Bildung und NO$_x$-Minderung bei Dampferzeugern für fossile Brennstoffe. VGB-B 301, VGB-Kraftwerkstechnik GmbH, Essen 1986
66 Asmuth, P.; Dettmann, P.: Aktivkohle-Katalysator hinter REA. In: VGB-Handbuch NO$_x$-Bildung und NO$_x$-Minderung bei Dampferzeugern für fossile Brennstoffe. VGB-301, VGB-Kraftwerkstechnik GmbH, Essen 1986
67 Weisweiler, W.; Hochstein, B.: Umweltverträgliche Katalysatoren zur Entstickung. Staub-Reinhalt. Luft 49 (1989) Nr. 2, S. 37—43
68 Hellstern, R.: NO$_x$-Minderung an Feuerungsanlagen mit Katalysatoren. Diplomarbeit Nr. 2197 am Institut für Verfahrenstechnik und Dampfkesselwesen der Universität Stuttgart 1986
69 Jordan, S.; Paur, H.-R.; Schikarski, W.: Simultane Entschwefelung und Entstickung mit dem Elektronenstrahlverfahren. Physik in unserer Zeit 19 (1988) Nr. 1, S. 8—15
70 Bussien, R. (Begr.); Goldbeck, G. (Hrsg.): Automobiltechnisches Handbuch. Ergänzungsband zur 18. Aufl. Berlin: de Gruyter 1978
71 Robert Bosch GmbH: Kraftfahrtechnisches Taschenbuch. 20. Aufl. . Düsseldorf: VDI 1987

72 Beier, R. u.a.: Verdrängungsmaschinen, Teil II Hubkolbenmotoren. Handbuchreihe Energie. München: Technischer Verlag Resch und Köln: TÜV Rheinland 1983
73 Berendes, H.; Eickhoff, H.: Entwicklung eines Regenerationssystems für Abgaspartikelfilter bei Dieselmotoren, Verbrennung und Feuerungen — 14. Dt. Flammentag. VDI-Ber. 765, S. 559—568. Düsseldorf: VDI 1989
74 Bernhardt, W.; Heitland, H.: Katalytische Abgasreinigung. Beitrag in [70], S. 1 296—1 311
75 Buck, D.: Der Abgaskatalysator — Aufbau, Funktion und Wirkung. Schriftenreihe der Adam Opel AG 42, Rüsselsheim 1984
76 Johnson Matthey Chemicals Limited: Kleine Katalysator-Kunde, Firmendruckschrift, Köln 1988
77 Ulrich, J.G.: Der Katalysator. Ingenieure berichten zum Thema Auto und Umwelt, Druckschrift der Fa. Dr. Ing. h.c. F. Porsche AG, Stuttgart 1985
78 Firma J. Eberspächer: Abgasreinigung—Katalysator—Rußfilter, Firmendruckschrift, Esslingen 1985
79 Robert Bosch GmbH (Hrsg.): Autoelektrik, Autoelektronik am Ottomotor. Düsseldorf: VDI 1987
80 Lies, K.-H.; Schulze, J.; Winneke, H.; Kuhler, M.; Kraft, J.; Hartung, A.; Postulka, A.; Gring, H.; Schröter, D.: Nicht limitierte Automobil-Abgaskomponenten. Volkswagen AG, Forschung und Entwicklung, Wolfsburg 1988
81 Goldschmidt, K.: Versuche zur Entschwefelung von Rauchgasen mit Weißkalkhydrat und Dolomitkalkhydrat bei Öl- und Kohlenstaub-Feuerungen. Fortschr.-Ber. VDI-Z., Reihe 6, Nr. 21 1968
82 Zentgraf, K.-M.: Beitrag zur SO_2-Messung in Rauchgasen und zur Rauchgasentschwefelung mit Verbindungen der Erdalkalimetalle. Fortschr.-Ber. VDI-Z., Reihe 3, Nr. 22, 1967
83 Michele, H.: Rauchgasreinigung mit trockenen Sorbentien — Möglichkeiten und Grenzen. Chem.-Ing. Tech. 56 (1984) Nr. 11, S. 819—829
84 Wildner, R.: Entstaubung und Entschwefelung kleiner bis mittelgroßer Kesselanlagen. Fachreport Rauchgasreinigung, Sonderteil der Zeitschriften BWK Brennstoff—Wärme—Kraft, Nr. 1/2 — 1986, und Umwelt, Nr. 1 — 1986
85 Hlubek, W.; Hein, K.: SO_2-Emissionsminderung bei Braunkohlefeuerungen — ein Beitrag zum Umweltschutz. VGB Kraftwerkstechnik 63 (1983), H. 4, S. 327—331
86 Reh, L.: Neue großtechnische Anwendungen des Reaktionsprinzips der zirkulierenden Wirbelschicht. Chem.-Ing.-Tech. 56 (1984) Nr. 3, S. 197—202
87 Chughtai, M.Y.; Michelfelder, S.: Schadstoffeinbindung durch Additiveinblasung um die Flamme. Brennstoff—Wärme—Kraft 35 (1983) Nr. 3, S. 75—83
88 Wickert, K.: Versuche zur Entschwefelung vor und hinter dem Brenner zur Verringerung des SO_2-Auswurfes. Mitteilungen der VGB (1963) Nr. 83, S. 74—82
89 Weisweiler, W.; Hoffmann, R.; Stein, R.: Trockene Rauchgasentschwefelung mit Kalkstein oder Dolomit: Verbesserung der Entschwefelungswirkung durch Optimierung der Feststoff-Porenstruktur, Pelletierung und chemische Aktivierung. Kernforschungszentrum Karlsruhe, Forschungsbericht KfK-PEF 31, Oktober 1987
90 L. & C. Steinmüller GmbH: Rauchgasreinigung. Firmendruckschrift P 8906-14-10/1. L, Gummersbach 1989
91 Bengtsson, S.; Ahman, S.; Kletsch, W.: Das Fläkt Drypac-Verfahren. Firmendruckschrift, Fläkt AG, Växjo, Schweden 1983
92 Schojan, N.: Alkali/CDAS, zwei Rauchgasentschwefelungsverfahren von Fläkt für kohlegefeuerte Kesselanlagen im Leistungsbereich 1—50 MW_{th}. Firmenbericht, Fläkt Industrieanlagen GmbH, Butzbach 1987
93 Babcock-BSH AG: Trockensorption von SO_2, HCl, HF aus Rauchgasen. Firmendruckschrift, Krefeld 1983
94 Gude, K.E.; Andreassen, J.: Das Sprühabsorptionsverfahren zur Rauchgasentschwefelung. Firmenbericht, Niro Atomizer, Söborg Dänemark 1984
95 Forck, B.: Entschwefelung von Rauchgasen — Stand und Entwicklung. VDI-Ber. Nr. 495, S. 67—72, Düsseldorf 1984
96 Voos, H.: Rauchgasreinigung — angewandte Verfahren und Entwicklungen. Technische Mitteilungen 78 (1985) H. 10, S. 498—504

97 Hüller, R.; Judersleben, P.; Hemming, H.: Betriebserfahrungen mit dem KRC/EVT Entschwefelungsverfahren. VDI-Ber. Nr. 495, S. 77–81, Düsseldorf 1984
98 Voos, H.; Makinejad; Mohn: Aktuelles zum Thema Rauchgasentschwefelung. L. & C. Steinmüller GmbH, Firmenbroschüre, Gummersbach, August 1983
99 Deutsche Babcock Anlagen AG: Rauchgasentschwefelung. Firmendruckschrift 33/6/87, Krefeld 1987
100 Grünewald, K.-G.; Clausen, J.: Abwasser aus Rauchgasentschwefelungsanlagen. Vorträge der VGB-Konferenz Kraftwerk und Umwelt 1989, S. 300–305, VGB-Kraftwerkstechnik GmbH, Essen 1989
101 Sieth, J.: Abwasser aus Rauchgasendreinigungsanlagen. Techn. Mitteilungen 78 (1985) H. 1/2, S. 71–73
102 Dettmann, P.; Luther-Goldmann, K.; Otterstetter, H.: Erfahrungen mit der Eindampfung von REA-Abwasser und Weiterbehandlung der Eindampfrückstände. Vorträge der VGB-Konferenz Kraftwerk und Umwelt 1989, S. 298–299, VGB-Kraftwerkstechnik GmbH, Essen 1989
103 Bundesverband der Gips- und Gipsbauplattenindustrie e.V.: Der Überschuß – Rauchgasgips – ein Substitutionsrohstoff für Naturgips? Energie 36 (1984), Nr. 6, S. 40–48
104 Täubert, U.: Verwertungskonzept für Reststoffe aus Kohlekraftwerken. VGB-Kraftwerkstechnik 68 (1988), H. 11, S. 1 172–1 179
105 Beckert, J.: Vergleich von Naturgips und REA-Gips. Vorträge der VGB-Konferenz Kraftwerk und Umwelt 1989, S. 271–274, VGB-Kraftwerkstechnik GmbH, Essen 1989
106 Baumüller, F.: Entschwefelung – Überblick über die Entschwefelungsverfahren. Anwenderreport Rauchgasreinigung der VDI-Zeitschriften, Staub-Reinhaltung der Luft und Brennstoff-Wärme-Kraft, Düsseldorf 1985
107 Leimkühler, J.; Weißert, H.: Betriebserfahrungen mit der Rauchgasentschwefelung Wilhelmshafen, Firmendruckschrift, Fa. Bischoff GmbH & Co. KG, Essen 1984
108 Herzog, G.W.; Kofler, G.; Priemer, H.; Veiter, E.: Das RCE-Entschwefelungsverfahren. Radex-Rundschau, H. 3 (1983) S. 260–281 (Österreichisch-Amerikanische Magnesit AG, Radentheim/Kärnten)
109 Uhde GmbH: Anlagen zur Rauchgasentschwefelung – Kobe Steel Verfahren. Firmendruckschrift BV I7 19 1000 83, Dortmund 1983
110 Thyssen Engineering GmbH: Entschwefelungsanlagen. Firmendruckschrift IWG 03474, Essen
111 S-H-L Saarberg-Hölter-Lurgi GmbH: Rauchgasentschwefelung und Abgasreinigung. Firmendruckschrift, Saarbrücken 1983
112 Hamm, H.; Müller, R.: Das zweistufige Knauf-Research-Cottrell-Verfahren zur Rauchgasentschwefelung am Beispiel des Kraftwerkes Franken. Zement-Kalk-Gips International 35 (1982) H. 6, S. 313–317 und Firmenprospekt KRC-EVT, Würzburg und Stuttgart
113 GEA Energietechnik GmbH & Co.: Abgasentschwefelung mit dem GEA CT-121 Verfahren. Bochum 18988
114 Walther Umwelttechnik: Rauchgasentschwefelung – Dünger aus Gas. Firmendruckschrift, Köln 1984
115 Neumann, U.: Regenerative Rauchgasentschwefelung nach dem Wellman-Lord-Verfahren. Chemie-Technik 12 (1983) H. 7, S. 21–24
116 Wahl, D.-J.; Rahm, J.; Grimm, H.H.: Erfahrungen mit der Rauchgasentschwefelung nach dem Wellman-Lord-Verfahren. Vorträge der VGB-Konferenz Kraftwerk und Umwelt 1989, S. 249–253, VGB-Kraftwerkstechnik GmbH, Essen 1989
117 Lange, M.: Fortschreibung von Regelwerken zur Emissionsbegrenzung bei Feuerungsanlagen. Fachveranstaltung „Aktuelle Fragen zur Rauchgasendreinigung" des Hauses der Technik, Essen, Vortragsveröffentlichungen 500, Essen: Vulkan 1985
118 Mannesmann Anlagenbau: Regenerativverfahren für die Entschwefelung von Rauchgasen. wlb „wasser, luft und betrieb" 3/86, S. 34–39
119 Haasis, H.-D.; Remmers, J.; Schons, G.; Rentz, O.: Ergebnisse eines Landesprogrammes über die Minderung von SO_2- und NO_x-Emissionen bei kleinen und mittleren Anlagen der Industrie. Die Industriefeuerung 40, Essen: Vulkan 1987
120 Energie-Versorgung Schwaben AG: Heizkraftwerk Heilbronn. Druckschrift, Stuttgart 5/89

7.8 Bibliography

121 Schönbucher, B.; Poensgen, Th.; Fahlenkamp, H.: Aufbau und Funktion der Rauchgasentschwefelungsanlage für Block 7 im Kraftwerk Heilbronn. VGB Kraftwerkstechnik 1987, Heft 3
122 IZE – Informationszentrale der Elektrizitätswirtschaft e.V.: Energiewirtschaft kurz und bündig. Frankfurt 1991
123 Umweltbundesamt: Luftreinhaltung '88 – Tendenzen – Probleme – Lösungen. Materialien zum 4. Immissionsschutzbericht der Bundesregierung an den Deutschen Bundestag, Berlin: Erich Schmidt 1989
124 Eigenberger, G.: Abluftreinigung – Schadgase und Gerüche. In: Arbeitsgruppe Luftreinhaltung der Universität Stuttgart, Jahresbericht 1988
125 Buonicore, A.J.; Davis, W.T. (Eds., Air & Waste Management Association): Air Pollution Engineering Manual. Van Nostrand Reinhold New York, 1992
126 VDI-Richtlinie 2280, Entwurf: Emissionsminderung – Flüchtige organische Verbindungen, insbesondere Lösemittel. Berlin: Beuth 1985
127 VDI-Richtlinie 3675 E: Abgasreinigung durch Absorption. Berlin: Beuth Mai 1981
128 VDI-Richtlinie 2443 E: Abgasreinigung durch oxidierende Gaswäsche. Berlin: Beuth Januar 1980
129 Umweltbundesamt (Hrsg): Handbuch Abscheidung gasförmiger Luftverunreinigungen. UMPLIS Information- und Dokumentationssystem Umwelt. Berlin: Erich Schmidt 1981
130 KT Kunststofftechnik KG: KT-Anlagen zur Luftreinhaltung – Gase, Dämpfe und Gerüche. Firmendruckschrift, Troisdorf 1985
131 Penzel, U.: Lösemittelhaltige Abluft – Entsorgung und Wertstoffrückgewinnung durch physikalische Absorption mit Desorption im Vergleich zum bio-catalytischen Abbau. VDI-Kolloquium „Aktuelle Probleme der Abluftreinigung und ihre Lösungswege", Düsseldorf 12./13.4.1989, Tagungsbericht, Düsseldorf: VDI
132 Fischer, K. et al.: Biologische Abluftreinigung. Kontakt & Studium, Band 212, Ehningen bei Böblingen Expert 1990
133 Fischer, K.: Biologische Reinigung lösemittelhaltiger Abluft. In: ALS Arbeitsgruppe Luftreinhaltung der Universität Stuttgart, Jahresbericht 1989
134 Kast, W.: Adsorption aus der Gasphase – Ingenieur-wissenschaftliche Grundlagen und technische Verfahren. Weinheim: VCH Verlagsgesellschaft 1988
135 VDI-Richtlinie 3674 E: Abgasreinigung durch Adsorption – Oberflächenreaktion und heterogene Katalyse. Berlin: Beuth Juni 1981
136 Kast, W.; Otten, W.: Der Durchbruch in Adsorptions-Festbetten: Methoden der Berechnung und Einfluß der Verfahrensparameter. Chem.-Ing.-Tech. 59 (1987) 8, S. 1–12
137 Krill, H.: Adsorption organischer Stoffe an Aktivkohle. Staub-Reinhalt. Luft 36 (1976) 7, S. 298–302
138 Lurgi: Anwendung von Aktivkohle zur Luftreinhaltung. Lurgi Information T 1 117, 8, 1–5, Frankfurt 1974
139 Kuschel, H.: Adsorptive Entfernung eines Lösemittelgemisches aus der Abluft eines folienverarbeitenden Betriebes. Staub-Reinhalt. Luft 36 (1976) 7, S. 303–306
140 Rotamill Maschinenbau GmbH: Aktivkohleanlagen zur Lösemittel-Rückgewinnung. Firmenprospekt, Siegen 1988
141 Davy Bamag GmbH: Entfernung und Rückgewinnung von flüchtigen Lösemitteln aus Abgas mittels Aktivkohle. Technische Informationen, Butzbach 1979
142 Günter, H.-G.: Industrielle Abluftreinigung nach dem Verfahren der Adsorption an Aktivkohle. Arbeitsbericht 24, Dürr Anlagenbau, Stuttgart 1982
143 Zimmermann, M.: Aktivkohle-Anlagen – Ihre Bedeutung im Einsatz mit Chemisch-Reinigungsmaschinen zur Rückgewinnung organischer Lösemittel aus der Luft. Wäscherei- und Reingungspraxis, Hefte 5 und 6/68
144 Bräuer, H.W.: Minderung der Geruchsstoff-Emissionen aus Lackierbetrieben durch adsorptive Reinigung; VDI-Berichte 416 (1982) 127–132
145 Grashof, J.: Erfahrungen mit einer Abluftreinigungsanlage an der Karosseriespritzkabine. Bericht, 1. Deutscher Automobilkreis, Porsche AG, Stuttgart 1987
146 Jeckel, A.: Untersuchungen zur Lösemittelabscheidung aus Spritzkabinenabluft. Diplomarbeit am Institut für Verfahrenstechnik und Dampfkesselwesen der Universität Stuttgart 1988
147 VDI-Richtlinie 2442: Abgasreinigung durch thermische Verbrennung. Berlin: Beuth Juni 1987

148 Kraftanlagen AG Heidelberg: TNV-Abhitzeanlage. Firmeninformation
149 VDI-Richtlinie 2441E: Katalytische Nachverbrennung. Berlin: Beuth Juli 1975
150 Schmidt, T.: Grundlagen und Anwendungen der katalytischen Nachverbrennung in der lösemittelverarbeitenden Industrie. VDI-Kolloquium „Aktuelle Probleme der Abluftreinigung und ihre Lösungswege", Düsseldorf 12./13.4.1989, Tagungsbericht, Düsseldorf: VDI
151 Nieken, U.: Katalytische Abluftreinigung mit autothermer Reaktionsführung. In: ALS Arbeitsgruppe Luftreinhaltung der Universität Stuttgart, Jahresbericht 1989
152 Strathmann, H.: Trennung von molekularen Mischungen mit Hilfe synthetischer Membranen. Darmstadt: Steinkopff 1979
153 Egli, S.; Ruf, A.; Buck, A.: Gastrennung mittels Membranen. Ein Überblick. Swiss Chem 6 (1984) Nr. 9, S. 3−24
154 Paul, H.; Dahm, W.; Rautenbach, R.; Strathmann, H.: Lösemittelrückgewinnung mit Hilfe von Membranen. Wasser, Luft und Betrieb (1988) Nr. 5, S. 39−45
155 Ripke, C.: Planung einer mobilen Meß- und Testanlage mit Gaspermeationssystem zur Rückgewinnung von Lösemitteln aus Abluft. Diplomarbeit, Fachhochschule für Technik, Esslingen 1989

8 Air Pollution Control Conventions, Laws and Regulations

The structure of air pollution control regulations on the international level is shown in Fig. 8.1.

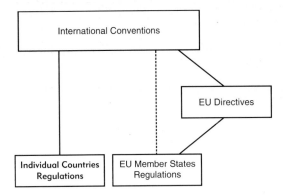

Fig. 8.1. Air pollution regulations within the international framework

8.1 International Conventions

As air pollutants do not refrain from crossing borders and as some effects of air pollution are of a global nature, e.g., the greenhouse effect of some gases or the depletion of the stratospheric ozone layer, it has been acknowledged the world over that international conventions are necessary to limit certain emissions. The first "Convention on Long-range Transboundary Air Pollution" was concluded in 1979 [1]. Further agreements followed, an overview of which is given in Table 8.1. The signatory states committed themselves to converting their assurances into national laws or regulations within a certain time frame.

8.2 EU Directives

The directives of the European Union are also meant to be a harmonization of the air pollution control regulations, at least in the European member states. In actual fact, important directive proposals have served to trigger off regulations in some EU

Table 8.1. Important international conventions for air pollution control

No.	Convention	Location	Year	Ref.
1	Convention on Long-range Transboundary Air Pollution	Geneva	Nov. 13, 1979	[1]
2	Protocol on long-term financing of the cooperative program for monitoring and evaluation of the long-range transmission of air pollutants in Europe (EMEP)	Geneva	Sept. 28, 1984	[2]
3	Convention for the Protection of the Ozone Layer	Vienna	March 22, 1985	[3]
4	Protocol on substances that deplete the ozone layer	Montreal	Sept. 16, 1987	[4]
5	Protocol on the reduction of sulfur emissions or their transboundary fluxes by at least 30 percent	Helsinki	July 8, 1985	[5]
6	Protocol concerning the control of emissions of nitrogen oxides or their transboundary fluxes	Sofia	Oct. 31, 1988	[6]
7	Protocol concerning the control of emissions of volatile organic compounds (VOC) or their transboundary fluxes	Geneva	Nov. 19, 1991	[7]
8	Convention on Climate Change	Rio de Janeiro	May 9, 1992	[8]

member countries which did not have any laws concerning the issues in the first place. In the member states the EU directives passed must be converted into national laws or regulations within a certain period and, in individual cases, they may contain requirements more stringent than the original directive, or requirements applying specifically to the partner state concerned. Also existing national law may be adapted to EU directives. If the directives are not converted within the agreed time-frame, then they will themselves become valid as regulations in the country concerned. This saves the country the legislative procedure, but the corresponding implementation regulations have to be passed nonetheless.

Table 8.2 lists some EU directives on emission restriction in industrial plants and on the introduction of air quality control standards. The regulations for vehicular emissions are not included in this table, they are treated separately in Chap. 8.4.

8.3 National Laws and Regulations

The following sections will give an overview of the laws and regulations of the two countries USA and Germany as examples of national air pollution and air quality control.

Table 8.2. Selection of EU directives for air pollution control and air quality standards in the member states (without vehicle emissions) [9]

Directive (75/716/EWG) on the sulfur content of certain liquid fuels	of Nov. 24, 1975 and of March 23, 1993	Sulfur contents of fuel oils and diesel fuel
Directive (80/779/EWG) on standards and guidelines of air quality for sulfur dioxide and suspended particulate matter	of July 15, 1980	
Directive (82/501/EWG) on severe accident hazards in certain industrial processes ("Seveso" Directive)	of Dec. 3, 1981	Regulations concerning national legislations for increasing the safety of certain industrial plants; amendment on Jan. 8, 1986
Directive (82/884/EWG) on a standard for the lead concentration in the air	of Dec. 3, 1982	Setting of a standard for the lead concentration in the air (annual mean value) of $2 \, \mu g/m^3$
Directive (84/360/EWG) for controlling air pollution caused by industrial plants ("Basic Directive Principle on Air Pollution Control")	of June 28, 1984	Introduction of a requirement of official permission for erecting, operating and modifying certain industrial plants: no harmful environmental effects, state-of-the-art preventive measures, compliance with air quality control standards
Directive (85/203/EWG) on air quality standards for nitrogen dioxide	of March 7, 1985	For the protection of human health; short-term air quality standard of $200 \, \mu g/m^3$; guidelines for more stringent regulations in special reserves: rules for measures and methods
Directive (85/210/EWG) on the lead content of gasoline	of March 20, 1985	
Directive (85/337/EWG) on the environmental impact assessment of certain public and private project plans	of June 27, 1985	Obligation to pre-examine pro-ject plans
Directive (88/609/EWG) on emission restrictions in large-scale furnaces	of Nov. 24, 1988	Emission restriction of particulate matter, SO_2, NO_x
Directive (89/369/EWG) on the prevention of air pollution emitted by incinerators for residential waste	of June 8, 1989	Emission regulations for arsenic, lead, cadmium, chrome, manganese, nickel, mercury, HCl, HF, SO_2 and particulate matter

8.3.1 US Air Pollution Laws and Regulations

In the United States the Clean Air Act of the year 1963 was the first modern environmental law enacted by the United States Congress. The Clean Air Act of 1970 reviewed and amended by Congress in 1975, 1977 and 1990 formed the basis of the federal air pollution control program of the United States. Health-based national ambient air quality standards were the strategic basis of the Clean Air Act. The standards were to be met through the application of control technologies to reduce emissions with the aim of improved air quality. Public health protection was to be given priority over cost and technological capability [10]. Fig. 8.2 gives an overview of the structure and the contents of the Clean Air Act. Fig. 8.3 gives information on the implementation of the regulations. The US Environmental Protection Agency (EPA) is responsible for implementing the Clean Air Act and for carrying out the program. There are, for instance, nationwide regulations directly concerning widespread products such as motor vehicles and gasoline or some major industries. US regulations limiting automobile emissions are described in Chap. 8.4. The EPA set up New Source Performance Standards (NSPS) that placed emissions limits on new industrial plants and on existing ones that were substantially modified. Other regulations contained in the State Implementation Plans (SIP's) are addressed to the executive authorities of the individual states. They are to observe them when, e.g., granting operating permits for industrial plants. SIP's are subject to EPA approval and if a

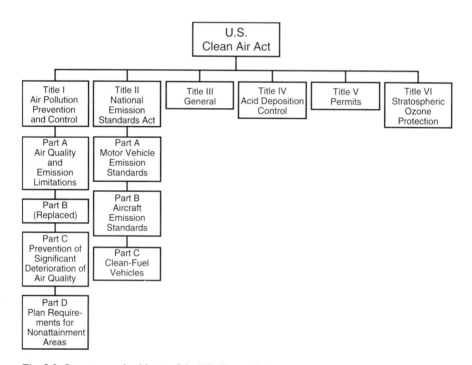

Fig. 8.2. Structure and subjects of the US Clean Air Act

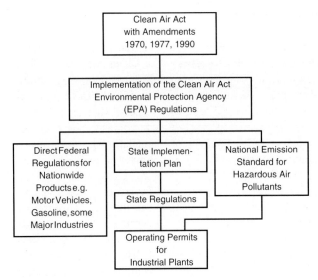

Fig. 8.3. Procedure for implementing the US Clean Air Act (from legislation to operating permits)

plan is not acceptable, the EPA is required to provide an implementation plan which the state is then required to enforce [10].

The National Ambient Air Quality Standards (NAAQS) set up by the EPA are mentioned in Table 4.6 (Chap. 4.6). The EPA is required ro review the standards every five years. Areas of the country already having high air quality have to meet prevention of significant deterioration (PSD) standards. Limitations are to be placed on new emissions in nonattainment areas [10].

8.3.2 German Laws and Regulations

The basis of the German air pollution regulations is the Bundes-Immissionsschutzgesetz (Federal Air Pollution Control Act) which was passed in 1974 [11]. Only motor vehicle exhaust gases and the lead content of gasoline are dealt with in separate laws, s. Chap. 8.4. The purpose of the Federal Air Pollution Control Act is to safeguard human beings, animals, plants and physical objects from detrimental environmental effects and from the hazards, significant disadvantages and irritations caused by industrial plants and to prevent environmentally harmful conditions in the first place.

The German Federal Air Pollution Control Act (BImSchG) consists of several parts, which can be basically divided into plant-, product- and area-related air quality control (s. Fig. 8.4). Besides this, some special areas are also dealt with. Vehicle emissions are also dealt with in the BImSchG but the details are regulated by the Road Safety Act and the EU directives, s. Chap. 8.4.

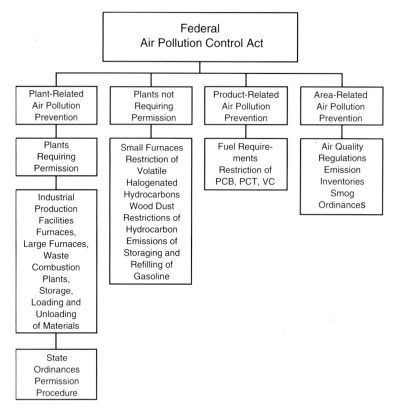

Fig. 8.4. Structure of the German Federal Air Pollution Control Act (Bundes-Immissionsschutzgesetz – BImSchG)

The BImSchG provides the framework for measures to be filled in by legal ordinances and administrative regulations. The latter are directed to administrative authorities, e.g, authorizing agencies, in order to uniformly regulate the proceedings for measures of air quality control. Thus, the law says that environment, human beings, animals, plants etc. are to be protected from harmful environmental effects, that industrial plants may not emit harmful pollutants etc.. However, what is to be considered a harmful polluting effect or an inadmissible and, seen in the context of the latest technology, avoidable emission is specified in the ordinances and administrative regulations which were passed by the federal government or the state governments after hearing the parties concerned.

8.3.2.1 Plants Requiring Official Permission

Since the last century the erection of plants whose emissions may cause significant disadvantages, hazards or irritations to residents in neighboring plots or the public in general has been permitted only with the official permission of the authority con-

Table 8.3. Types of plants requiring official permission acc. to the 4th BImSchV [12]; the individual plants only require permission when they exceed a certain size, size ranges are listed in the 4th BImSchV

Number of plant type	Type of plant	Examples
1	Heat generation, mining, energy	Power plants with fossil fuels and furnaces in general, briquette factories
2	Rocks and earths, glass, ceramics, building materials	Cement factories, glass melting plants, brickworks
3	Steel, iron and other metals including processing	Raw iron and steel production, foundries
4	Chemical products, pharmaceutics, mineral oil refineries and processing	Chemical plants for various products, fertilizer factories, detergent production, mineral oil refineries
5	Surface treatment with organic substances, production of sheet-shaped materials made of plastic, other processing of resins and plastics	Spray-painting and drying plants, printing shops
6	Wood, cellulose	Paper factories, chipboard production
7	Foodstuffs, semi-luxuries and feed, agricultural products	Mass breeding, animal carcass disposal plants, gelatin and glue production, grain mills, oil mills, sugar factories, coffee roasters, chocolate and milk powder factories
8	Utilization and disposal of residual substances	Waste incinerators, compost works, chemical processing
9	Storage, loading and unloading of products	Gas and mineral oil storage facilities and those of various chemicals
10	Others	Plants for the production of pyrotechnical products, rubber products, glue and cleaning products, dyeing and bleaching plants

cerned. The 4th ordinance for the BImSchG defines which plants require official permission and lists the plants with their individual volume ranges. Table 8.3 gives an overview of the plant types listed in the 4th BImSchV [12]. As early as 1869 a major part of today's plants required official permission according to the rulebooks of the time [13].

The federally uniform requirements which the authorizing agency must prescribe for the erection and operation of plants are contained in the Technische Anleitung zur Reinhaltung der Luft (Technical Directive on Air Pollution Control) – TA Luft [14]. An overview of the contents of the TA Luft is given in Table 8.4.

Table 8.4. Overview of the content of the German Technical Directive on Air Pollution Control (Technische Anleitung zur Reinhaltung der Luft – TA-Luft) [14]

1	Field of application	Valid for plants requiring official permission acc. to the 4th BImSchV
2	General regulations on air pollution and air quality control	
2.1	Definitions and units of measurement	Emisson standards for 3 substance classes
2.2	General principles for permission and pre-permission	
2.3	Carcinogenic substances:	Methods for calculating minimum stack heights
2.4	Releasing of flue gases	
2.5	Immission values	Fixing of air quality values for the protection of health and for the protection against considerable disadvantages and annoyances
2.6	Determination of air quality characteristics	Measuring plan for determining background concentrations and additional pollution (establishment of calculation method for pollutant dispersion – Gaussian model)
3	Restricting and measuring emissions	
3.1	General regulations for limiting emissions	Process optimization, optimization of starting and closing-down processes, taking into account flue gas dilution; standards for particulate materials and for heavy metals. Handling of goods causing particulate emissions; standards for vaporous and gaseous substances: inorganic and organic substances in 3 classes
3.2	Measuring and monitoring emissions	Measuring sites, single measurements: measurement planning, measurement methods, evaluation, measurement of odorous substances. Continuous measurements: measuring programs, measurement of particulate matter and gases, measuring equipment (certified instruments), evaluation and discussion of results, calibration and function test.
3.3	Special regulations for certain plant types	Emission standards are staggered acc. to plant size

For special types of plants requiring official permission, e.g., large furnaces (> 50 MW oil- and solid fuel-fired, > 100 MW gas-fired) and waste-combustion plants special ordinances were passed in Germany because of their importance for clean air. Emission threshold values for large furnaces are listed in Table 8.5, a selection of the most important threshold values for waste combustion plants are listed in Table 8.6.

8.3 National Laws and Regulations 469

Table 8.5. Emission standards of the Ordinance for Large-scale Furnaces [15]

Fuel	Solid fuels					Liquid fuels			Gaseous fuels	
Range of capacity	50-100	100-300	>300	up to 300	>50	50-100	100-300	>300	100-300	>300
Special characteristics				Fluidized bed furnace	Slag-tap furnace					
Pollutant mg/m³										
Particulate matter	50	50	50	50	50	50	50	50	5[a]	5[a]
Arsenic, lead, cadmium, chrome, cobalt, nickel and their compounds as particulate matter, total[b]	0.5	0.5	0.5	0.5	0.5	2[c]	2[c]	2[c]	–	–
Carbon monoxide CO	250	250	250	250	250	175	175	175	100	100
Nitrogen oxides NO+NO₂ as NO₂	800	800	800	800	1800	450	450	450	350	350
Dynamized threshold values[d] for NO+NO₂ as NO₂	400	400	200	400	no exception: 400 or 200	300	300	150	200	100
Sulfur oxides SO₂+SO₃ as SO₂	2000	2000	400	400	400	1700[e]	1700[e]	400 (650)	35 100[f] 5[g]	35 100[f] 5[g]
max. degree of sulfur emission in %	–	40	15	25	15	–	40	15	–	–
Halogen compounds										
HCl	200	200	100	200	200	30	30	30	–	–
HF	30	30	15	30	30	5	5	5	–	–

For different types of furnaces different reference O₂ concentrations in the flue gases apply.
[a] for blast furnaces 10 mg/m³ particulate matter apply, for industrial gases from steel production 100 mg/m³.
[b] Is only valid for solid fuels other than coal or wood.
[c] For fuel oils with a nickel content higher than 12 mg/kg fuel.
[d] Possibilities of reducing emissions by furnace-related or other state-of-the-the-art measures are to be fully utilized. The conference of the Ministers of the Environment of 1984 defined how far NO$_x$ reduction is feasible: these values are called dynamized threshold values.
[e] Corresponds to 1 % in fuel oil.
[f] Coking plant gas.
[g] Liquid gas.

Table 8.6. Emission standards of the 17th BImSchV for waste incineration plants (related to 11 % O_2) [16]

Pollutant	Emission standards in mg/m³		
	Daily mean value	Half-hourly mean value	90 % of all values
CO	50	100	150
Total particulate matter	10	30	-
Total organic carbon	10	20	-
Inorganic chlorine compounds as HCl	10	60	-
Inorganic fluorine compounds as HF	1	4	-
Sulfur oxides SO_2+SO_3 as SO_2	50	200	-
Nitrogen oxides $NO+NO_2$ as NO_2	200	400	-

Metals and dioxins	Mean value over sampling time
Cadmium, thalium and their compounds	0.05 mg/m³ total
Mercury and its compounds	0.05 mg/m³
Antimony, arsenic, lead, chrome, cobalt, copper, manganese, nickel, vanadium, tin and their compounds	0.5 mg/m³ total
Polychlorinated dibenzodioxins and -furans (PCDD/PCDF)	0.1 ng/m³

8.3.2.2 Plants Exempted from Official Permission

Plants not requiring official permission are mainly small plants which are not covered by the 4th BImSchV. Most of such plants are domestic furnaces. Requirements concerning these plants are laid down in the 1st ordinance for the Federal Air Pollution Control Act [17], with an excerpt given in Table 8.7. This ordinance also sets limits on the plants' energy losses. These small plants are checked by chimneysweeps once a year.

8.3.2.3 Product-Related Air Pollution Prevention

The most important ordinance in this field concerns restriction of the sulfur content of light fuel oil and diesel fuel to 0.2 % [18].

Other products whose use is regulated by a special law, the 10th BImSchG [19], are polychlorinated biphenyls (PCB), polychlorinated terphenyls (PCT) and mixtures of both and vinyl chloride (1-chloroethene; VC). PCB, PCT and VC as propellants for aerosols are practically forbidden, apart from a few exceptions, as for instance in transformers, condensors, heat exchangers and certain hydraulic plants.

The lead content of gasoline is not regulated in the Federal Air Pollution Control Act but in the gasoline lead law [20] with its ordinances and administrative regulations. According to this law, regular-grade gasoline may no longer contain any lead compounds (< 0.013 g/l) and premium-grade gasoline no more than 0.15 g/l.

8.3 National Laws and Regulations   471

Table 8.7. Selected emission standards of the 1st BImSchV for small furnaces [17]

Fuel	Light fuel oil			Mineral coal, lignite, peat	Wood, natural, in chunks or shavings, meal, bark or straw				Varnished or coated wood, plywood, chipboards or fiberboards without wood preservatives		
	Evaporation burner		Atomizer burner								
Pollutant	Capacity in kW										
	11	>11	up to 5000	15···1000	15···50	50···150	150···500	500···1000	50-100	100-500	500-1000
Particulate emissions in mg/m³	–	–	–	150	–	150	150	150	150	150	150
Blackening acc. to Bacherach (soot number)	3	2	1	–	–	–	–	–	–	–	–
CO g/m³	–	–	–	–	4	2	1	0.5	0.8	0.5	0.3
SO₂: max. sulfur content of the fuel in %	0.2	0.2	0.2	1	–	–	–	–	–	–	–

NO_x emissions are to be minimized in keeping with state-of-the-art technology. In gas furnaces only heat losses in the exhaust gas are limited.

8.3.2.4 Area-Related Air Pollution Prevention

To be able to recognize level and development of air pollution in the Federal Republic and to acquire the bases for remedial and preventive measures, the agencies authorized by state laws have the task of constantly determining, in the areas to be examined, type and extent of certain air pollutants in the atmosphere and of examining the conditions leading to their formation and dispersion.

The authorities given jurisdiction by state law must also set up emission inventories for the areas to be examined. These inventories must contain data on the type, amount, spatial and temporal distribution and discharge conditions of pollutants from certain plants and vehicles. These emission inventories are to be regularly updated.

Identification of air quality and emission inventories are to be evaluated with regard to meteorological conditions.

If the evaluation suggests that harmful environmental conditions due to air pollution have occurred or are expected to occur in the entire polluted area or in parts of it the authorities concerned are to set up a clean air plan for this area. This clean air plan contains type and extent of the air pollution measured or expected and the harmful environmental effects caused by it, statements on the causes of the air pollution and measures for the reduction of air pollution and for their prevention.

For areas needing special protection it has been determined that
1. non-stationary plants may not be erected,
2. stationary plants may not be erected,
3. non-stationary and stationary plants may only be operated at certain times or must comply with more stringent operational requirements or
4. fuels in plants may not be used or only to a restricted extent.

In addition, the state governments have been authorized to pass so-called smog ordinances for cases of low-exchange weather conditions with high-pollution situations. It has been prescribed that in highly polluted areas
1. non-stationary (e.g. traffic) or stationary plants may only be operated at certain times or
2. fuels causing air pollution to a particularly high degree may not be used in plants at all or only to a restricted extent.

Trigger criteria for the smog pre-warning and for the smog alarm were determined uniformly in the federal German states according to a model smog ordinance. These smog alarm values are listed in Table 4.6.

8.4 Vehicle Exhaust Emissions

Legislation on vehicle exhaust emission control has existed since 1964 in the United States, since 1966 in Japan and since the seventies in Europe [21]. Meanwhile exhaust emission legislation has been passed in many countries. Some of them have their own regulations but generally existing test procedures and standards have been adopted as a rule. Concerning permits for vehicles with regard to air pollutants such as CO, CO_2, hydrocarbons (HC), NO_x and particulate matter the regulations for countries of the European Community are coupled with European Economic Community (EEC) directives. German requirements concerning vehicle exhaust emissions are regulated by Section 47 of the Straßenverkehrs-Zulassungs-Ordnung (StVZO = Automobile Safety Act) with all its numerous decrees [22].

Table 8.8. Three-Stage-Plan for European Exhaust Emission Standards (Conformity of Production) for Passenger Cars [1] [23,24]

		91/441/EEC 1st Stage from 1992 [2]	93/116/EC 2nd Stage from 1996 [3]	German Proposal 3rd Stage from 1999
Gasoline	CO	3.16 g/km	2.2 g/km	1.5 g/km
	HC+NO_x	1.13 g/km	0.5 g/km	0.2 g/km
Diesel	CO	3.16 g/km	1.0 g/km	0.5 g/km
	HC+NO_x	1.13 g/km	0.7 g/km	0.5 g/km
	PM	0.18 g/km	0.08 g/km	0.04 g/km
Direct-Injecting Diesel	CO	3.16 g/km	1.0 g/km	[6]
	HC+NO_x	1.58 g/km	0.9 g/km	[6]
	PM	0.25 g/km	0.10 g/km	[6]
Evaporation [4]		2.0 g/Test	2.0 g/Test	[6]
Fuel Economy (CO_2)		–	manufacturer's standard [5]	[6]

Remarks:
[1] Exhaust emission limitations are applicable to passenger cars ≤ 6 persons (incl. driver) and ≤ 2500 kg max. The Fuel Economy limit is applicable to passenger cars ≤ 9 persons (incl. driver) and ≤ 3500 kg max.
[2] New type approval standard with effect from July 1, 1992;
New vehicle registration with effect from Dec. 31, 1992.
For direct-injecting diesel engines:
New type approval standard with effect from July 1, 1994;
New vehicle registration with effect from Dec. 31, 1994.
The exhaust emission standards correspond to ECE R 83/01 regulation from Dec. 30, 1992 for Stage B (unleaded gasoline) and Stage C (diesel fuel) respectively.
[3] New type approval standard with effect from Jan. 1, 1996;
New vehicle registration with effect from Jan. 1, 1997.
Voluntary application for CO_2 from Apr. 1, 1994.
[4] According to the SHED (Sealed Housing for Evaporative Determinations) procedure.
[5] For passenger cars up to 3500 kg including light-duty vehicles and off-road vehicles. Manufacturer's standard (= proposal of manufacturer for a certain vehicle series) may be exceeded at the type approval test for max. 4%; in series control however, the statistical average must be below the manufacturer's standard. Declaration of fuel economy in l/100 km for City-Cycle (part 1), Extra-Urban Driving Cycle, EUDC (part 2), MVEG (part1+part2).
[6] No regulations fixed so far.

Table 8.8 gives an overview of the so-called Three-Stage-Plan which is to reduce emission standards in three stages 1992, 1996 and 1999 and which lays down the emission standards to be observed for newly registered cars with Otto and Diesel engines, starting from different dates [23]. The internationally scheduled reductions of emission standards until the year 2005 are shown in Fig. 8.5. Unlike earlier EEC

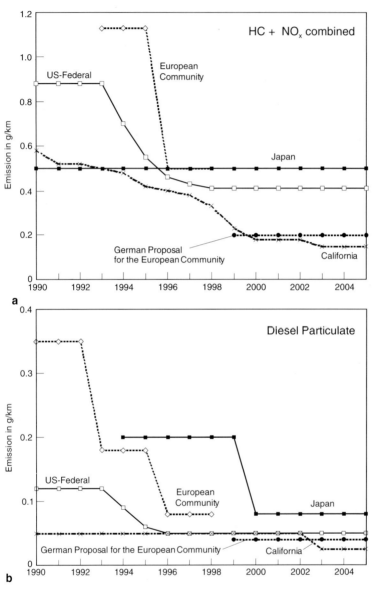

Fig. 8.5. International exhaust emission standards for passenger cars until 2005. **a.** HC plus NO_x combined; **b.** Diesel particulate [23, 26]
Note: the US, Europe and Japan use different test procedures which can influence the actual stringency of the requirements

directives where emission standards were initially classified in vehicle weight classes and later in displacement classes, there have been uniform standards since 1992 which have made European exhaust legislation much clearer. As a new aspect with effect from 1996, CO_2-emission standards will be considered. For the time being the manufacturer's proposal value will be considered valid as CO_2 emission standard; in series control, however, the statistical average value must lie below this standard value [24].

Compliance with emission standards is established by emission measurements according to defined test procedures which are based on the driving cycles mentioned in Chap. 5.4.2. Since July 1992 a new combined driving cycle consisting of city-cycle and extra-urban driving cycle – called MVEG-Cycle (Motor Emission Vehicle Group) – has been used in Europe.

Apart from European and Japanese exhaust emission regulations international legislation is mostly based on the American FTP-75 driving cycle. The exhaust emission standards of the USA (California) and Japan are listed in Table 8.9 and 8.10 respectively. In California vehicles are classified according to their exhaust emission levels. A continuous lowering of vehicle exhaust emissions is to be achieved by the mandatory NMOG fleet average (Non-Methane Organic Gases) from the model year 1994 onwards. The progressively decreasing NMOG limit is forcing manufacturers to sell more and more clean categories. In the other US-states emission standards from 1996 on basically correspond to the Californian Tier 1 level. But the Clean Air Act on which US exhaust emission legislation is based provides for the adoption of Californian standards by any other US state if air quality standards are not met. Besides, there are fuel economy regulations in the USA based on the FTP-75 cycle and the highway-test which have remained unchanged since 1991 at 27.5 miles per gallon or 8.55 liters per 100 km. In case of non-compliance a fine has to be paid. Many countries, even in Europe, have followed American exhaust emissions legislation because for years American standards were more stringent than European ones and the driving cycle used was more appropriate to determine vehicle exhaust emissions than the European city-cycle which was used until 1992. Table 8.11 gives an overview of internationally applied test procedures to which emission limits are referred. While emission limits vary from country to country, there is a general need for all the countries mentioned to use 3-way catalytic converters to meet emission limits.

In addition to exhaust emission standards for new vehicles there are In-Use Surveillance Tests to check the emission behaviour of in-use vehicles, as they are finally decisive for the actual emission. Generally in-use testing regulations for Otto vehicles provide, among other things, the measurement of CO while idling and at about 3000 rpm. However, by means of such short tests only outsider vehicles can be reliably identified. Much more far-reaching are the Californian regulations that prescribe OBD II (On-Board-Diagnosis, Stage 2) from 1996 on. This emission control monitoring system which is a part of the motor management of the vehicle performs a continuous functioning control of all exhaust components. This requirement has represented a big challenge for the manufacturers. But the advantages are obvious. Unlike the German short test (AU) which since end of 1992 provides that all vehicles with catalytic converters be checked every two years, OBD II performs a continuous control. If any fault occurs, the driver is advised. At the same time it is stored and

Table 8.9. Passenger Car Exhaust Emission Regulations – USA (California) [24]

Vehicle Category	Pollutant	Standard	Year	NMOG-Fleet Average [g/km] [3]	ZEV-percentage [%] [9]
Tier 1	NMHC [1]	0.155 g/km			
	CO	2.11 g/km	1994	0.155	-
	NO$_x$	0.25 g/km	1995	0.144	-
	PM	0.05 g/km	1996	0.140	-
TLEV [4]	NMOG [2]	0.078 g/km	1997	0.126	-
	CO	2.11 g/km	1998	0.098	2
	NO$_x$	0.25 g/km	1999	0.070	2
	PM	0.05 g/km	2000	0.045	2
	HCHO [10]	0.009 g/km	2001	0.043	5
LEV [5]	NMOG [2]	0.047 g/km	2002	0.041	5
	CO	2.11 g/km	2003	0.039	10
	NO$_x$	0.12 g/km			
	PM	0.05 g/km			
	HCHO [10]	0.009 g/km			
ULEV [6]	NMOG [2]	0.025 g/km			
	CO	1.06 g/km			
	NO$_x$	0.12 g/km			
	PM	0.025 g/km			
	HCHO [10]	0.005 g/km			
ZEV [7]	all pollutants	0			
	Evaporation [8]	2 g/Test			

Remarks:

[1] NMHC (Non-Methane-Hydrocarbon) = standard for hydrocarbon emissions without methane.

[2] NMOG (Non-Methane Organic Gases) = total mass of oxygenated and non-oxygenated hydrocarbon emissions, i.e. it also includes all ketones, aldehydes, alcohols and esters with 5 or fewer carbon atoms as well as all known alkanes, alkenes, alkines and aromatics containing up to 12 carbon atoms. NMOG measurement is adjusted by a reactivity correction factor, to take into account the emission's potential towards ozone formation. The reactivity depends on the fuel type as well as on the emission control system.

[3] With effect from model year 1994 the required NMOG-fleet average becomes mandatory. There are no limitations about how many cars of the different vehicle categories – except the production of ZEV vehicles, which is mandated from 1998 on – may be sold. However, the NMOG-average must be met.

[4] TLEV (Transitional Low Emission Vehicle) = vehicle planned as transition towards lower emissions.
[5] LEV (Low Emission Vehicle) = vehicle with very low emissions.
[6] ULEV (Ultra Low Emission Vehicle) = vehicle with extremely low emissions.
[7] ZEV (Zero Emission Vehicle) = vehicle causing practically no emissions (e.g. battery vehicle without a vehicle heating or cooling system using fossil fuels).
[8] Beginning with model year 1995 a new Evaporative Emission Test procedure will gradually be introduced.
[9] Production of ZEVs related to annual sales in California is mandated according to predetermined percentages.
[10] HCHO (formaldehyde) standards are applicable for methanol vehicles as well as for vehicles of the categories TLEV, LEV and ULEV.

Table 8.10. Passenger Car Exhaust Emission Regulations – Japan [24,27]

		1978 Standard	1994 Standard	2000 Standard
Gasoline [1] Hot Start Test (10·15-mode)	HC	0.25 g/km		
	CO	2.1 g/km		
	NO_x	0.25 g/km		
Gasoline [1] Cold Start Test (11-mode)	HC	1.7 g/km		
	CO	14.7 g/km		
	NO_x	1.1 g/km		
Diesel [2] Hot Start Test (10·15-mode)	HC		0.4 g/km	0.4 g/km
	CO		2.1 g/km	2.1 g/km
	NO_x		0.5 g/km	0.4 g/km
	PM		0.2 g/km	0.08 g/km
Evaporation		2.0 g/Test		

Remarks:
[1] Covers vehicles (no mass limitation) which serve exclusively for transportation of passengers up to a max. transportation capacity of 10 persons.
[2] Covers vehicles up to a reference mass of ≤1265 kg. For heavier vehicles NO_x limits are about 20% higher.

Table 8.11. International Exhaust Emission Regulations for Passenger Cars with Otto Engines [24]

Country	Test Procedure
Australia (Commonwealth) Australia (New South Wales und Victoria) Brazil Chile Canada Norway [1] Switzerland [2] South Korea Taiwan USA	FTP-75 SHED (Sealed Housing for Evaporative Determinations)
Costa Rica Hong Kong Iran Mexico	FTP-75
European Community	MVEG (Motor Vehicle Emission Group)
Guatemala	[3]
India	India-Cycle
Israel	[4]
Japan	10·15-mode, 11-mode, Evap.
Saudi Arabia	[5]
Singapore	[6]

Remarks:
[1] With effect from 1995 adoption of EC regulations scheduled.
[2] With effect from 10/95 adoption of EC regulations scheduled.
[3] There are no stipulated exhaust emission limits. Vehicle registration is possible with US, Canadian, Mexican or EU certificate.
[4] Prescribes fulfilment of exhaust emission standards according to US-regulation (CFR 40.86) or 91/441/EEC.
[5] From 1996 on catalytic converters are mandatory.
[6] With effect from 1.7.1994 91/441/EEC (SHED-Certification voluntary), alternatively Japan 78 Standards. Self-Certification by the manufacturer: "First-of-its-kind" Exhaust Emission Test.

can be retrieved later in the garage. Thus, by controlling the emission system an identification of the faulty part is achieved as well.

For heavy trucks there have been in the European Community since the mid-eighties exhaust emission limits for the components CO, hydrocarbons and NO_x and since 1992 for particulate matter (PM) as well. Table 8.12 shows that by means of the decrease of exhaust emission limits decided on in 1992 a considerable reduction of emissions will be achieved by 1999. As trucks are used for purposes of work, their emission standards are, for practical reasons, assigned to the work done, and indicated in grams/kWh. Unlike the certification procedure for passenger cars and light-duty vehicles where the whole vehicle is tested on a chassis dynamometer, the certification of trucks is based on a motor dynamometer test. In Europe the test procedure is based on regulation ECE-R 49 which provides for a 13-point-test. Exhaust emissions are measured at 13 stationary test points and the average emission is calculated applying weighting factors to these test points. This 13-point-test is to be modified for the third stage of exhaust emission reduction from 1999 on, yet the industry proposed that the test will remain a stationary one. In the USA, however, since model year 1985 there has been instead of the former 13-point-test a new test procedure called the Transient Test, which simulates dynamic inner- and extra-urban driving behavior. Compared with the stationary European 13-point-test the Transient Test produces on an average slightly higher hydrocarbon and particulate emissions while the NO_x emissions are slightly lower [25]. As in Europe there are also in the USA further reductions of exhaust emission standards for heavy trucks and busses scheduled for the following years (Figure 8.6).

Also for motorcycles, largely ignored so far, there will be in the European Community from 1995/96 on much more stringent emission standards which are to be even further reduced from 1999 on. This means that motorcycles will have to be equipped with catalytic converters too.

Table 8.12. Three-Stage-Plan for European Exhaust Emission Standards for Commercial Vehicles [23]

	88/77/EEC from 1988/90	91/542/EEC 1st Stage (EURO I) from 1992/93	91/542/EC 2nd Stage (EURO II) from 1995/96	German Proposal 3rd Stage (EURO III) from 1999
CO	12.3 g/kWh	4.9 g/kWh	4.0 g/kWh	2.0 g/kWh
HC	2.6 g/kWh	1.23 g/kWh	1.1 g/kWh	0.6 g/kWh
NO_x	15.8 g/kWh	9.0 g/kWh	7.0 g/kWh	below 5.0 g/kWh
PM	–	0.4 g/kWh [1]	0.15 g/kWh	below 0.1 g/kWh

Remarks:
[1] For engines with a maximum power of less than 85 kW a limit of 0.68 g/kWh has to be applied.

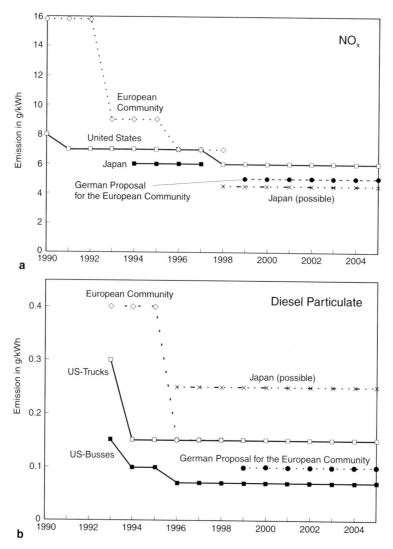

Fig. 8.6. International exhaust emission standards for commercial vehicles until 2005. **a.** NO$_x$; **b.** Diesel particulate [23, 26]
Note: the US, Europe and Japan use different test procedures which can influence the actual stringency of the requirements

8.5 Bibliography

1. Convention on Long-range Transboundary Air Pollution, of 13 November 1979; text in: International Legal materials, vol. 18 (1979), p 1442
2. Protocol on long-term financing of the cooperative programme for monitoring and evaluation of the long-range transmission of air pollutants in Europe (EMEP), of 28 September 1984; text in: International Legal Materials, vol. 24 (1985), p 484

8.5 Bibliography 481

3   Convention for the protection of the ozone layer, of 22 March 1985; text in: International Legal Materials, vol. 26 (1987), p 1529
4   Protocol on substances that deplete the ozone layer, of 16 September 1987; text in: International Legal Materials, vol. 26 (1987), p 1550
5   Protocol on the reduction of sulphur emissions or their transboundary fluxes by at least 30 per cent, of 8 July 1985; text in: International Legal Materials, vol. 27 (1988), p 707
6   Protocol concerning the control of emissions of nitrogen oxides or their transboundary fluxes, of 31 October 1988; text in: International Legal Materials, vol. 28 (1989), p 214
7   Protocol concerning the control of emissions of volatile organic compounds (VOC) or their transboundary fluxes, of 19 November 1991; text in: International Legal Materials, vol. 31 (1992)
8   Convention on Climate Change, of 9 May 1992; text in: International Legal Materials, vol. 31 (1992)
9   Becker, B.: Fundstellen- und Inhaltsverzeichnis Umweltschutzrecht der Europäischen Union (EU). Luftreinhaltung; Klimaschutz; Lärmbekämpfung (No. 2000 ff.), Verlag R. S. Schulz, 1. April 1995
10 CLEAN AIR ACT – Law and Explanation. Commerce Clearing House, Inc., Chicago, Illionois, 1990
11 Gesetz zum Schutz vor schädlichen Umwelteinwirkungen durch Luftverunreinigungen, Geräusche, Erschütterungen und ähnliche Vorgänge (Bundes-Immissionsschutzgesetz – BImSchG) vom 15.3.1974, BGBl. I, S. 721, ber. S. 1193, in der Fassung vom 29.11.1986, BGBl. I, S. 2089
12 Vierte Verordnung zur Durchführung des Bundes-Immissionsschutzgesetzes (Verordnung über genehmigungsbedürftige Anlagen – 4. BImSchV) vom 24.7.1985, BGBl. I. S. 1586
13 GewO des Ndd. Bundes vom 21.6.1989, Teil II, Nr. 1 S. 16
14 Erste Allgemeine Verwaltungsvorschrift zum Bundes-Immissionsschutzgesetz (Technische Anleitung zur Reinhaltung der Luft – TA Luft -) i.d.F. vom 27.2.1986, GMBl., S. 95
15 Dreizehnte Verordnung zur Durchführung des Bundes-Immissionsschutzgesetzes (Verordnung über Großfeuerungsanlagen – 13. BImSchV) vom 22.6.1983, BGBl. I, S. 7189
16 Siebzehnte Verordnung zur Durchführung des Bundes-Immissionsschutzgesetzes (Verordnung über Verbrennungsanlagen für Abfälle und ähnliche brennbare Stoffe – 17. BImSchV) vom 23.11.1990, BGBl. I, S. 2545, 2832
17 Erste Verordnung zur Durchführung des Bundes-Immissionsschutzgesetzes (Verordnung über Kleinfeuerungsanlagen – 1. BImSchV) vom 15.7.1988, BGBl. I, S 1059
18 Dritte Verordnung zur Durchführung des Bundes-Immissionsschutzgesetzes (Verordnung über Schwefelgehalt von leichtem Heizöl und Dieselkraftstoff – 3. BImSchV) vom 15.1.1975, BGBl. I, S. 264
19 Zehnte Verordnung zur Durchführung des Bundes-Immissionsschutzgesetztes (Beschränkung von PCB, PCT und VC – 10. BImSchV) vom 26.7.1978, BGBl. I, S. 1138
20 Gesetz zur Verminderung von Luftverunreinigungen durch Bleiverbindungen in Ottokraftstoffen für Kraftfahrzeugmotore (Benzinbleigesetz – BzBlG) vom 5.8.1971, BGBl. I, S. 1234
21 Klingenberg, H.: Automobil-Meßtechnik Band C: Abgasmeßtechnik. Berlin, Heidelberg: Springer 1995
22 Beck'sche Textausgaben: Straßenverkehrsrichtlinien. Textsammlung. München: C. H. Beck, laufende Ergänzungen
23 Schadstoffminderung bei Kraftfahrzeugen. Umwelt, Informationen des Bundesministers für Umwelt, Naturschutz und Reaktorsicherheit 9 (1992), S. 342 – 344
24 Mercedes-Benz: Abgas-Emissionen – Grenzwerte, Vorschriften und Messung der Abgas-Emissionen sowie Berechnung des Kraftstoffverbrauchs aus dem Abgastest (PKW). Druckschrift Nr. 14 EP/TZ, Stuttgart Mai 1994
25 Mercedes-Benz: Umweltschutz beim Nutzfahrzeug – Das Konzept von Mercedes-Benz. 2. Auflage GBN/ES, Stuttgart Juli 1991
26 Walsh, M.P.: Motor Vehicle Pollution Control: A Local, Regional and Global Challenge. Proceedings of the 10th World Clean Air Congress, Volume 4. Espoo, Finland: The Finish Air Pollution Prevention Society 1995
27 Schäfer, F.; van Basshuysen, R.: Schadstoffreduzierung und Kraftstoffverbrauch von PKW-Verbrennungsmotoren. Wien: Springer 1993

Subject Index

Absorption bands   226
Accuracy   304
Acid precipitation   191
Acid rain   109, 119
Adiabatic
- lapse rate   80
- layers   81

Aerosol   128, 150
- components   130
- size distribution   129

Air   24
- excess   15, 17
- lack of   15, 18
- natural composition   1

Air pollutants   10
Air pollution
- area-related   472
- effects   174
- history   2
- natural   1, 174
- product-related   470
- standards   463

Air quality
- concentration   12
- in large cities   139

Aldehydes   25, 117
- effects on human health   204

Ambient air measurement   292
- forest station   295
- sampling system   293, 297
- set-up of measuring stations   295
- significance of locations   292

Ammonia   133
- effects on plants   191

Analytical function   303
Asbestos, impacts   204
Atmospheric
- layers   82
- lifetime   103

Barrier layer   88
Benzene
- concentration   135
- effects   204, 205
- life-span   127

Bergerhoff   273
Bronchi   201
BTX-Benzene, toluene, xylenes   20
Bundesimmissionsschutzgesetz   465
Bush burning   131

Calibration   298
- basic   298
- dynamic methods   299, 301
- function   303
- gas production   300
- gases   299
- routine and control calibrations   298
- static methods   299

Carbon compounds
- organic   127
- inorganic   127

Carbon dioxide (CO_2)   8, 142
- emission   64
- increase   179

Carbon emission   45
Carbon monoxide   18, 24, 53, 54, 70, 127, 132
- effects on human health   203, 204

Subject Index

- trends 70
- Cascade impactor 270
- Catalytic converter
  - desulfurization 366
  - selective catalytic reduction (SCR) 404
  - three-way 413
    (for automotive exhaust gases)
- Chemical transformation 100
  - modeling 166
- Chemiluminscence 236
- Chlorinated hydrocarbons 135
- Chlorine
  - effects on plants 191
  - on human health 203
- Clean air 1
- Clean Air Act 464
- Climatic change 175
- Clinical sudies 200
- Coal 27
  - ash 42
  - combustion 41
- Combustion
  - incomplete 18
  - nitrogen oxides 36
  - process 15, 16
- Composition of air 1
- Conventions, international 461
- Correlation 350
  - pollutant-wind 350
- Corrosion
  - of metals 184
  - of stones 183

- Dead time 305
- Degree of removal 367
- Demage 187
  - forest 194
  - of plants 188
  - profiles 194
- Desulfurization 417
  - chemical reactions 428
  - citrate process 435
  - dry 418
  - effluents 433
  - gypsum 433
  - lime scrubbing 426
  - mass flows 437
  - principle of process 432
  - residues 423
  - semi-dry 420, 423
  - spray tower 427
  - Wellman-Lord process 434
  - wet 426
- Diesel engine
  - emissions 56
  - particulate 474
  - standards 480
- Dioxin 46f
  - sources 50
  - toxicity 48
- Directives, EU 461, 463
- Dispersion 79
  - in valley 97
  - widespread 91
- Dispersion models 156
  - box models 162
  - chemical transformation 166
  - diagnostic 158
  - Gaussian 164
  - k-models 163
  - Lagrange 161
  - prognostic 159
  - statistical 167
- Distribution
  - spatial 130f
  - temporal 142f
  - vertical 123
- Diurnal profile, course 90, 112, 122
- Domestic furnace 43, 470
- Dose
  - effect relation 188
  - wind rose 354

Driving
- behavior 58
- cycle 57, 290, 475

ECE test 60, 479
Eddy 82
EEC regulations 474
Effects
- harmful 199
- long term 337
- medium term 337
- on climate 175
- on humans 337
- on human health 203
- on material 181, 337
- on plants 187, 190
- on vegetation 186, 337
- short term 337

Efficiency 367
Electrostatic precipitator 383
- collecting electrodes 388
- constructions 390
- corona discharge 386
- electric wind 387
- electrodes 391
- equation 389

EMEP 462
Emission 67f
- annual 62, 331
- carbon 45
- carbon dioxide 64
- concentration 12
- control (furnaces) 363
- development 60
- diesel engines 56
- domestic furnaces 328, 333, 363
- factor 12, 322
- household 67f
- industry 328, 363
- industrial processes 61
- inventory 326
- mass flow 12
- mass flow (rate) 322
- particulate 364
- per capita 63
- power plant 67, 333, 363
- products of incomplete combustion 15, 364
- rates 25, 31
- rate (mass flow) 323
- solvent 362
- sources 11
- temporal course 69
- transportation 67f
- traffic 327, 333
- trends 67
- vehicles 51f

Emission measurement 217, 220, 221f
- constandt volume sampling (CVS) 289
- location for sampling 282
- motor vehicles 288
- sampling techniques 282

Engine performance characteristics 52
EPA 464
Epidemiology 200
Evaluation
- averaging 336
- diurnal course 338, 334
- half-hour value 336
- mean diurnal course 348
- mean value 335
- monthly profile 338
- of measurements 334

Excess Air 15, 17, 24, 35, 52, 54
Exhaust gas 11, 52, 55
- purification 367

Extincition 226
Extra-Urban driving cycle 291

FITNAH 161
Flame ionization 239
Flue gas 11, 284
- flow rate 323

Fluorochlorinated hydrocarbons 135, 144
- effects on plants 191
Forest demage 9, 193, 194
Formaldehyde 22, 24, 25, 118, 234
- effects 204
Frequency distribution 344
FTP-75 478
Fuel 16
- /air ratio 52
- change 366
- coal 26f
- desulfurization 364
- gas 28
- NO 35
- oil 16, 23, 27
- rate 323
- wood 21, 29
Furan 46f
- Sources 50
- Toxity 48
Furnaces
- domestic 470
- industrial 37
- nitrogen oxides 37
- ordinances 468

Gas 28
Gas Chromatography 254
- GC/MS 256
- as chromatogram 23
Gauss-Krueger grid 331
Gaussian model 164
German Federal Air Poll Control Act 465
Graphic representation 346
Greenhouse effect 179
Guidelines 202, 205
Gypsum 182

Halocarbons 144
Harmattan 131

Heavy metals 130, 325
- effects 191
Henry's law 440
- oefficients 284
Hydrocarbon 18, 23, 25, 54, 55, 127, 135
- chemical transformation 104, 115
- chlorinated 135
Hydrochloric acid (HCl) 133
Hydrogen chloride
- effects on human health 210
- effects on plants 191
Hydrogen fluoride 189
- effects on plants 191
- guidelines 210
Hydrogen sulfide 28, 30, 132
- effects 206
Hydroxyl radicals 115

Impact 187
- on human health 198
Incinerator
- regulation 463
Industrial processes 61
Infrared (IR)
- absorption 221
- non-dispersive analyzer 229
- photometer 227
Interference 306
Inversion 83f, 92
- formation 85
- frontal 87
- radiational 85
- subsidence 86
- types 84
Isoplethes 124

Lambda 53, 54
- calculation 17
- control 59
Lambert-Beer law 226
lapse rate 80

Laws 461
- German 465
- national 462
- US 464
- vehicles 472
Lead
- standard 463
Life spans 127
Limited Components 55, 290
Long-range transport 91
Long-range transboundary pollution 461
Los Angeles smog 139

Magnetic susceptibility 249
Maintanance interval 305
Measuring range 303
Measuring techniques 216
- accuracy 302
- air quality, components 223
- aldehydes 254
- amperometry 243
- chemiluminescence 236
- CO 229, 253
- CO_2 249
- colorimetric methods 234, 251
- conductometry 241
- continous 217
- detection limit 303
- difficulties 301
- discontinous 219
- emission, components 221, 222
- errors 313
- fault consideration 312
- fields of application 217
- flame photometry 238
- flame ionization 239
- formaldehyde 234
- gas chromatography 254
- GC/MS 256
- H_2 249
- H_2S 253
- HCl 246, 253
- HF 246
- HPLC 257
- infrared absorption 227
- interference 308
- ion chromatography 257
- linearity 307
- manual analysis 251
- molecular excitation 220
- NH_3 253
- NO 231, 294,
- NO_x 236, 253, 294
- O_2 246, 249
- olfactometry 261
- overview 225
- ozone 232, 238
- PAH 254, 258
- performance characteristics 303
- polychlorinated dibenzodioxins 254, 258
- potentiometry 245
- precision 304
- reference methods 219
- repeatability 304
- ring tests 310
- sampling 252
- sensitivity 307
- significance 302
- SO_2 235, 241, 253, 267, 285, 294
- SO_3 253
- student factor 307
- thermal conductivity 249
- titration 251
- UV absorption 230
- volumetry 251
Mercaptane 27
Metals 40
- heavy 130
Meteorology 79
Methane 127, 144
- climate effect 176
MIK 212

- values 210
Mixing layer 87
Models, mathematical 156
- box 163
- chemical transformation 166
- diagnostic 158
- dispersion 160
- flow and turbulence 158
- Gaussian 164
- K-models 162
- Lagrange 161
- prognostic 159
Motor vehicle 51
- diesel 56
- emissions 51f
- exhaust gas 51
- standards 472f
Motor vehicle traffic 327

Nickel 326
Nitrate 133
Nitric acid 119
Nitrogen oxide (NO_x) 32, 53, 58, 63, 111, 133, 150, 330, 363
- abatement 355
- calculation 324
- chemical transformation 104, 110
- concentration 325
- control 462
- emissions 38
- emission control 364
- flue-gas recirculation 400
- fuel 35
- in furnace measures 397
- low-NO_x burners 401
- measuring 285
- pollutant wind rose 354
- prompt 33
- reducing air excess 397
- reducing combustion chamber load 399
- removal 397

- stage combustion
- thermal 32
- trends 68
- values 141
Nitrogen dioxide (N_2O) 133
- horizontal profile 142
- oxidation 118
- photolytic degradation 111
- ringtest 311
- spatial distribution 140
- standards 463
- trends 154
- values 141
- vertical profile 123
Nitrous oxide (N_2O) 133, 176
Nutrient
- leaching 196
- intake 196

Occupational medicine 200
Odor 24
Oil 16, 23, 27
Olfactometry 261
On-board measurements 60
Opacimeter 267
Oxides
- nitrogen 32
- sulfur 29
Oxygen
- excess 15, 35
Ozone 112, 136, 150f
- climatic effects 176
- diurnal profile 122
- effects on human health 203
- effects on materials 185
- effects on plants 190
- historical concentration 151
- standards 207
- stratospheric 119
- trends 154
- tropospheric 121

- vertical distribution 123
Ozone layer 177
- destruction 178
- protection 462

Painting 361
Paramagnetisms 249
Particles 37f, 38, 55, 131
- components 130
- composition 42
- deposition in lung 201
- emissions 41
- formation 38
- in the atmosphere 128
- light scattering 180
- size distribution 39, 45, 129
- Soot 40
Particulate matter 63, 364
Particulate matter measuring methods 262
- atomic absorption spectometry
- cascade impactor 270
- chemical composition 280
- chemical reaction 276
- condensate 286
- filter cartridge 265
- gravimetric 263
- high-volume sampler 274
- in-situ determination 168
- isokinetic sampling 263
- leakages 286
- low-volume sampler
- mass spectometry and laser microprobe mass spectometry (LMMS) 281
- photometric measurement 267
- PM 10
- β radiation attenuation 268
- scattered-light photometer 278
- sedimentation 271
- setting up 282
- soot number 269
- X-ray fluorescence 281

Particulate removal 369
- baghouse filters 396
- centrifugal force collector 371
- cyclone 372
- electrostatic precipitator 383
- filters 392
- filter fabrics 394
- inertial collectors 370
- multi-cyclone 374
- rotary flow particle collector 373
- sedimentation 378
- tube filters 393
- venturi scrubber 379
- wet scrubber 375, 379
- whirl scrubber 379
Permission 466
Percentile 347
Performance characteristics 303
Peroxyacetyl nitrate (PAN) 116, 133
- effects on plants 190
- guidelines 211
Photometry
- long path 233
Photostationary equilibrium 112
Pitot tube 264
Plants
- requiring permission 467
Pollutants 12
- combustion 15
- concentration 12, 323
- deposition 12
- dose 12, 322
- flux 322
- wind rose 350
Polychlorinated
- biphenyle 46
- dioxine, furane 46
Polycyclic Aromatic Hydrocarbons (PAH) 20, 22, 130
Potentimetry 245
Precipitation 128

Precision 304
Process modification 361
Products of incomplete combustion 364
Product related prevention 470
Prompt NO 33

Radiation, solar 101
Reaction rates 105, 113
Reaction time
- of effects 337
Reduction potential 403
- activated coke 411
- dependency of temperatur 407
- electron-beam process 412
- oxidation processes 411
- SCR reactor 409
- selective catalytic reduction (SCR) 405
- selective non-catalytic reduction (SNCR) 404
Regulations 461
- California 476
- EEC 474
- international 478
- Japan 474, 477
- national 462
- ordinances 468
- passenger cars 476
- state 465
- US 464, 474
- vehicles 473
Removal
- absorption, physical 439
- absorption, chemically 442
- adsorbers 448, 449
- adsorption 445
- adsorption isothermes 446
- biofilter 443, 448
- biological 442
- catalytic afterburning 551
- combustion 450
- condensation 439
- membrane process 552

- of organic substances 437
- of vehicle emissions 413
- process for particulate 369
- thermal afterburning 450
Repeatibility 304
Reproducibility 305
Response time 305
REWIMET 160
Ring test 310
Rise time 305

Sampling system 284
Sensitivity 303
Short Range Dispersion 95
Smog 6, 7, 10, 44
- alarm values 213
- photochemical 116, 139
- warning 338
Smoke
- blue and white 57
- demage 193
- plume 82
- visibility 199
Smoking
- effects 199
Soil acidification 195
Solubility 282
Soot 19, 23, 39f, 57, 204
- number 269
Sources 15
Spatial distribution 327
Standards 203, 205
- busses 479
- commercial vehicles 479
- emission (German) 470
- European, exhaust emission 473
- intern, exhaust emission 474
- motorcycles 479
- US National Ambient Air Quality 465
Standard deviation 303
Stove 20

Subject Index

Student factor 304, 307
Sulfate 108, 133
Sulfur 26f
- compounds 132
- dioxide 29f, 63, 66
- dioxide trends 67
- emission rate 31
- mission reduction 462
- oxides 65
- trioxide 30
Sulfur dioxide (SO_2) 132, 330, 363
- calculation 324
- chemical transformation 104
- concentration course 341
- deposition 183
- development 145
- dispersion 91
- effects on human health 203
- effects on plants 190
- emission control 364
- guidelines 210
- oxidation 105
- standards 463
Suspended Particultae Matter (SPM) 146

TA Luft 467
(Technical Directive on Air Poll. Control)
Temperature increase 177
Temporal resolution (of measurements) 335
- distribution 332
Thermal conductivity 249
Thermal NO 32
Thirteen-point-test 479
Three
- reactions 414
- way catalytic converter 413
Three-Stage-Plan 473
Titration 251
Topographic influence 83

Toxicity 49
- toxicity equivalent 49
Toxicology 200
Trace compounds 132
Trace elements 45, 130
Transmission 10
Turbulence 80

Ultraviolett (UV)
- absorption spectra 230
- fluorescence 235
- measuring technique 231
Uncertainty 034, 347
Units 12
Urea 405
US-Test 291

Valley 97
Vanadium oxides 326
Variance 303
Variation coefficient 303
VDI
- guidelines 201
Vehicle exhaust emissions 472
Vent air 11
VOC
- control 462
Volatile Organic Compounds (VOC) 63, 72, 135

Wind rose 350
Windspeed 79
Wood
- combustion 21, 43
- emissions 45
World Health Organization 199
- guidelines 209

Zero point 303
Zirconium dioxide measuring sensor 246

Printing: COLOR-DRUCK DORFI GmbH, Berlin
Binding: Buchbinderei Lüderitz & Bauer, Berlin